GLOBAL ISSUES IN WATER, SANITATION, AND HEALTH

Workshop Summary

Rapporteurs: Eileen R. Choffnes and Alison Mack

Forum on Microbial Threats
Board on Global Health

INSTITUTE OF MEDICINE
OF THE NATIONAL ACADEMIES

THE NATIONAL ACADEMIES PRESS
Washington, D.C.
www.nap.edu

THE NATIONAL ACADEMIES PRESS 500 Fifth Street, N.W. Washington, DC 20001

NOTICE: The project that is the subject of this report was approved by the Governing Board of the National Research Council, whose members are drawn from the councils of the National Academy of Sciences, the National Academy of Engineering, and the Institute of Medicine.

This project was supported by contracts between the National Academy of Sciences and the U.S. Department of Health and Human Services: National Institutes of Health, National Institute of Allergy and Infectious Diseases, the Centers for Disease Control and Prevention, and the Food and Drug Administration; U.S. Department of Defense, Department of the Army: Global Emerging Infections Surveillance and Response System, Medical Research and Materiel Command, and the Defense Threat Reduction Agency; U.S. Department of Veterans Affairs; U.S. Department of Homeland Security; U.S. Agency for International Development; the American Society for Microbiology; sanofi pasteur; Burroughs Wellcome Fund; Pfizer; GlaxoSmithKline, Infectious Diseases Society of America; and the Merck Company Foundation. Any opinions, findings, conclusions, or recommendations expressed in this publication are those of the author(s) and do not necessarily reflect the view of the organizations or agencies that provided support for this project.

International Standard Book Number-13: 978-0-309-13872-7
International Standard Book Number-10: 0-309-13872-8

Additional copies of this report are available from the National Academies Press, 500 Fifth Street, N.W., Lockbox 285, Washington, DC 20055; (800) 624-6242 or (202) 334-3313 (in the Washington metropolitan area); Internet, http://www.nap.edu.

For more information about the Institute of Medicine, visit the IOM home page at: **www.iom.edu.**

Printed in the United States of America

The serpent has been a symbol of long life, healing, and knowledge among almost all cultures and religions since the beginning of recorded history. The serpent adopted as a logotype by the Institute of Medicine is a relief carving from ancient Greece, now held by the Staatliche Museen in Berlin.

Cover credit: Copyright Ralph A. Clevenger/Corbis.

Suggested citation: IOM (Institute of Medicine). 2009. *Global issues in water, sanitation, and health.* Washington, DC: The National Academies Press.

"Knowing is not enough; we must apply.
Willing is not enough; we must do."

—Goethe

INSTITUTE OF MEDICINE

OF THE NATIONAL ACADEMIES

Advising the Nation. Improving Health.

THE NATIONAL ACADEMIES

Advisers to the Nation on Science, Engineering, and Medicine

The **National Academy of Sciences** is a private, nonprofit, self-perpetuating society of distinguished scholars engaged in scientific and engineering research, dedicated to the furtherance of science and technology and to their use for the general welfare. Upon the authority of the charter granted to it by the Congress in 1863, the Academy has a mandate that requires it to advise the federal government on scientific and technical matters. Dr. Ralph J. Cicerone is president of the National Academy of Sciences.

The **National Academy of Engineering** was established in 1964, under the charter of the National Academy of Sciences, as a parallel organization of outstanding engineers. It is autonomous in its administration and in the selection of its members, sharing with the National Academy of Sciences the responsibility for advising the federal government. The National Academy of Engineering also sponsors engineering programs aimed at meeting national needs, encourages education and research, and recognizes the superior achievements of engineers. Dr. Charles M. Vest is president of the National Academy of Engineering.

The **Institute of Medicine** was established in 1970 by the National Academy of Sciences to secure the services of eminent members of appropriate professions in the examination of policy matters pertaining to the health of the public. The Institute acts under the responsibility given to the National Academy of Sciences by its congressional charter to be an adviser to the federal government and, upon its own initiative, to identify issues of medical care, research, and education. Dr. Harvey V. Fineberg is president of the Institute of Medicine.

The **National Research Council** was organized by the National Academy of Sciences in 1916 to associate the broad community of science and technology with the Academy's purposes of furthering knowledge and advising the federal government. Functioning in accordance with general policies determined by the Academy, the Council has become the principal operating agency of both the National Academy of Sciences and the National Academy of Engineering in providing services to the government, the public, and the scientific and engineering communities. The Council is administered jointly by both Academies and the Institute of Medicine. Dr. Ralph J. Cicerone and Dr. Charles M. Vest are chair and vice chair, respectively, of the National Research Council.

www.national-academies.org

FORUM ON MICROBIAL THREATS

[*]Until June 9, 2009. Dr. Hamburg is currently the Commissioner of the Food and Drug Administration.

†Current Vice Chair.

---

[*]Until January 16, 2009. Kent Kester, Commander of Walter Reed Army Institute of Research, is the current U.S. Army Medical Research and Materiel Command representative on the Forum.

BOARD ON GLOBAL HEALTH

Margaret Hamburg (*Chair*), Consultant, Nuclear Threat Initiative, Washington, DC
Jo Ivey Boufford (*IOM Foreign Secretary*), President, New York Academy of Medicine, New York
Claire V. Broome, Adjunct Professor, Division of Global Health, Rollins School of Public Health, Emory University
Jacquelyn C. Campbell, Anna D. Wolf Chair, and Professor, Johns Hopkins University School of Nursing, Baltimore, Maryland
Thomas J. Coates, Professor, David Geffen School of Medicine, University of California, Los Angeles, Los Angeles, California
Valentin Fuster, Director, Wiener Cardiovascular Institute, Kravis Cardiovascular Health Center, Professor of Cardiology, Mount Sinai School of Medicine, Mount Sinai Medical Center, New York, New York
Sue Goldie, Associate Professor of Health Decision Science, Department of Health Policy and Management, Center for Risk Analysis, Harvard University School of Public Health, Boston, Massachusetts
Richard Guerrant, Thomas H. Hunter Professor of International Medicine and Director, Center for Global Health, University of Virginia School of Medicine, Charlottesville
Peter J. Hotez, Professor and Chair, Department of Microbiology, Immunology, and Tropical Medicine, The George Washington University, Washington, DC
Gerald T. Keusch, Assistant Provost for Global Health, Boston University School of Medicine, and Associate Dean for Global Health, Boston University School of Public Health, Massachusetts
Michael Merson, Director, Duke Global Health Institute, Duke University, Durham, North Carolina
Fitzhugh Mullan, Professor, Department of Health Policy, George Washington University, Washington, DC
Philip Russell, Professor Emeritus, Bloomberg School of Public Health, Johns Hopkins University, Baltimore, Maryland

Staff

Patrick Kelley, Director
Allison Brantley, Senior Program Assistant

IOM boards do not review or approve individual reports and are not asked to endorse conclusions and recommendations. The responsibility for the content of the report rests with the authors and the institution.

Reviewers

This report has been reviewed in draft form by individuals chosen for their diverse perspectives and technical expertise, in accordance with procedures approved by the National Research Council's Report Review Committee. The purpose of this independent review is to provide candid and critical comments that will assist the institution in making its published report as sound as possible and to ensure that the report meets institutional standards for objectivity, evidence, and responsiveness to the study charge. The review comments and draft manuscript remain confidential to protect the integrity of the deliberative process. We wish to thank the following individuals for their review of this report:

Rima Khabbaz, Centers for Disease Control and Prevention
Bud Rock, Arizona State University
Mary Wilson, Department of Population and International Health, Harvard University

Although the reviewers listed above have provided many constructive comments and suggestions, they were not asked to endorse the final draft of the report before its release. The review of this report was overseen by **Dr. Melvin Worth**. Appointed by the Institute of Medicine, he was responsible for making certain that an independent examination of this report was carried out in accordance with institutional procedures and that all review comments were carefully considered. Responsibility for the final content of this report rests entirely with the authoring committee and the institution.

Preface

The Forum on Emerging Infections was created by the Institute of Medicine (IOM) in 1996 in response to a request from the Centers for Disease Control and Prevention (CDC) and the National Institutes of Health (NIH). The purpose of the Forum is to provide structured opportunities for leaders from government, academia, and industry to meet and examine issues of shared concern regarding research, prevention, detection, and management of emerging or reemerging infectious diseases. In pursuing this task, the Forum provides a venue to foster the exchange of information and ideas, identify areas in need of greater attention, clarify policy issues by enhancing knowledge and identifying points of agreement, and inform decision makers about science and policy issues. The Forum seeks to illuminate issues rather than resolve them; for this reason, it does not provide advice or recommendations on any specific policy initiative pending before any agency or organization. Its value derives instead from the diversity of its membership and from the contributions that individual members make throughout the activities of the Forum. In September 2003, the Forum changed its name to the Forum on Microbial Threats.

ABOUT THE WORKSHOP

In the early days of space exploration, the first images taken of our home planet showed the Earth to be a bright blue marble in the vastness of space. The striking blue in these images—covering more than 70 percent of the planet's surface—represents our planet's water resources. Yet, despite this seemingly endless supply of water, only about 2.5 percent is fresh water, two-thirds of which is trapped as ice in glaciers.[1] Over 90 percent of the fresh water that is not ice may be found in underground aquifers that, once drained, may take hundreds if not thousands of years to recharge.

During the past century, the human population has more than tripled, and water consumption has more than quadrupled, placing ever-increasing demands on the world's limited freshwater resources. Approximately one-third of the world's population now lives in areas with scarce water resources. A U.N. report estimates that water scarcity will affect two-thirds of the population by 2025.[2] In addition, increasing amounts of pollution from domestic, industrial and agricultural runoff is contaminating an ever-shrinking water supply.

The lack of access to and availability of clean water and sanitation has had devastating effects on many aspects of daily life. Areas without adequate supplies of freshwater and basic sanitation carry the highest burdens of disease which disproportionately impact children under five years of age. Lack of these basic necessities also influences the work burden, safety, education, and equity of women. While poverty has been a major barrier to gaining access to clean drinking water and sanitation in many parts of the developing world, access to and the availability of clean water is a prerequisite to the sustainable growth and development of communities around the world.

Worldwide, over one billion people lack access to an adequate water supply; more than twice as many lack basic sanitation.[3] Unsafe water, inadequate sanitation, and insufficient hygiene account for an estimated 9.1 percent of the global burden of disease and 6.3 percent of all deaths, according to the World Health Organization.[4] This burden is disproportionately borne by children in developing countries, with water-related factors causing more than 20 percent of deaths of people under age 14. Nearly half of all people in developing countries have infections or diseases associated with inadequate water supply and sanitation.[5]

The effects of water shortages and water pollution have been felt in both industrialized and developing countries, and it will be necessary to transcend international and political boundaries to meet the world's water needs in a sus-

---

[1]UNESCO (United Nations Educational, Scientific, and Cultural Organization). 2006. The state of the resource. In *Water, a shared responsibility: the United Nations world water development report 2*. New York: UNESCO/Berghahn Books.

[2]United Nations. 2006. *Factsheet on water and sanitation*, http://www.un.org/waterforlifedecade/factsheet.html (accessed August 11, 2008).

[3]WHO/UNICEF (World Health Organization/United Nations Children's Fund). 2006. *Meeting the MDG drinking water and sanitation target: the urban and rural challenge of the decade*. Geneva: WHO/UNICEF.

[4]Prüss-Üstün, A., R. Bos, F. Gore, and J. Bartram. 2008. *Safer water, better health: costs, benefits and sustainability of interventions to protect and promote health*. Geneva: World Health Organization.

[5]Bartram, J., K. Lewis, R. Lenton, and A. Wright. 2005. Focusing on improved water and sanitation for health. *Lancet* 365(9461):810-812.

tainable manner that will conserve and preserve this common resource. In the past few decades, national and international organizations from both the public and private sectors have come together to tackle global issues in water and sanitation.

Recognizing water availability, water quality, and sanitation as fundamental issues underlying infectious disease emergence, the Forum on Microbial Threats of the Institute of Medicine held a two-day public workshop in Washington, DC, on September 23 and 24, 2008. Through invited presentations and discussions, participants explored global and local connections between water, sanitation, and health; the spectrum of water-related disease transmission processes as they inform intervention design; lessons learned from water-related disease outbreaks; vulnerabilities in water and sanitation infrastructure in both industrialized and developing countries; and opportunities to improve water and sanitation infrastructure so as to reduce the risk of water-related infectious disease.

ACKNOWLEDGMENTS

The Forum on Microbial Threats, and the IOM, wish to express their warmest appreciation to the individuals and organizations who gave their valuable time to provide information and advice to the Forum through their participation in this workshop. A full list of presenters may be found in Appendix A.

The Forum is indebted to the IOM staff who contributed during the course of the workshop and the production of this workshop summary. On behalf of the Forum, we gratefully acknowledge the efforts led by Dr. Eileen Choffnes, director of the Forum; Kate Skoczdopole, senior program associate; Sarah Bronko, research associate; K. C. Ostapkovich, research associate; and Kenisha Peters, senior program assistant, for dedicating much effort and time to developing this workshop's agenda and for their thoughtful and insightful approach and skill in planning for the workshop and in translating the workshop's proceedings and discussion into this workshop summary. We would also like to thank the following IOM staff and consultants for their valuable contributions to this activity: Alison Mack, Heather Phillips, Bronwyn Schrecker, Jackie Turner, and Jordan Wyndelts.

Finally, the Forum wishes to recognize the sponsors that supported this activity. Financial support for this project was provided by the U.S. Department of Health and Human Services: National Institutes of Health, National Institute of Allergy and Infectious Diseases, the Centers for Disease Control and Prevention, and the Food and Drug Administration; U.S. Department of Defense, Department of the Army: Global Emerging Infections Surveillance and Response System, Medical Research and Materiel Command, and the Defense Threat Reduction Agency; U.S. Department of Veterans Affairs; U.S. Department of Homeland Security; U.S. Agency for International Development; the American Society for Microbiology; sanofi pasteur; Burroughs Wellcome Fund; Pfizer; GlaxoSmithKline, Infectious Diseases Society of America; and the Merck Company Foundation. The views presented in this workshop summary report are those of the workshop participants and rapporteurs and are not necessarily those of the Forum on Microbial Threats or its sponsors.

David A. Relman, *Chair*
Margaret A. Hamburg, *Vice Chair*[*]
Forum on Microbial Threats

[*]Until June 9, 2009. Dr. Hamburg is currently the Commissioner of the Food and Drug Administration.

Contents

Tables, Figures, and Boxes

TABLES

FIGURES

BOXES

Workshop Overview

GLOBAL ISSUES IN WATER, SANITATION, AND HEALTH

Water is a fixed commodity. At any time in history, the planet contains about 332 million cubic miles of it. Most is salty. Only 2 percent is freshwater and two-thirds of that is unavailable for human use, locked in snow, ice, and permafrost. We are using the same water that the dinosaurs drank, and this same water has to make ice creams in Pasadena and the morning frost in Paris. It is limited, and it is being wasted. . . . But usage is only part of the problem. We are wasting our water mostly by putting waste into it.

Rose George (2008)

Worldwide, over one billion people lack access to an adequate water supply; more than twice as many lack basic sanitation (WHO/UNICEF, 2006). Unsafe water, inadequate sanitation, and insufficient hygiene account for an estimated 9.1 percent of the global burden of disease and 6.3 percent of all deaths, according to the World Health Organization (Prüss-Üstün et al., 2008). This burden is disproportionately borne by children in developing countries, with water-related factors causing more than 20 percent of deaths of people under age 14. Nearly half of all people in developing countries have infections or diseases associated with inadequate water supply and sanitation (Bartram et al., 2005).

The effects of water shortages and water pollution have been felt in both industrialized and developing countries, and it will be necessary to transcend international and political boundaries to meet the world's water needs in a sustainable manner that will conserve and preserve this common resource. In the last few decades, national and international organizations from both the public and private sectors have come together to tackle global issues in water and sanitation.

The lack of access to and availability of clean water and sanitation has had devastating effects on many aspects of daily life. Areas without adequate supplies of freshwater and basic sanitation carry the highest burdens of disease which disproportionately impact children under five years of age. Lack of these basic necessities also influences the work burden, safety, education, and equity of women. While poverty has been a major barrier to gaining access to clean drinking water and sanitation in many parts of the developing world, access to and the availability of clean water is a prerequisite to the sustainable growth and development of communities around the world.

As the human population grows—tripling in the past century while, simultaneously, quadrupling its demand for water—Earth's finite freshwater supplies are increasingly strained, and also increasingly contaminated by domestic, agricultural, and industrial wastes (UNESCO, 2006). Today, approximately one-third of the world's population lives in areas with scarce water resources (UN, 2009). Nearly one billion people currently lack access to an adequate water supply, and more than twice as many lack access to basic sanitation services (Prüss-Üstün et al., 2008). It is projected that by 2025 water scarcity will affect nearly two-thirds of all people on the planet (Figure WO-1).

The majority of these people live in rural areas without community infrastructure. With the rise of "megacities," urban population growth may overtake the ability of local communities and governments to meet their residents' water needs through infrastructure creation and improvements, adding to the estimated 3.6 million people who die each year from inadequate access to safe water, sanitation, and hygiene (Prüss-Üstün et al., 2008). Nearly one in four deaths among children under the age of 14 result from inadequate access to safe water, sanitation, and hygiene. Lack of these necessities also establishes a vicious cycle, for poverty bars many in the developing world from obtaining the safe drinking water and sanitation needed to drive sustainable community growth and development.

Recognizing that water availability, water quality, and sanitation are fundamental issues underlying infectious disease emergence, the Forum on Microbial Threats of the Institute of Medicine held a two-day public workshop in Washington, DC, on September 23 and 24, 2008. Through invited presentations and discussions, participants explored global and local connections between water, sanitation, and health; the spectrum of water-related disease transmission processes as they inform intervention design; lessons learned from water-related disease outbreaks; vulnerabilities in water and sanitation infrastructure in both industrialized and developing countries; and opportunities to improve water

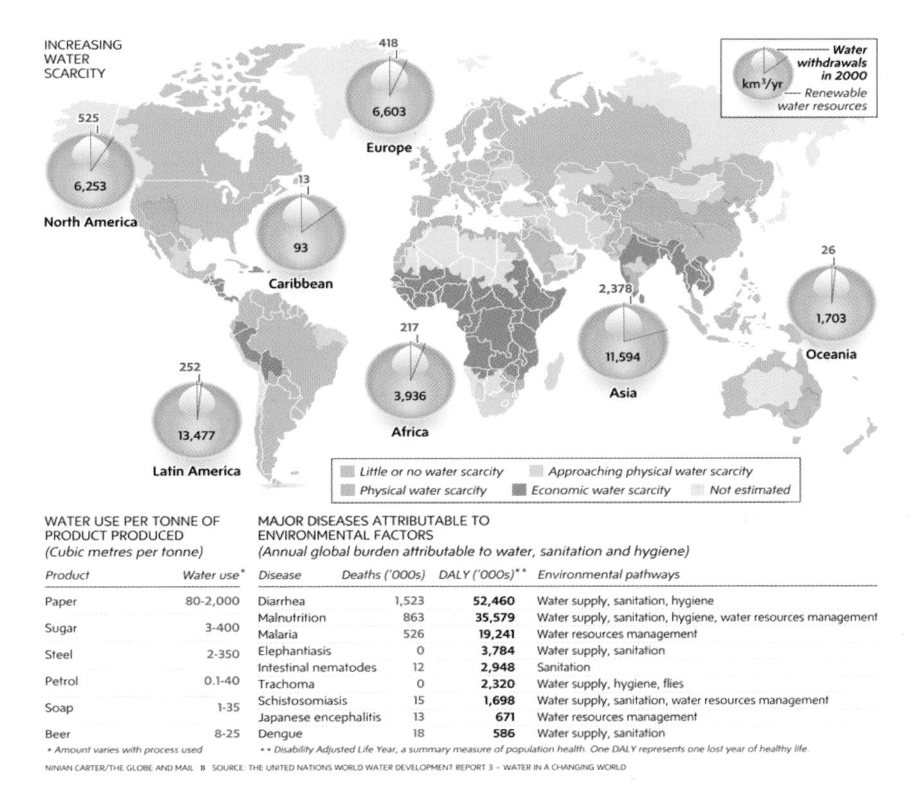

WATER USE PER TONNE OF PRODUCT PRODUCED
(Cubic metres per tonne)

Product	Water use*
Paper	80-2,000
Sugar	3-400
Steel	2-350
Petrol	0.1-40
Soap	1-35
Beer	8-25

* Amount varies with process used

MAJOR DISEASES ATTRIBUTABLE TO ENVIRONMENTAL FACTORS
(Annual global burden attributable to water, sanitation and hygiene)

Disease	Deaths ('000s)	DALY ('000s)**	Environmental pathways
Diarrhea	1,523	**52,460**	Water supply, sanitation, hygiene
Malnutrition	863	**35,579**	Water supply, sanitation, hygiene, water resources management
Malaria	526	**19,241**	Water resources management
Elephantiasis	0	**3,784**	Water supply, sanitation
Intestinal nematodes	12	**2,948**	Sanitation
Trachoma	0	**2,320**	Water supply, hygiene, flies
Schistosomiasis	15	**1,698**	Water supply, sanitation, water resources management
Japanese encephalitis	13	**671**	Water resources management
Dengue	18	**586**	Water supply, sanitation

** Disability Adjusted Life Year, a summary measure of population health. One DALY represents one lost year of healthy life

FIGURE WO-1 Population growth, climate change, reckless irrigation, and chronic waste are placing the world's water supplies in danger.
SOURCE: Reprinted from Mittelstaedt (2009) with permission from *The Globe and Mail*; based on UN (2009).

and sanitation infrastructure so as to reduce the risk of water-related infectious disease.[1]

Some topics important to water quality and health were either not covered at the workshop, covered only in passing, or were explored in greater detail in other National Research Council (NRC) reports. These topics included desalination,[2] bioterrorism,[3] conflicts over water and the implications for global security,[4]

[1]For a discussion about Bradley's four categories of water-related disease (water-borne, water-washed, water-based, and water-related insect vectors), see page 18 and his paper in Chapter 1.

[2]See NRC (2004a,b,c, 2005, 2006, 2007, 2008a,b).

[3]See NRC (2002).

[4]See *Running Dry* DVD (enclosed) and MacPherson in IOM (2008).

pharmaceuticals,[5] heavy metals,[6] and issues related to runoff from farms and pollution of water supplies.[7]

Organization of the Workshop Summary

This workshop summary was prepared for the Forum membership by the rapporteurs and includes a collection of individually authored papers and commentary.[8] Sections of the workshop summary not specifically attributed to an individual reflect the views of the rapporteurs and not those of the Forum on Microbial Threats, its sponsors, or the Institute of Medicine (IOM). The contents of the unattributed sections are based on the presentations and discussions at the workshop.

The workshop summary is organized into chapters as a topic-by-topic description of the presentations and discussions that took place at the workshop. Its purpose is to present lessons from relevant experience, to delineate a range of pivotal issues and their respective problems, and to offer potential responses as discussed and described by the workshop participants.

Although this workshop summary provides an account of the individual presentations, it also reflects an important feature of the Forum's philosophy. The workshop promotes a dialogue among representatives from different sectors and allows them to present their beliefs about which areas may merit further attention. The reader should be aware, however, that *the material presented herein expresses the views and opinions of the individuals participating in the workshop* and not the deliberations and conclusions of a formally constituted IOM consensus study committee. These proceedings merely summarize the statements of participants at the workshop and are not intended to be an exhaustive exploration of the subject matter nor a representation of a consensus evaluation.

Global and Grassroots Perspectives

The workshop opened with a screening of the film *Running Dry* (Thebaut, 2005; CD included on the inside front cover of report volume), introduced and followed by remarks from its writer, producer, and director, James Thebaut. The documentary explores the growing global water crisis and its staggering toll of some 14,000 "quiet preventable deaths" per day. Focusing on China, the Middle East, Africa, India, and the United States, *Running Dry* presents compelling argu-

[5]See Davies in IOM (2009).

[6]See NRC (1993a, 2008c).

[7]See NRC (1993b); see also Caravati et al. in Chapter 4.

[8]Speakers Mark Sobsey, Thomas Clasen, and Vahid Alavian did not submit manuscripts for this summary report. To ensure that their contributions to this meeting were captured in this summary report we have supplemented the overview section of the chapter in which their material would have appeared.

TABLE WO-1 Estimation of Mortality Due to Diarrhea in India

Crude death rate (India, rural) *Sample Registration System Bulletin*. 2001; 32	9.3 per 1000 population
Total number of deaths	6,897,441
Total deaths in 0-6 years*	1,517,437 (22% of total rural deaths)
Total deaths in >6 years*	5,380,004 (78% of total rural deaths)
Crude death rate (India, urban) *Sample Registration System Bulletin*. 2001; 32	6.3 per 1000 population
Total number of deaths	1,797,736
Total deaths in 0-6 years*	221,122 (12.3% of total urban deaths)
Total deaths in >6 years*	1,576,614 (87.7% of total urban deaths)
Total deaths (all ages; rural + urban)*	8,695,177
Total 0–6 years deaths (rural + urban)*	1,738,559
Proportionate mortality due to diarrhea (all ages)*	5.23% [SBHI, 2002]
Total diarrheal deaths (all ages)	454,758
Proportionate mortality due to diarrhea* (0-6 years)	9.1% [SBHI, 2002]
Total diarrheal deaths among 0-6 years*	158,209
Total diarrheal deaths among 6+ years*	296,549

*Age-specific death rates: Sample Registration System, 1998.
NOTE: The estimated total deaths due to diarrhea are less than the estimation of 576,480 deaths by Zaidi et al. (2004).
SOURCE: National Institute of Cholera and Enteric Diseases, Kolkata (NICED, 2005) with permission from the Ministry of Health and Family Welfare, Government of India.

ments for international cooperation on water issues and highlights some promising grassroots programs to improve access to safe water (see Chapter 1).

In China and India, rapid economic expansion has intensified demand for increasingly polluted water. China, the film notes, contains more than one-fifth of the world's people but only seven percent of its fresh water (Thebaut, 2005). Industrial consumption of water drains the storied Yellow River to such an extent that in drought years, it runs dry. Seventy percent of China's rapidly growing cities lack a sewage treatment plant, and agricultural and industrial waste pollutes the country's major reservoirs.[9] Another fifth of the world's population—and approximately half of the world's poor—live in India, where more than 100 cities release their untreated human, animal, and industrial wastes directly into the sacred river Ganges, transforming it into an open sewer. This entirely preventable environmental catastrophe, coupled with widespread groundwater contamination, undoubtedly contributes to India's heavy burden of death (Table WO-1) and disability (Table WO-2) from diarrheal disease.

Rampant over-consumption and misuse of water occurs in the United States, with consequences that reach well beyond our borders. Despite the existence of a

[9]For a good summary article, see http://www.dailywealth.com/archive/2008/aug/2008_aug_11.asp and http://www.msnbc.msn.com/id/17704190/.

TABLE WO-2 Data Used for Estimation of Burden Due to Diarrhea in India

Indices	Current Values (2001)	Projected Values 2001-2006	2006-2011	2011-2016
Total Population (in crore*)	102.7	109.41 ('06)	117.89 ('11)	126.35 ('16)
Life expectancy at birth (years)				
Male	62.30 (projected)	63.87	65.65	67.04
Female	65.27 (projected)	66.91	67.67	69.18

*Crore (Hindi: करोड़) (often abbreviated cr) is a unit in the Indian numbering system and was formerly a unit in the Persian numbering system, still widely used in Bangladesh, India, Maldives, Nepal, Pakistan, and Sri Lanka, and formerly in Iran. An Indian crore is equal to 100 lakh or 10,000,000.
SOURCE: National Institute of Cholera and Enteric Diseases, Kolkata (NICED, 2005) with permission from the Ministry of Health and Family Welfare, Government of India. Data retrieved from the Registrar General of India, 1996.

bilateral treaty—the U.S.-Mexico Water Treaty of 1944[10]—that stipulates that the two countries will share the Colorado River's waters, the demands of upstream users of the Colorado River—a primary source of water for seven states in the western United States—are now so great that its waters rarely reach the Sea of Cortez in Mexico (Cohen and Henges-Jeck, 2001). This is but one example among the growing number of social, political, and economic conflicts arising over access to water, according to Peter Gleick, cofounder and president of the Pacific Institute for Studies in Development, Environment, and Security (Gleick, 2001; Thebaut, 2005).

Where political tensions already exist—as in the arid Middle East—competition for and access to clean water, a natural resource more valuable than oil,[11] may intensify them. On the other hand, as both Shimon Peres, former Prime Minister of the State of Israel, and Nabil Sharif, Chairman of the Palestinian Water Authority, observed in *Running Dry*, the process of making policy to meet water needs may also offer adversaries an opening for resolving other conflicts. Extended to a global level, the necessity for international cooperation on water issues may thus be viewed as an opportunity for regional conflict resolution.

Water and Health in Africa

Africa poses particular challenges to providing safe, accessible water for its rapidly growing population. Although the continent, particularly in the Congo Basin, possesses abundant water resources, the majority of Africans lack *access* to safe water, primarily as a consequence of poverty and armed conflicts (UNICEF,

[10]For more information on this treaty, please see http://www.usbr.gov/lc/region/g1000/pdfiles/mextrety.pdf.

[11]Urban Water Conference, http://www.urbanwaterconference.be/db (accessed March 13, 2009).

2006). Outbreaks of cholera and other water-related diseases have been frequent occurrences, affecting the health and well-being of thousands of individuals. In sub-Saharan Africa, water resources are scarce and water availability may be seasonal. According to the World Health Organization (WHO) and United Nations Children's Fund (UNICEF) Joint Monitoring Programme for Water Supply and Sanitation (JMP), 28 percent of the population of sub-Saharan Africa defecates in the open, and an additional 23 percent use "unimproved" sanitation facilities that "do not ensure hygienic separation of human excreta from human contact" (JMP, 2008).

Moreover, even where clean water and flush toilets are available in Africa, lack of hygiene awareness continues to result in outbreaks of water-related diseases (Thebaut, 2005). Clearly, in order to benefit from advances in sanitation, people must appreciate the connection between water, sanitation, and health—a link that keynote speaker Donald Hopkins, Vice President of Health Programs at the Carter Center in Atlanta, Georgia, and his colleagues have tried to forge at the grassroots level in African communities (see Hopkins in Chapter 1). His description of two such programs with very different outcomes—one addressing trachoma in Ethiopia and the other targeting dracunculiasis (Guinea worm disease) in Ghana—revealed the importance of social factors as both catalysts and barriers to efforts to improve health in under-resourced communities by improving sanitation.

Trachoma in Ethiopia Trachoma, a chronic infection of the cornea[12] and conjunctiva[13] caused by the bacterium *Chlamydia trachomatis*, is the world's leading preventable cause of blindness (Hopkins et al., 2008). Ten percent of the global population is considered to be at risk for developing this disease, which disproportionately affects females. Trachoma is transmitted through multiple routes, including contaminated fingers, flies, and fomites[14] such as dirty face cloths (Figure WO-2). The disease is currently managed through a multipronged strategy: surgery to prevent blindness in people with severe infections; antibiotics to fight infection in its early stages; educating people about the importance of proper face-washing to prevent the accumulation of discharge around the eyes; and environmental interventions such as improved sanitation to reduce populations of flies that spread the disease, and which breed mainly in human feces (Emerson et al., 2006; Hopkins, 2008; Hopkins et al., 2008).

[12]The transparent, dome-shaped window covering the front of the eye (http://www.stlukeseye.com/anatomy/cornea.asp).

[13]The thin, transparent tissue that covers the outer surface of the eye. It begins at the outer edge of the cornea, covering the visible part of the sclera, and lining the inside of the eyelids. It is nourished by tiny blood vessels that are nearly invisible to the naked eye (http://www.stlukeseye.com/Anatomy/Conjunctiva.asp).

[14]Inanimate objects or substances that can transmit infectious organisms from one host to another.

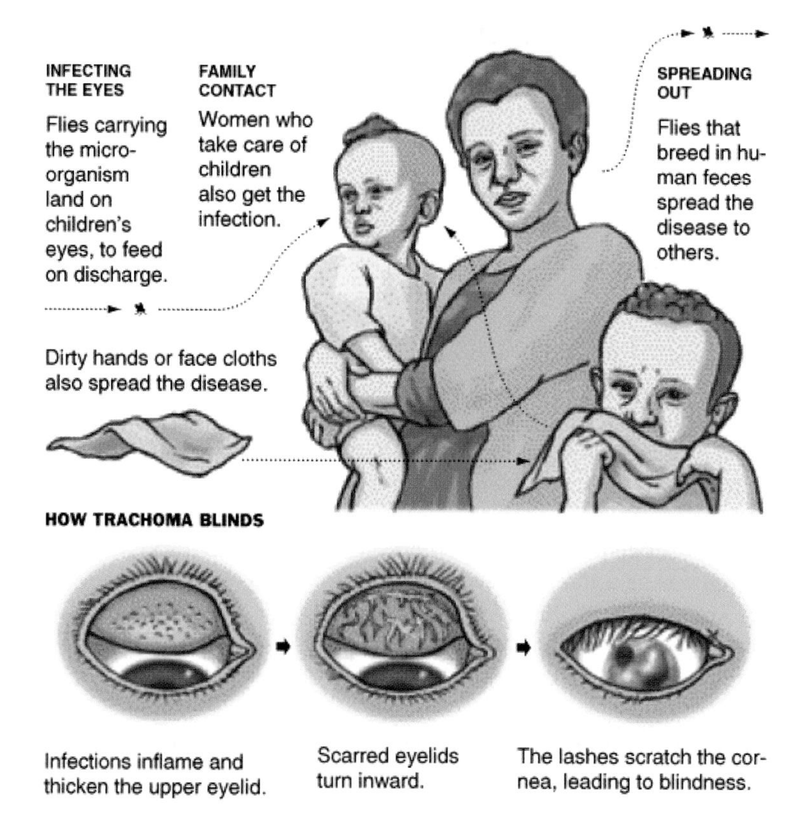

FIGURE WO-2 The life cycle of trachoma.
SOURCE: Reprinted from Dugger (2006). Copyright 2006 New York Times Graphics.

Of the approximately 50 countries where trachoma is endemic, Ethiopia is thought to have the largest number of cases; about one-third of these occur in the country's impoverished Amhara region (Hopkins, 2008). Because this region's geology and ecology favored the construction of latrines from abundant wood, Amhara provided a promising target for sanitation-based interventions to combat trachoma, Hopkins explained. Working with the Amhara Regional Health Bureau, and with local Lions Clubs, the Carter Center mobilized residents to build latrines. "There came to be a competition between villages, between families, in who could build the better latrine or the faster latrine," he recalled, with much of the momentum provided by women. They welcomed the project because women without access to latrines were self-described "prisoners of the daylight" due to cultural taboos against women defecating in the open during daylight hours, when they might be seen by a man.

"This feminist aspect of this problem did not become apparent until we began this intervention . . . [and] it was not primarily to prevent trachoma," Hopkins observed. Women's demand for convenience provided "enormous energy" for the project, he said, and it was efficiently harnessed by organizing tightly knit kin groups to perform work that benefited their relatives, leaving little room for corruption. As a result, the program surpassed its initial goal to build 10,000 latrines in Ethiopia in 2004 by more than eightfold, and has continued to build a cumulative total of more than 600,000 latrines, as of mid-2008 (Figure WO-3). Even with subsequent declines due to political unrest and a focus on other diseases by the Carter Center, Ethiopia is well on its way to meeting the Millennium Development Goal (MDG; Box WO-1) for providing latrines to at least half of the population by 2015 that did not have latrines in 2000, he reported. Existing latrines in Ethiopia, and also those constructed in a similar project in Niger, are being used and in some cases upgraded, Hopkins reported. "The most important thing that has happened [as a result of this initiative] has been, not so much the physical act of building the latrine, but the change in the mindset," he concluded. The behavioral foundations of this success should be studied so it can be replicated elsewhere, he added: "What we haven't done, and can't really afford to do, is to get some proper anthropologists and others to go into that area and really understand, better than we do, what happened there, to see how that can be applied more broadly, perhaps even beyond Ethiopia."

Dracunculiasis in Ghana Dracunculiasis[15] is caused by the nematode *Dracunculus medinensis* and is limited to humans, who typically ingest the parasite in water containing copepods[16] that carry the worm's larvae (Hopkins et al., 2008). After a year-long asymptomatic incubation period, worms as long as one meter emerge through the skin on any part of the host's body, causing extreme pain that, along with frequent secondary bacterial infection, typically incapacitates patients for two to three months (see Figure WO-6).

Half the population of a village may simultaneously experience these symptoms, drastically reducing school attendance and agricultural productivity. While there is no immunity to or cure for dracunculiasis, Hopkins noted, the disease could be eradicated by providing safe drinking water to vulnerable populations—a goal approached over the course of the past two decades, during which the number of infected individuals has declined from 3.5 million to 10,000 (Hopkins et al., 2008; Figures WO-7 and WO-8). Interventions included preventive health education and instruction in the use of cloth filters to remove copepods from drinking water; preventing contamination of surface water by

[15]Dracunculiasis, or Guinea worm disease, is the only infectious disease that is caused exclusively by ingestion of contaminated drinking water.

[16]Copepods are a type of crustacean which may live in both salt and freshwater. They are one of the most abundant animals on the planet (see http://jaffeweb.ucsd.edu/pages/celeste/Intro/index.html).

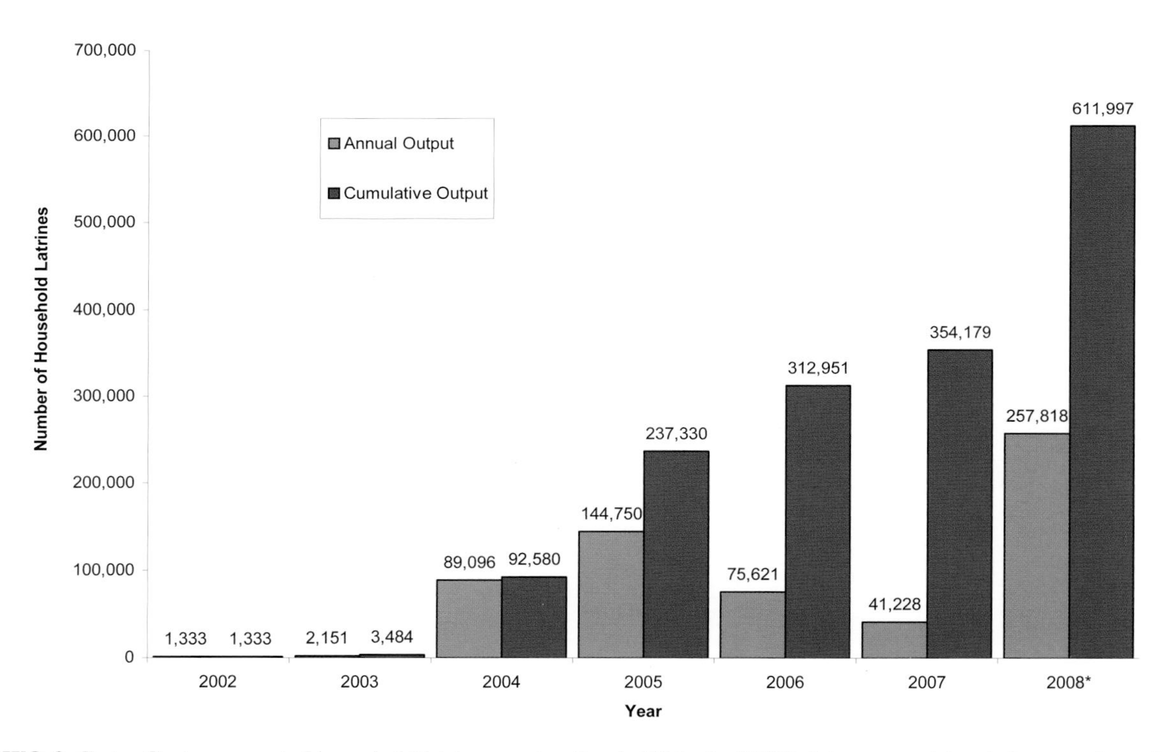

FIGURE WO-3 Carter Center-supported household latrine construction in Ethiopia. *2008 data are provisional, January-August.
SOURCE: Courtesy of The Carter Center.

BOX WO-1
Millennium Development Goals

The Millennium Development Goals (MDGs) are based on the actions and targets contained in the Millennium Declaration, which was adopted by 189 nations and signed by 147 heads of state and governments during the UN Millennium Summit in September 2000 (UNDP, 2009). In response to the world's primary development challenges, the MDGs are comprised of eight goals to be achieved by 2015. On water and sanitation specifically, the MDGs aim to halve the proportion of people without sustainable access to safe drinking water and basic sanitation. Basic sanitation is defined as access to, and use of, excreta and sullage* disposal facilities and services that provide privacy and dignity while at the same time ensuring a clean and healthful living environment both at home and in the immediate neighborhood of users (UN Millennium Project, 2005).

The indicators of progress toward this target are

- the proportion of population with sustainable access to an improved drinking water source, urban and rural (Figure WO-4); and
- the proportion of the urban and rural population with access to improved sanitation (Figure WO-5).

continued

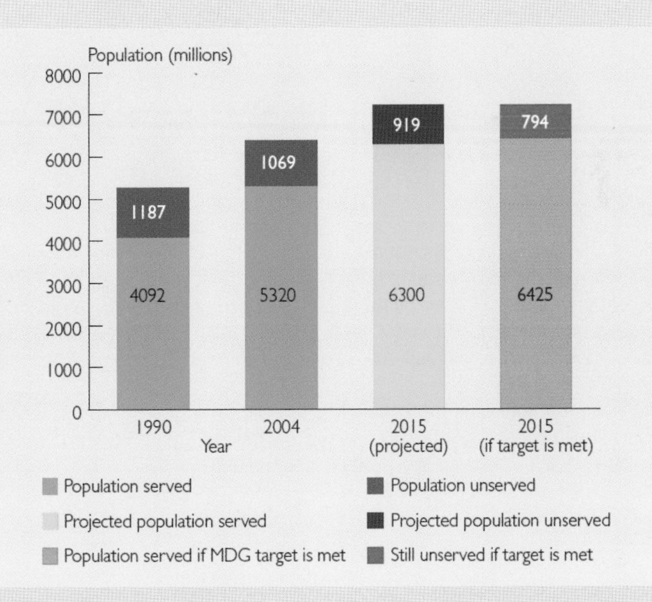

FIGURE WO-4 World population with and without access to an improved drinking water source in 1990, 2004, and 2015.
SOURCE: Reprinted with permission from WHO/UNICEF (2006).

BOX WO-1 Continued

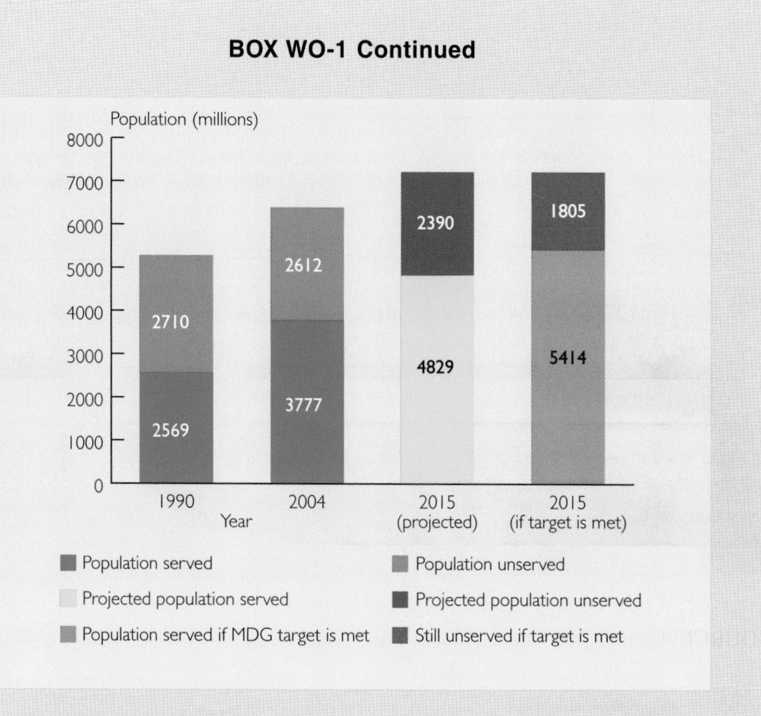

FIGURE WO-5 World population with and without access to improved sanitation in 1990, 2004, and 2015.
SOURCE: Reprinted with permission from WHO/UNICEF (2006).

There are two main challenges to achieving the MDG's drinking water and sanitation target: the rapid pace of urbanization requires a major effort to maintain the current coverage levels and a huge number of rural people do not have basic sanitation and safe drinking water, which calls for an intensive mobilization of resources to reduce the vast gap in coverage between urban and rural populations (WHO/UNICEF, 2006).

The following figures show the world's population with and without access to improved drinking water (Figure WO-4) and sanitation (Figure WO-5) in 1990 and 2004, respectively, and what it is projected to be in 2015.

*Sullage (or grey water) is dirty water from the laundry, kitchen, and bathroom. Grey water contains chemicals such as dish detergent and soap as well as fats, grease, and whatever washes off our body while bathing. Sullage does not usually contain sewage but can be equally contaminated and can cause infections (Northern Territory Government, Department of Health and Families, 2009).

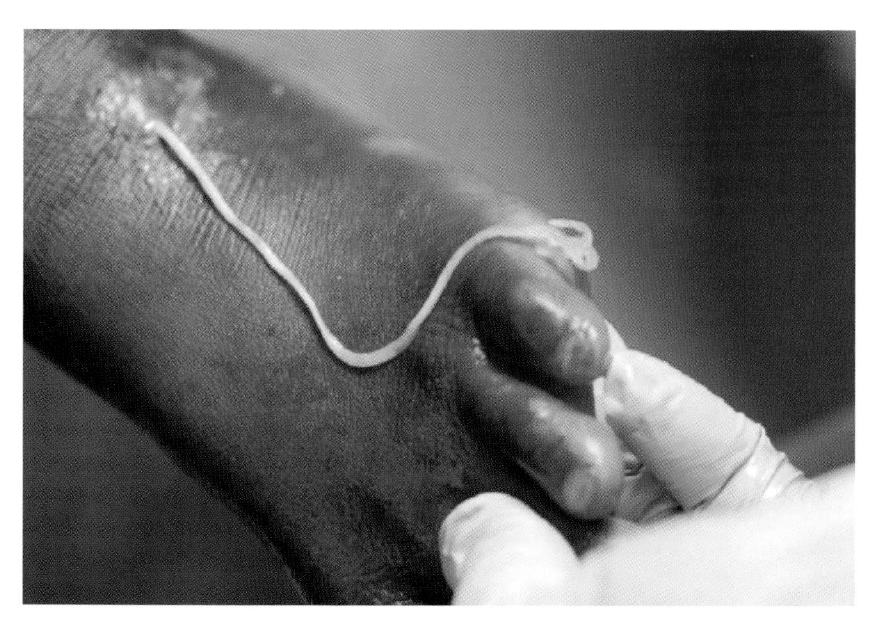

FIGURE WO-6 Guinea worm disease.
SOURCE: Courtesy of The Carter Center/L. Gubb.

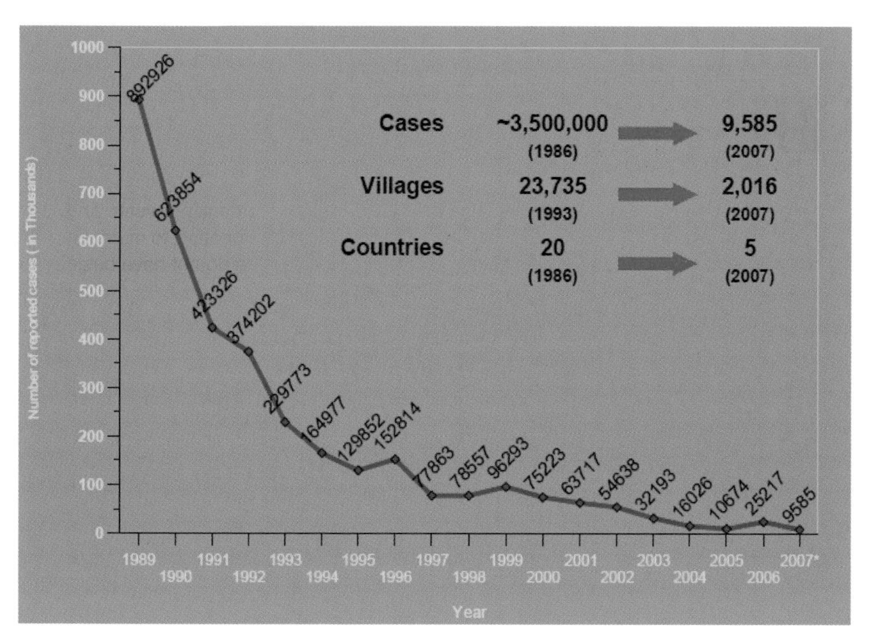

FIGURE WO-7 Number of reported cases of dracunculiasis by year, 1989-2007.
SOURCE: Courtesy of The Carter Center.

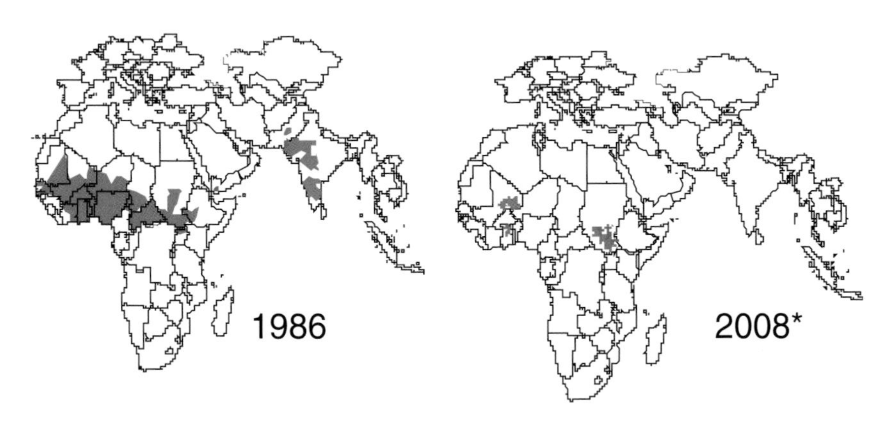

FIGURE WO-8 Guinea worm reduction over time.
SOURCE: WHO Collaborating Center for Research, Training, and Eradication of Dracunculiasis, CDC.

infected persons discharging larvae from skin lesions; treating water reservoirs with a mild insecticide that kills copepods but does not harm humans, fish, or vegetation; and, most effective of all, providing clean water from underground borehole wells.

Despite the overall success of the dracunculiasis eradication campaign, some countries, including Ghana, continue to report thousands of cases per year, Hopkins reported. The highest rates of disease occur in the impoverished north of that country, where access to safe water is extremely limited due to scarce rainfall and geology that makes well-drilling difficult. As a result, residents are often forced to drink contaminated surface water. Hopkins observed, however, that the greatest obstacles to progress against dracunculiasis in northern Ghana were sociopolitical in nature. The country has been torn by often violent conflict among its several constituent ethnic groups. At the same time, official corruption at all levels of government and administrative "red tape" have rendered funded programs ineffective.

Because dracunculiasis is not fatal, and because its victims are mainly poor rural residents distant from the country's southern power base, the political will to eradicate this disease is often lacking. Recalling his efforts to gain cooperation from the Ghanian government to address dracunculiasis in their country, Hopkins remarked that while he expected to encounter political fraud and corruption, he "was not prepared for the kind of indifference of people in agencies, in ministries, and the lack of a sense of urgency in getting help to these people, who were growing their food out in those rural areas." Today, wells dug in northern Ghana through the efforts of the Carter Center and other nongovernmental organizations are largely useless, having fallen into disrepair or because they were inappropriately sited, he reported. Cloth filtration of water remains the best source of protection against dracunculiasis in northern Ghana. Since this method of water "purification" fails to remove bacterial or viral contaminants, it is far from optimal. He observed that "if it's bad for children to have diarrhea, it's worse for them to have diarrhea and Guinea worm disease."

In reviewing the challenges and successes of the programs in Ethiopia and Ghana, Hopkins identified the following four lessons learned:

- Priorities for clean water and sanitation—and thereby, health—must incorporate considerations of economics, politics, and geology.
- Fraud, corruption, and indifference must be challenged in order to implement interventions to improve access to clean water.
- Outcomes of water and sanitation interventions must be monitored to gauge and support their effectiveness over the long term.
- It is better to implement limited interventions immediately than delay action until perfection is achievable.

Transmission and Prevention of Water-Related Diseases

The range of water-related microbial infectious diseases is vast, encompassing pathogens transmitted by diverse—and often nonexclusive—routes. While water quality affects transmission rates of many water-related diseases, water availability (or lack thereof) also plays a significant role in the spread of infection. Changes in water flow or quality, which can influence the population dynamics of vector species that transmit infectious diseases and intermediate hosts for microbial pathogens, also influence the prevalence and transmission dynamics of infectious diseases. Workshop presentations demonstrated how an analysis of transmission processes guides disease prevention efforts, and how such analyses may reveal the importance of the household as a target for clean water interventions.

Classification of Disease Transmission Processes

The first effort to classify the various routes by which water-related diseases may be transmitted, and thereby to relate transmission processes to potential interventions for disease reduction, was undertaken by workshop presenter David Bradley, of the London School of Hygiene and Tropical Medicine, and coworkers (White et al., 1972). In order to determine how best to reduce the burden of water-related disease using the limited resources available in East Africa in the 1960s, the researchers attempted "to disaggregate the various components of water-related disease and the way in which water affected them," Bradley said (see Bradley in Chapter 1). They described four key categories of disease transmission processes:[17]

- ***Water-borne***: The pathogen is acquired through consumption of contaminated water, as occurs in diarrheal diseases, dysenteries and typhoid fever.
- ***Water-washed***: The pathogen is spread from person to person due to lack of water for hygiene, as occurs in diarrheal diseases, scabies, and trachoma.
- ***Water-based***: The pathogen is transmitted to humans through contact with and infection, multiplication in, and excretion from aquatic intermediate hosts, as occurs in the diseases schistosomiasis and dracunculiasis (Guinea worm).
- ***Water-related insect vectors***: The pathogen is carried and transmitted by insects that breed in or bite near water, as occurs in dengue fever, malaria, and trypanosomiasis (sleeping sickness).

This classification scheme structures and clarifies information critical to effectively target interdisciplinary intervention efforts to the health effects of water and sanitation, which tend to be planned and executed by engineers, Bradley explained. Where once such efforts were limited almost exclusively to providing households with piped water, recognition of the importance of water-washed diseases led to an appreciation that, under some circumstances, simply making more water available—even without improving its quality—could provide significant health benefits. Bradley also adapted this classification scheme in order to estimate the relative burden of disease attributable to each type (Cairncross and Valdmanis, 2006); a modified version of the now well-known Bradley Classification of Water-Related Infections provides a framework for describing the spectrum of water-related diseases in Box WO-2.

Bradley and coworkers have also identified six sanitation-related disease transmission categories, based on a combination of three key characteristics: latency of infectivity[18] following excretion, pathogen persistence[19] in the environment, and the ability of pathogens to multiply[20] in the environment (see Table WO-3). Various combinations of traits require different types of sanitary

[17]Bradley has since identified a fifth category of water-related disease transmission processes—water aerosols—which transmit respiratory pathogens such as *Legionella* (see Chapter 1).

[18]Bradley defines "latency" as the length of time until a pathogen becomes infectious after being released from its host. It has nothing to do with infection in the human host (enterobiasis/pinworm, for example, takes 2-6 weeks of development in the human host before infectious eggs can be excreted, but eggs are infectious as soon as excreted; see Chapter 1).

[19]Bradley defines "persistence" as the capacity to survive outside the human body (see Chapter 1).

[20]By "multiply," Bradley is referring to whether an organism multiples in excreta or in the environment (see Chapter 1).

BOX WO-2
Spectrum of Water-Related Diseases

Water-related infectious diseases are typically classified into the following four groups, based on their primary routes of transmission, in order to connect water supply variables with the burden of disease (Cairncross and Feachem, 1993). These categories are not exclusive, however, as many water-related diseases are spread through multiple routes of exposure.

Waterborne Diseases

Waterborne diseases may result when pathogenic organisms including but not limited to viruses (hepatitis A and hepatitis E), parasites (e.g., giardia, cryptosporidium), and bacteria (e.g., *Shigella, Campylobacter jejuni, Escherichia coli, Salmonella spp.*, and *Vibrio cholerae*) present in feces or urine from human and animal waste contaminate water supplies, and this water is subsequently used for drinking or food preparation without adequate treatment. These are largely diarrheal diseases, which as a group cause more than 1.5 million deaths per year, of which more than 90 percent occur in children under the age of 14 (Prüss-Üstün et al., 2008).

Water-Washed Diseases

Person-to-person transmission of certain skin and eye infections is favored by inadequate hygiene conditions, which frequently result from lack of access to water (Cairncross and Valdmanis, 2006). These water-washed diseases include shigellosis, a bacterial infection affecting the intestinal tract; trachoma (discussed previously and in Chapter 1); and scabies, a skin infection caused by the microscopic mite *Sarcoptes scabei.*

Water-Based Diseases

Aquatic intermediate hosts are involved in the transmission of water-based diseases such as schistosomiasis and dracunculiasis (Guinea worm; discussed previously and in Chapter 1). Schistosomiasis infects an estimated 200 million people worldwide and causes 280,000 deaths per year in sub-Saharan Africa alone (van der Werf et al., 2003). Although associated with inadequate sanitation, schistosomiasis is limited to environments favorable to the snail in which the larval pathogen develops into a stage that is released from the snail into fresh water and can penetrate intact human skin. Schistosomiasis was introduced into both Mauritania and Senegal when the damming of the Senegal River created a less salty environment that allowed snails to flourish (WHO, 2008c).

continued

BOX WO-2 Continued

Water-Related Vector-Borne Disease

Water-related vector-borne diseases include mosquito-borne diseases such as malaria, which causes more than 500 million cases of severe illness and 100 million deaths per year (WHO, 2008e), and dengue fever. Dengue infects approximately 50 million people per year, a small percentage of whom develop dengue hemorrhagic fever, a severe and sometimes fatal form of the infection (WHO, 2008a). Flies that breed in water transmit the parasitic worm that causes onchocerciasis (river blindness) (WHO, 2008f); the tsetse fly, which bites near water, spreads African trypanosomiasis (sleeping sickness) (WHO, 2008b).

barriers in order to prevent disease transmission, as shown in Figure WO-8. Bradley said, for example, that while some transmission processes can be interrupted by providing latrines, this will not reduce transmission of "water-washed" pathogens that spread directly from person to person in the absence of adequate personal hygiene. In a given setting, preventive measures at the level of the individual (e.g., face washing), the household (e.g., latrine construction), or the community (e.g., sewage treatment) may be most effective, depending upon the sanitation-related transmission processes contributing to the local disease burden.

As Bradley discusses in his contribution to Chapter 1, the categories are related to the relative efficacy of various sanitary interventions. Human hygienic behavior (and particularly hand washing with soap) is most effective in the first two categories of disease transmission, the sanitary infrastructure affects categories II-IV the most, and the last two categories are most dependent on rather specific measures unless the general sanitary standard is relatively high.

These concepts, depicted in the cartoon for Figure WO-9, illustrate the anthropogenic components of human hygiene behaviors that are relevant to breaking the cycle of disease transmission and targeted prevention efforts.

To further advance the understanding of disease processes provided by water- and sanitation-based transmission classifications, Bradley proposed the systematization of two additional transmission-related variables: hygiene behavior and spatial components of water and sanitation services. In the first case, echoing Hopkins' earlier remarks, Bradley advocated an anthropological analysis of human behaviors relevant to disease transmission and prevention (Figure WO-9). In the second, he recommended the analysis of disease control on multiple spatial scales using geographical information systems, based on his observation that effective interventions to address water- and sanitation-related infectious diseases are often

TABLE WO-3 Excreta-Related Transmission

Type	Latency	Persistence	Multiplication	Biology, predominantly:	Examples	Routes of entry and egress*
I	No	Short	No	Viruses, Protozoa, Helminths	*Enterobius vermicularis,*	or, pa
					Rotavirus	or, fe
II	No	Longer	Yes	Bacteria	*Salmonella typhi,*	or, ur/fe
					Leptospira interrogans	pc, ur
III	Yes	Long	No	Helminths	*Ascaris lumbricoides,*	or, fe
					Necator americanus	pc, fe
IV	Yes	[Long]	In cow/pig	Helminths	*Taenia solium, T*	or, fe
					aenia saginata	or, fe
V	Yes	[Long]	In aquatic organisms	Helminths	*Schistosoma haematobium,*	pc, ur
					Schistosoma mansoni	pc, fe
VI	Spread by excreta-related insects			Some mosquitoes		

*For entry: or = oral; pc = percutaneous or through mucosa; for egress: fe = in feces; pa = perianal region; ur = in urine.
SOURCE: David Bradley.

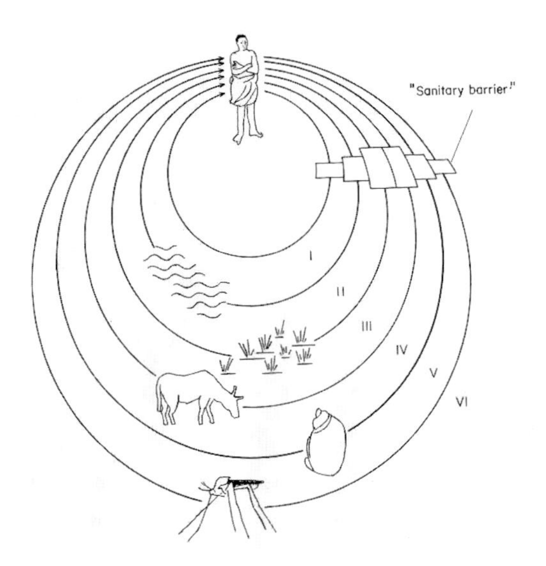

FIGURE WO-9 Length and dispersion of transmission cycles of excreted infections.
SOURCE: Reprinted from Feachem et al. (1983) with permission from the International Bank for Reconstruction and Development/The World Bank.

implemented at the local level but can achieve significant impact when replicated, with appropriate adaptations, in large numbers of other communities.

Disease Prevention at the Household Level

Despite significant progress over the past three decades in extending coverage in low-income countries, approximately one billion people lack access to improved water supplies, and many more rely on sources that are microbiologically contaminated (Clasen, 2008). While the provision of safe, piped water to every home is a distant goal, household-based water treatment and storage interventions represent important interim solutions for the prevention of water-related disease. Speaker Thomas Clasen, of the London School of Hygiene and Tropical Medicine, noted that a systematic review and meta-analysis of 57 studies comparing the bacterial content of drinking water at its source with water stored in the home (Wright et al., 2004) found that in many settings, the bacteriological quality of drinking water declined significantly after collection. The authors concluded, therefore, that policies that aim to improve water quality at the source may be compromised unless they are accompanied by corresponding measures to ensure safe household water storage and treatment.

The profound disease burden attributed to diarrhea makes it the most important target for waterborne disease prevention, Clasen said. Following respiratory infections diarrheal diseases are the second leading cause of deaths from infectious disease, and the second leading cause of childhood mortality, exceeded only by neonatal conditions (Black et al., 2003). Following a systematic review of intervention trials to improve water quality, Clasen and his coauthors concluded that household-based interventions were nearly twice as effective as source-based measures for preventing diarrheal disease (Clasen et al., 2007b).[21] Similarly, workshop speaker Pete Kolsky, of the World Bank, and coauthors concluded that "the householder's perspective and priorities are similar to those that emerge from a public health perspective. As most of the victims of poor environmental health are children under five, it makes sense to focus attention on where they spend the most time, which is at home" (Bostoen et al., 2007).

Workshop presentations by Clasen and Robert Tauxe, of the Division of Foodborne, Bacterial and Mycotic Diseases at the U.S. Centers for Disease Control and Prevention (CDC), described a range of household-based (also known as point-of-use) strategies to improve water quality and thereby reduce the burden of water-related disease. Clasen (see Chapter 4) noted that the most common method of household water treatment is boiling. Although highly effective in reducing microbiological contamination, boiled water can be readily recontaminated. Furthermore, the physical process of boiling water is relatively costly in terms of time and energy, is associated with a small but measurable risk for burn accidents, and contributes to indoor air pollution as well as carbon emissions depending on the carbon source(s) used as fuel to boil water (Clasen, 2008; Clasen et al., 2008). Despite these shortcomings, the widespread use of boiling indicates good potential for the adoption of household water treatment methods that are more effective, more convenient, less costly, more appealing, less hazardous, and more sustainable.

Tauxe described strategies to improve the quality of the water obtained by individual households for drinking, washing, and food preparation, and to maintain its purity during storage in the home (see Tauxe et al. in Chapter 1). He and coworkers have studied the specific health effects of a variety of interventions to make unsafe water safer to drink, as well as ways to increase the uptake of these interventions through social marketing. Their combined water chlorination and storage "safe water system," sold for a low price in Africa under the brand name WaterGuard®,[22] has proven to be scalable, sustainable, and measurable in its health benefits, Tauxe reported. In a study conducted in households in Uganda containing at least one member with HIV, use of the safe water system for one year was associated with a 20 percent reduction in diarrheal episodes among all family members, and a 25 percent reduction among those with HIV (Lule et al.,

[21] Handwashing with soap and water is important for the prevention of diarrheal disease and respiratory infections.

[22] WaterGuard® is the chlorine solution made in some countries for water purification. The "safe water system" involves safe water storage combined with health education.

2005). This finding, coupled with the results of 20 additional published studies on point-of-use water chlorination, provides strong evidence that this intervention reduces the risk of childhood diarrhea, among other health benefits (Arnold and Colford, 2007).

Tauxe observed that in addition to providing measurable protection against diarrheal diseases, safe water programs can serve as a springboard for additional public health interventions to tackle water-related infectious diseases. Combined initiatives promoting hand washing and the use of the safe water system, conducted in maternal and child health clinics and in schools in both Kenya (O'Reilly et al., 2008; Parker et al., 2006) and China (Bowen et al., 2007), have led to measurable improvements in hygiene behavior, as well as in health measures such as reduced school absenteeism. Moreover, researchers found that children in school-based hand hygiene and safe water programs positively influenced their parents' behavior. "The parents reported that they picked up more WaterGuard® [and used it] . . . more often," Tauxe said. "There was more soap in the home, and there was a lot of hand washing going on." Based on these successes, similar interventions are currently being expanded to serve 1,500 Kenyan schools and are being assessed in schools in Pakistan and the Philippines.

The availability of safe water also supports a safe food supply. At the local level, Tauxe reported, safe water systems have been shown to reduce contaminants such as *E. coli* in food prepared in the home and by street vendors. However, as global markets expand, the quality of water used to grow, harvest, and process produce and other foodstuffs for worldwide export and consumption also has increasingly far-reaching consequences. These risks—which were explored in-depth in a recent Forum workshop summary report entitled *Addressing Foodborne Threats to Health* (IOM, 2006)—are illustrated by recurrent outbreaks of foodborne illnesses traced to imported produce, Tauxe noted. To address this situation, Tauxe suggested that companies importing produce into the United States should be compelled to meet water quality standards "both in the packing sheds and in the workers' homes." Yet, in light of recent foodborne disease outbreaks linked to produce grown in the United States—spinach laced with *E. coli* O157:H7 (Calvin, 2007) and *Salmonella* in alfalfa sprouts (CDC, 2009)—coupled with the observation that "sanitary standards at U.S. food production facilities are vastly inferior to others in foreign sites" (Osterholm, 2006), it appears that more stringent domestic water quality standards and improved sanitary standards at food production/processing facilities may also be in order (see Tauxe in Chapter 4 and IOM [2006]).

Lessons from Waterborne Disease Outbreaks in the Americas

The following three case studies comprised a workshop session that focused on connections between climate and weather, human demographics, land use, infrastructure, and infectious disease outbreaks. Each presentation featured an outbreak chronology, an analysis of contributing factors, and a consideration of lessons learned (see Chapter 2).

Cholera in Peru, 1991

Cholera, a diarrheal disease caused by *Vibrio cholerae* that is endemic to the Ganges Delta in India, has also emerged in seven distinct pandemic waves since the early 1800s (Figure WO-10). When cholera reemerged in the Western Hemisphere in 1991, after being absent for more than a century, it first appeared in Peru. There, three simultaneous outbreaks in three coastal cities produced hundreds of thousands of cases in a matter of months (Seas et al., 2000), igniting an eight-year epidemic that spread across South America. In his workshop presentation, Forum member Eduardo Gotuzzo of the Universidad Peruana Cayetano Heredia in Lima, Peru, described social and environmental contributors to the epidemic, the public health response (in which he played a leadership role), and the potential use of early warning systems, based on environmental conditions, to anticipate future cholera outbreaks (see Seas and Gotuzzo in Chapter 2).

Cholera arrived in Peru during a time of hyperinflation, political unrest, episodes of terrorism, and high rates of unemployment, Gotuzzo recalled. Water quality was poor throughout the country and especially in rural areas, where sanitation coverage reached only about one in five people. As the number of cases mounted, epidemiological investigations revealed a number of underlying risk factors for infection, all of which would be expected to raise the risk of water- or foodborne infections via the fecal-oral route: having a family member with diarrhea, consuming water and food from street vendors, and being unemployed (a marker for poverty). Orders to boil water were issued, street vending of food and drinks was prohibited, and consumption of the national dish, ceviche (seafood "cooked" with acidic juices, rather than with heat), was prohibited. The rapid establishment of free treatment centers, where cholera victims received oral and intravenous rehydration therapy, helped to prevent high mortality rates, Gotuzzo noted.

Public health officials also investigated the presence of *V. cholerae* in the environment and were able to detect the pathogen's presence in communities before the first human cases appeared, Gotuzzo said. Subsequent investigations of the complex, climate-dependent relationship between *V. cholerae* and the zooplankton species that serve as its reservoir host, as depicted in Figure WO-11, suggested that El Niño conditions, which increase sea surface temperatures, raised the concentration of the pathogen in Peruvian coastal waters (Gil et al., 2004; Seas et al., 2000).

Once introduced into coastal communities in concentrations large enough for human infection to occur, cholera could have spread readily via contaminated water and food. The ability to measure sea surface temperature and zooplankton

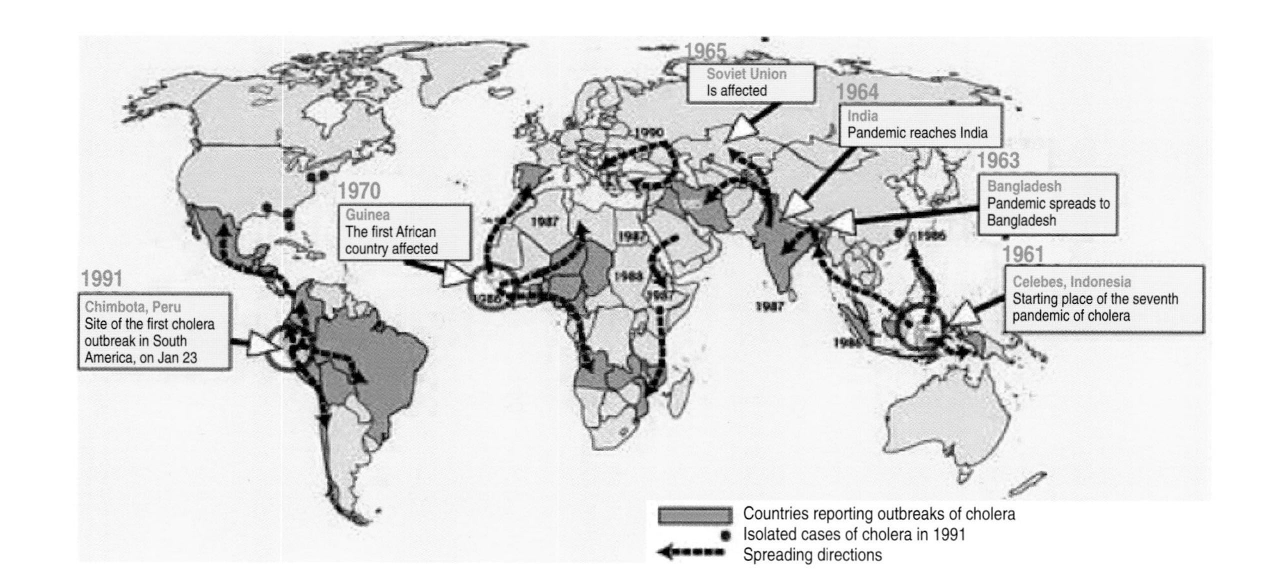

FIGURE WO-10 The seventh cholera pandemic.

SOURCE: Carlos Seas. Cólera. Medicina Tropical. CD-ROM. Version 2002. Instituto de Medicina Tropical, Príncipe Leopoldo. Amberes, Bélgica. Instituto de Medicina Tropical Alexander von Humboldt; Lima, Peru. Universidad Mayor de San Simón, Cochababmba, Bolivia.

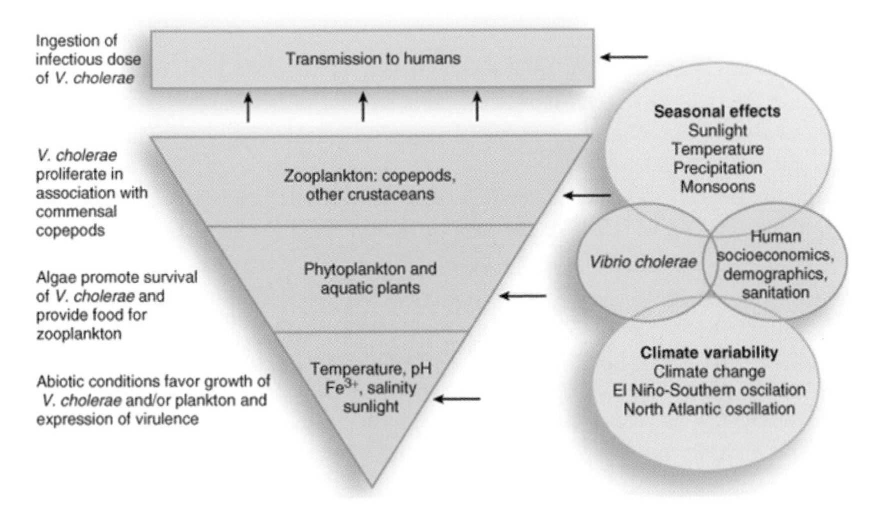

FIGURE WO-11 A hierarchical model for cholera transmission.
SOURCE: Reprinted from Lipp et al. (2002) with permission from the American Society for Microbiology.

concentrations via remote sensing now enables the prediction of cholera outbreaks based on environmental conditions (Colwell, 2004; Gil et al., 2004).[23]

Cryptosporidiosis in Milwaukee, 1993

While poverty and lack of sanitation coverage fueled the cholera "tsunami" in Peru and its subsequent spread across South America in the early 1990s, an epidemic of cryptosporidiosis two years later in a U.S. city, which sickened over 400,000 people and caused at least 50 deaths, demonstrated that even "modern" water treatment and distribution facilities are susceptible to contamination. This latter event resulted in a massive waterborne epidemic—the largest ever recorded in the continental United States. Developed countries bear only a fraction—about 1 percent—of the global burden of diarrheal diseases (Prüss-Üstün et al., 2008). Nonetheless, there have been several serious incidents of waterborne disease outbreaks in some of the world's wealthiest countries, including the United States and Canada. According to the CDC, a total of 36 waterborne disease outbreaks were reported in 19 states between 2003 and 2004; 30 of these outbreaks were

[23]For a detailed discussion of the ecological basis of cholera epidemiology, see the recent Forum workshop summary, *Global Climate Change and Extreme Weather Events: Understanding the Contributions to Infectious Disease Emergence* (IOM, 2008).

associated with drinking water and were estimated to have caused more than 2,700 illnesses (mainly gastroenteritis, as well as acute respiratory illness associated with water aerosols) and four deaths (Liang et al., 2006).

Like the cholera outbreak in Peru, the epidemic in Milwaukee became clear when unusually large numbers of people sought treatment for gastrointestinal illness and were absent from work and school (see Davis et al. in Chapter 2). "We felt that this was a waterborne outbreak until proven otherwise," recalled speaker Jeffrey Davis of the Wisconsin Division of Public Health. When their initial, cursory review of local finished water quality data revealed a spike in turbidity just prior to the outbreak's onset, they began an in-depth investigation of the city's water works. Information on infection rates among various "fixed populations" (e.g., nursing home residents) in different locations pointed to one of the city's two water treatment plants as the source of the outbreak, as depicted in Figure WO-12. Meanwhile, after stool samples from some patients with gastrointestinal illness (that had tested negative for enteric pathogens) were further tested for cryptosporidium and found to be positive, Milwaukee's mayor issued a "boil water" advisory.

Davis and colleagues pursued the question of how cryptosporidium came to contaminate finished water from the city's southern treatment plant from several angles (Mac Kenzie et al., 1994). Eventually, they concluded that the outbreak resulted from a "perfect storm" of conditions, which included the contamination of a storm sewer that emptied into Lake Michigan with animal wastes from an abattoir, record-setting rainfall while the ground remained frozen, unusual wind conditions and directions that served to funnel contaminated water toward the southern water plant's intake grid in Lake Michigan, and inadequate removal of particulate matter due to problems associated with a change in the coagulants used in the treatment plant. Amplification of the pathogen in humans,[24] coupled with delays in diagnosing the infectious agent (due to the fact that patients with gastrointestinal illness were not routinely tested for the presence of cryptosporidium oocytes in their stools), served to intensify the outbreak and its consequences for public health, Davis explained.

Among the lessons learned from this harrowing experience, Davis noted

- The importance of stringent water quality standards, and in particular that turbidity is more than an aesthetic property of finished water;
- The demand for advances in water sampling that maximize safety and minimize response time;
- The necessity of good public health surveillance for determining the scope, source, and progress of an outbreak;

[24]Following a minimal infective dose of approximately 130 cryptosporidium oocysts, humans excrete billions of oocysts per day in their stools and continue to do so after disease symptoms resolve.

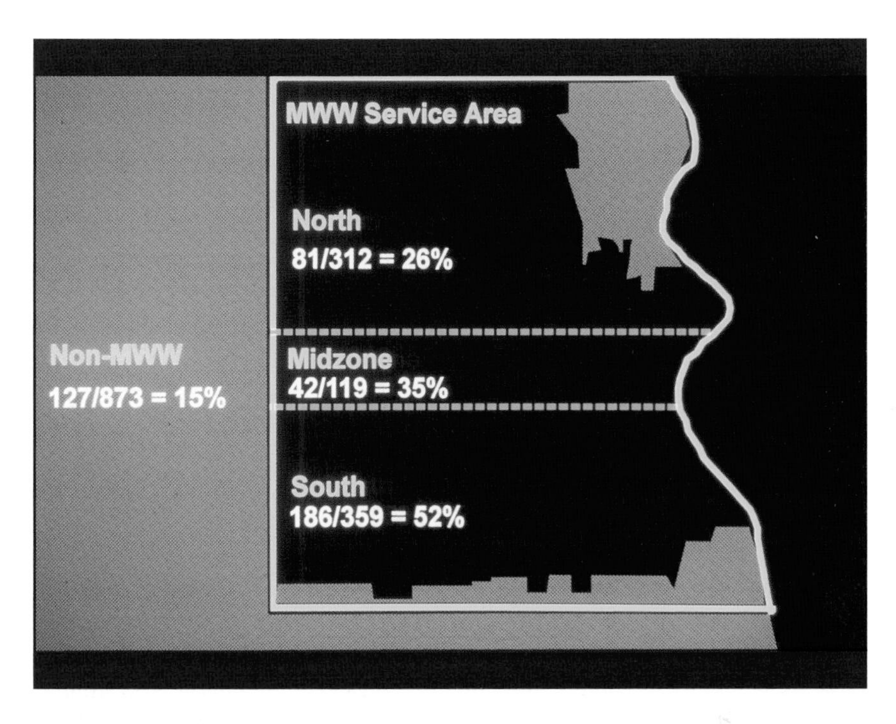

FIGURE WO-12 Rate of watery diarrhea from March 1 through April 28, 1993, among respondents in a random-digit telephone survey of households in the five-county Greater Milwaukee area, by Milwaukee Water Works region.
SOURCE: Adapted from Mac Kenzie et al. (1994).

- The need for communication between public health and water authority agencies to prevent and address waterborne disease outbreaks; and
- The critical role of the media in disseminating public health messages.

Perhaps the most enduring impact of the 1993 outbreak was increased public awareness of cryptosporidium, which at the time was a relatively novel microbial threat, and which led to improvements in disease surveillance, water testing, and enhanced water treatment. "Once these modifications were made in [Milwaukee's] treatment facilities, people from all over the world have been visiting, looking at the changes that were made," Davis said, "It has become very, very instructive."

E. coli *O157:H7 and Campylobacter, Walkerton, Ontario, 2000*

Bacterial contamination of the water supply in Walkerton, Ontario, in the spring of 2000 sickened nearly half of the town's 5,000 residents and caused 7 deaths, as well as 27 cases of hemolytic uremic syndrome, a serious kidney disease with potential lifelong complications (Hrudey and Walker, 2005). The outbreak also resulted in severe economic consequences for the town and province, and a loss of confidence in the public trust. As speaker Steve Hrudey, of the University of Alberta, observed, prevention of this tragedy—which occurred largely as a result of negligence, ignorance, and mendacity—was "painfully easy in hindsight."

The outbreak began on May 18, 2000, six days after an extremely heavy rainfall (a 60-year event) flooded the area, washing manure from a barnyard into a nearby shallow well (Figure WO-13; see Hrudey and Hrudey in Chapter 2). Although this well had previously been identified as vulnerable to agricultural contamination, it was not maintained to achieve prescribed chlorination standards, nor was chlorination increased in response to the flooding event.

FIGURE WO-13 Location of Walkerton Well 5 near farms to south and west.
SOURCE: Adapted from original photo taken for the Walkerton Inquiry by Constable Marc Bolduc, Royal Canadian Mounted Police.

The resulting fatal outbreak was precipitated by failures at every level of oversight, from the operators of the Walkerton water system to the government and people of Ontario—a chain of blame so long that one wondered, as Hrudey did, how 22 years managed to elapse between the well's ill-conceived creation and its deadly outcome.

In order to find ways to prevent the occurrence of similar tragedies, Hrudey and coauthor Elizabeth Hrudey (Hrudey and Hrudey, 2007) analyzed 73 published case studies of drinking water outbreaks that occurred in 15 affluent countries between 1974 and 2007, in order to determine "what failed and why." While each outbreak had unique features, Hrudey noted that the single most common factor was complacency among officials responsible for water safety, and indeed on the part of society at large. This in turn led to failure to recognize problems within water systems and to address conditions (such as flooding) that pose challenges to system capacity. "Prevention of future outbreaks does not demand perfection, only a commitment to learn from past mistakes and to act on what has been learned," the authors concluded.

Water Infrastructure: Recognizing Vulnerabilities and Opportunities for Improvements

Several workshop speakers offered diverse perspectives on the multifaceted issue of water infrastructure vulnerabilities and strategies to address them. Despite the vast differences between water infrastructures in developing and developed country settings, participants acknowledged common elements underlying effective public health strategies for preventing water-related diseases.

Threats to Safe Water Availability

Drinking water distribution systems Most of the approximately one million miles of pipes that comprise the U.S. drinking water distribution system's infrastructure are due for replacement within the next 30 years (NRC, 2006). According to presenter Michael Beach of the CDC, over the next 20 years this "routine maintenance" of the water distribution/sanitation infrastructure may cost the country between $300 billion and $1 trillion. Deficiencies in the water distribution system, according to Beach, are estimated to account for between 13 and 17 percent of all U.S. waterborne disease outbreaks. As the system's infrastructure continues to deteriorate, such outbreaks are likely to become more frequent.

A recent report by the National Research Council (2006), *Drinking Water Distribution Systems: Assessing and Reducing Risks*, concluded that addressing the threat of waterborne disease associated with water distributions systems would require a greater understanding of the ecological underpinnings of these threats, as well as specifically targeted epidemiologic studies. Beach noted, however, that pathogen surveillance informed by such efforts would be limited to regulated water distribution systems within the jurisdiction of a utility and would therefore exclude the increasing numbers of wells and other small, private water distribution systems in the United States—which now serve about 12 percent of the population—as well as "premise plumbing": pipes that lie on private property, within houses and buildings (see Beach et al. in Chapter 3).

Premise plumbing is of particular concern, according to Beach, given the rise of *Legionella*-associated disease in the United States, which in 2005-2006 accounted for approximately half of all reported waterborne disease outbreaks. Although the pathogen may be present at undetectable levels in public water systems, "this is a premise plumbing issue," Beach explained. "*Legionella* is colonizing buildings [such as hospitals and nursing homes] and then causing disease, especially in vulnerable populations."

Another potential, and largely uncharacterized, source of waterborne disease in the United States is recreational water[25] use. Over recent decades, Beach noted, the number of outbreaks of cryptosporidiosis and other diarrheal diseases associated with swimming pools, water parks, and other "disinfected venues" has increased dramatically. As discussed earlier, after cryptosporidiosis contaminated Milwaukee's drinking water in 1993 (see Davis in Chapter 2), "satellite" recreational outbreaks occurred in swimming pools. Today, this pathogen is increasingly linked to community-wide outbreaks that stem from inadequate chlorination in individual swimming pools.

The measurable reduction in waterborne disease outbreaks in the United States over the past 30 years attests to the effectiveness of water quality regulations to improve public health, Beach concluded. Further improvements will require a better understanding of the circumstances surrounding current waterborne disease outbreaks, and therefore "where the gaps in regulation exist and where we need to step in again."

Climate change effects Several phenomena associated with climate change—including extreme weather events (e.g., flooding and droughts), altered patterns of precipitation and river runoff, and decreased snow cover—are expected to present challenges to the consistent availability of safe water (IPCC, 2008). In her workshop presentation speaker Joan Rose, of Michigan State University, observed that in addition to directly affecting the size of water supplies, temperature and precipitation interact with other factors such as land use patterns to produce indirect effects on water quality through agricultural runoff and industrial pollution.

Research concerning the effects of climate on water quality tends to measure changes in three main indicators: concentrations of fecal bacteria (typically *E. coli*) in water supplies, the overall pathogen load present in water from various sources, and the burden of waterborne disease. In her contribution to Chapter 3,

[25]Recreational water is that which is used for water-based activities in marine, freshwater, hot tubs, spas, and swimming pools (Pond, 2005).

Rose describes the basis for each of these types of studies and presents examples of their findings. Among the 540 waterborne disease outbreaks reported to the CDC between 1948 and 1994, for example, about 50 percent were preceded (immediately, in the case of surface waters; after a two-month lag, in the case of ground waters) by episodes of rain that exceeded the 90th percentile for local precipitation (Curriero et al., 2001). As was previously noted, similar heavy precipitation events preceded major waterborne disease outbreaks in both Milwaukee, Wisconsin (see Davis et al. in Chapter 2) and Walkerton, Ontario (see Hrudey and Hrudey in Chapter 2). Heavy rains—as well as high winds—also preceded the widespread contamination of groundwater by sewage on South Bass Island in Lake Erie in 2005, Rose reported (see Chapter 3). Hundreds of residents and tourists were sickened by a combination of pathogens that included *Campylobacter* and adenoviruses.

Floods affect more people than droughts, windstorms, and geological disasters (e.g., earthquakes) combined (Hoyois et al., 2007). Floods are also associated with a range of waterborne illnesses, including diarrhea, cholera, typhoid, hepatitis, and leptospirosis, Rose noted. However, she added, because there is no widely accepted definition of "flooding," it is difficult to estimate its contribution to the burden of waterborne disease. Flooding must also be defined so that different flooding events can be recorded and compared in order to make actionable predictions of future risks to vulnerable water supplies in advance of climate variations such as El Niño, Rose said.

Disease Surveillance and Pathogen Detection

U.S. waterborne disease surveillance The spectrum of waterborne diseases in the United States, as described in Box WO-3, derives from reports made by local health departments to the CDC's Waterborne Disease and Outbreak Surveillance System (WBDOSS). Due to the passive nature of this surveillance, and its inability to track endemic waterborne disease, WBDOSS registers only a fraction of the true burden of disease associated with water exposure, Beach said. He noted that estimates of the actual number of acute gastrointestinal illnesses associated with public water supplies range from 4 to 33 million cases per year (see Beach et al. in Chapter 3).

Pathogen monitoring In the United States, pathogen monitoring is generally limited to water treatment facilities, where measures such as fecal and total coliform bacterial counts are used to gauge treatment efficacy. Not only do these methods fail to detect a broad spectrum of human pathogens (including some known to resist standard treatments), they also miss microbes that enter the distribution system beyond the treatment plant, such as when floodwaters breach distribution systems or when deteriorating pipes allow contaminated groundwater to enter the drinking water supply, according to workshop pre-

BOX WO-3
Spectrum of Water-Related Disease in the United States

1. Acute gastroenteritis
   - *Cryptosporidium*, toxigenic *E. coli*, *Giardia*, *Shigella*, norovirus, chemicals
2. Skin infections
   - *Pseudomonas* dermatitis/folliculitis, fungal infections
3. Ear infections
   - *Pseudomonas*
4. Eye infections and irritation
   - Adenoviruses, chloramines
5. Respiratory infections
   - *Legionella*, *Mycobacterium*
6. Neurologic infections
   - Echovirus, *Naegleria*
7. Wound infections
   - *Vibrio*
8. Hepatitis
   - Hepatitis A virus
9. Other
   - Leptospirosis
10. Nosocomial urinary tract infections
    - *Pseudomonas*

SOURCE: Courtesy of Michael Beach, Ph.D., CDC.

senter Kelly Reynolds of the University of Arizona (see Reynolds and Mena in Chapter 3).

In order to evaluate these inconsistencies, while avoiding the logistical difficulties associated with "testing at the tap" in private homes and businesses, Reynolds and coworkers examined pathogens present on filters in water vending machines (Miles et al., 2008). Twenty-seven percent of the 45 filters, which were not specifically designed to trap pathogens, tested positive for total coliform bacteria, including several fecal indicator species, suggesting that contamination had occurred post-treatment. The investigators also identified several types of culturable viruses. Reynolds advocated more extensive real-time monitoring in order to survey exposure to waterborne disease, as well as long-term analyses to determine exposure levels for different populations.

Water testing The United States, like many other industrialized countries, does not routinely require drinking water testing for the presence of specific microbial contaminants, according to speaker Mark Sobsey of the University of

North Carolina at Chapel Hill. While more accurate and comprehensive water testing could reduce waterborne disease risks in industrialized countries, testing is essential to providing safe water in developing and developed countries. In these settings, water quality data are needed to select promising sources for drinking water and appropriate treatments to ensure its safety, as well as to classify existing sources for the purposes of studying their health effects. Unfortunately, Sobsey observed, most water tests are not accessible, too complicated, or too costly for routine use in developing countries.

The "ideal" water test for microbial contamination in low-resource settings would be low cost, portable, self-contained, lab-free, electricity-free, and globally available, Sobsey said. It would also support data communication and serve as a tool for educating and mobilizing stakeholders, especially youth, to improve public health. Tests currently being developed to achieve these goals include a variety of culture methods for *E. coli* and other pathogens that can be conducted at ambient temperatures, or which contain internal temperature controls (see Chapter 3). Sobsey went on to predict that culture-free and direct methods for detecting waterborne pathogens would predominate in the future.

Risk Assessment

Estimates of the true impact of waterborne disease vary widely because they rely on extrapolations of exposure, rather than on surveillance based on accurate monitoring and testing of pathogens in the water supply, Reynolds observed. She noted that risk assessments for water-related diseases are also hampered by the dearth of epidemiological data on pathogen infectivity rates and dose responses, particularly in vulnerable populations such as infants, immunocompromised individuals, and the elderly.

Notwithstanding these limitations, Reynolds and coworkers have used mathematical modeling techniques and a predetermined standard of "acceptable risk" to evaluate various pathogens in water supplies (see Reynolds and Mena in Chapter 3). "In a particular water source, we can evaluate either the level of pathogens that can be present to meet the acceptable risk level, or which treatment measures need to be initiated to meet [the defined acceptable risk level] at the point of consumption," Reynolds explained. Through an iterative process, the models' plausibility is judged against epidemiological data, while the modeled risk estimates can be used to target future epidemiological studies. According to Reynolds, it is currently possible to make "reasonable assumptions" about how to extrapolate disease risk to different populations at different levels of exposure to waterborne pathogens. Rose insisted, however, that accurate exposure and dose-response measures could only be obtained from animal models and properly designed and conducted outbreak investigations.

The risk modeling process also provokes further questions, Reynolds observed, including

- What is our goal of acceptable risk?
- What is the realized risk?
- How bad is exposure to these different types of organisms or a mixture of these organisms? Can we evaluate that?
- What are our risk-reduction potentials?

Water Treatment

Methodology Despite his emphasis on the importance of water testing, Sobsey—along with other participants at this workshop—indicated that, given limited resources, it may be better to treat and protect water first, then test if possible to ensure that the treatment was effective. But, as his University of North Carolina colleague Philip Singer made clear in a presentation he called "Sanitary Engineering 101," determining the "best" treatment regime for a given water supply is far from simple (see Singer in Chapter 4). Having reviewed a broad range of disinfectants, Singer focused on the use of chlorine, describing water quality factors including reduced inorganic material, dissolved organic carbon, and microbial contents that must be addressed in order to achieve disinfection by this method. Singer also discussed the parameters and limitations of various approaches to water treatment, emphasizing the significant barriers to disinfection posed by particulate matter and describing different methods and approaches for its removal by filtration and flocculation.[26]

Effectiveness and cost-effectiveness Based on a systematic review of intervention trials to improve microbiological water quality levels at both source (dug wells, boreholes, or stand posts) and point of use (improved storage, chlorination, solar disinfection, filtration, or combined flocculation-disinfection using a water-purifying product) for their ability to prevent diarrhea, Clasen and coworkers (Clasen et al., 2007b; see Chapter 4) concluded that all were effective for preventing diarrhea in children younger than five years of age, but were inconsistent in their effectiveness for reasons unexplained by research. For example, filtration appeared to be more effective than other methods in preventing diarrhea, but this difference could not be attributed to its ability to remove microbes from water. Instead, Clasen suggested, filtration may have been used more routinely than other methods due to its superior ability to improve how water looks, tastes, and smells as compared with other disinfection methods such as chlorination or solar (ultra-violet [UV]) exposure.

[26]The separation of a solution. Most commonly, flocculation is used to describe the removal of a sediment from a fluid. In addition to occurring naturally, flocculation can also be forced via agitation or the addition of flocculating agents. Numerous manufacturing industries use flocculation as part of their processing techniques, and it is also extensively employed in water treatment (http://www.wisegeek.com/what-is-flocculation.htm).

Given this result, Clasen's group conducted a cost-effectiveness analysis to determine the cost per disability-adjusted life year (DALY, a measure of disease burden) averted for a similar range of interventions against diarrhea at the water source as well as chlorination, filtration, solar disinfection, and flocculation at the household level (Clasen et al., 2007a). "These aren't cost-benefit analyses," he explained. "We aren't looking at improvements in productivity . . . we are just looking at . . . actual savings from health care expenditures as a result of reduced diarrheal disease." The researchers found that, upon reaching 50 percent of a country's population, interventions involving household chlorination and solar disinfection paid for themselves, and that all interventions were cost-effective.

Scaling up household water treatment and storage interventions Among the various definitions of "scaling up," Clasen has chosen the most expansive to describe efforts to increase the availability, uptake, and correct, consistent use of household water treatment and safe storage systems (Clasen, 2008). These efforts are spearheaded by the International Network to Promote Household Water Treatment and Safe Storage, a consortium of interested UN agencies, bilateral development agencies, international nongovernmental organizations, research institutions, international professional associations, and private sector and industry associations (WHO, 2008d).

Presently, only a tiny fraction of the millions of people who could benefit from household water treatment and safe storage interventions—far more than the nearly one billion who use "unimproved" water sources—are being served, and those who need them most are the most difficult to reach. This uptake problem is not restricted to safe water initiatives, Clasen added: it is common to a host of public health interventions targeting individual households, including oral rehydration salts, insecticide-treated bed nets, and improvements in sanitation services. Examining uptake successes for other such products, Clasen discussed several initiatives that are currently being employed to boost public acceptance and use of household safe water and storage (see Chapter 4). Prominent among these strategies is the promotion of nonhealth benefits such as cost saving, convenience, and aesthetic appeal. Clasen also noted several challenges to these efforts, including the popular notion that diarrhea is not a disease, but merely an annoyance—or worse, a way to purify one's body.

Water Access

Solutions for the developing world As Kolsky and coauthors observe in Chapter 4, "water access *at the household scale* is critical to increasing the quantity of water used to improve hygiene . . . [but] the perspective of the service provider . . . is often different from that of the householder and the public health specialist. . . . The highest priority of the water engineer is often the intake and central treatment works. Their attitude is, if this link in the chain fails, the rest will fail" (Bostoen et al., 2007). While this engineer's worldview largely defines water infrastructure in the industrialized world, according to workshop speaker Joseph Hughes, of the Georgia Institute of Technology, it represents a barrier to solving water needs in developing countries.

Hughes described the diverse and complex elements of the U.S. civil water system, which range from water supply creation to resource protection to the "heavy infrastructure" of treatment plants and systems for water conveyance, storage, and the management of residual (post-treatment) waste (see Caravati et al. in Chapter 4). "We have a system that is poorly translatable" to the developing world, he concluded. "So if the system won't work where we need it, we need a new system." He proceeded to describe small, distributed, simple solutions, as well as technologies that lie somewhere between the disparate ends of the spectrum currently occupied by the developing (low-tech) and industrialized (high-tech) worlds. "Functionally we have nothing in the middle [of this spectrum] right now," he observed, "and certainly no research."

While acknowledging the many and formidable barriers to innovation under these circumstances, Hughes offered several reasons for optimism. Over recent decades, he noted, the combined efforts of engineers and medical researchers have vastly advanced medical instrumentation; they could do the same for safe water technologies. As the growth of distributed energy systems reduces energy costs to levels supportable by microfinance, more people will be able to invest in small-scale infrastructure appropriate to the developing world, he added. Microfinancing can also support businesses based on safe water or sanitation, such as the use of fecal matter to generate electricity. With little more than "a piece of wire and two pieces of cloth—[and] the bacteria [present in] feces—you can make electricity," Hughes said. "Not enough to light up a city, but you can charge a lot of cell phones."

Successful development of innovative solutions for water and sanitation in developing countries will require careful attention to matters of scale, Hughes continued. These extend from the rate and area of deployment of an intervention—in the form of a product or service, governed by a business plan—to educational requirements necessary for its implementation. "There is technology being developed all over the world . . . that could be brought into the [water and sanitation] sector if there is a business opportunity," he concluded.

The toilet-to-tap conundrum As previously noted herein, and highlighted in *Running Dry* (Thebaut, 2005; see enclosed DVD), water supplies—as depicted in Figure WO-14, which shows the consumption of water in the United States as compared to other countries—are becoming increasingly limited. Drought conditions that have exacerbated pressure on dwindling freshwater resources in the southeastern United States and the Colorado River basin have encouraged development of systems for wastewater reuse, Beach said, including so-called "toilet-to-tap" renewal (see Beach et al. in Chapter 3). This is a misnomer, he added, because the majority of water consumed in U.S. households is used for cooking and bathing.

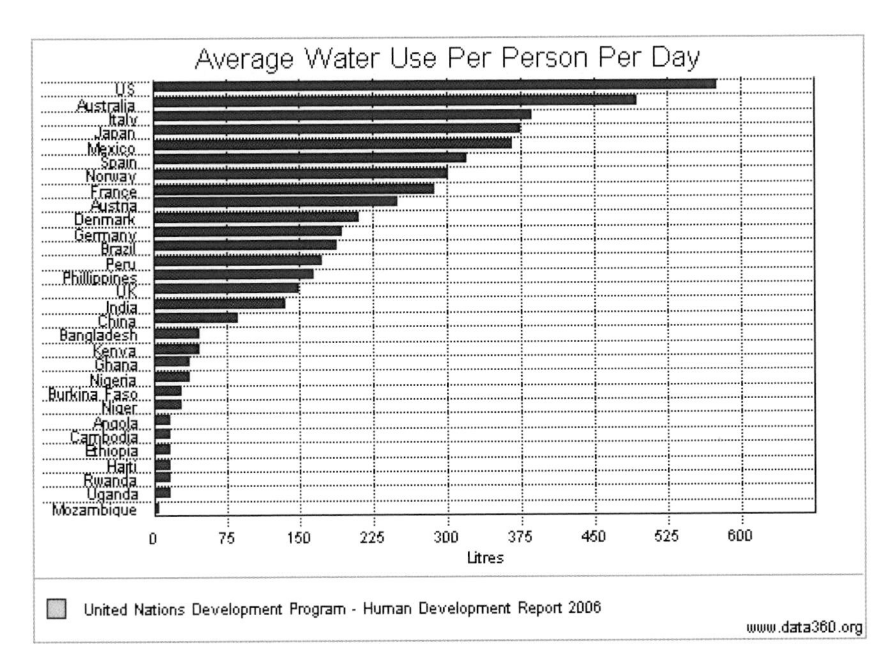

FIGURE WO-14 Water consumption in the United States compared with other countries. SOURCE: Courtesy of Data360, http://www.data360.org/dsg.aspx?Data_Set_Group_Id=757 (accessed June 4, 2009).

We need to investigate the health effects associated with such increasingly necessary adaptive strategies for water scarcity, Beach advised. Understanding how to use different grades of water safely will help to address potentially conflicting needs for water quality and quantity.

> Improving Sanitation. Diarrhea—usually caused by feces-contaminated food or water—kills a child every fifteen seconds. That means more people die of diarrhea than all the people killed in conflict since the Second World War. Diarrhea, says the UN children's agency UNICEF, is the largest hurdle a small child in a developing country has to overcome. Larger than AIDS, or TB, or malaria. 2.2 million people—mostly children—die from an affliction that to most westerners is the result of bad takeout. Public health professionals talk about water-related diseases, but that is a euphemism for the truth. These are shit-related diseases (George, 2008).

A gram of feces can contain 10 million viruses, 1 million bacteria, 1,000 parasite cysts, and 100 worm eggs; one sanitation specialist has estimated that people who live in areas with inadequate sanitation ingest 10 grams of fecal matter every day (George, 2008). Perhaps it is not surprising, then, that approximately 80 per-

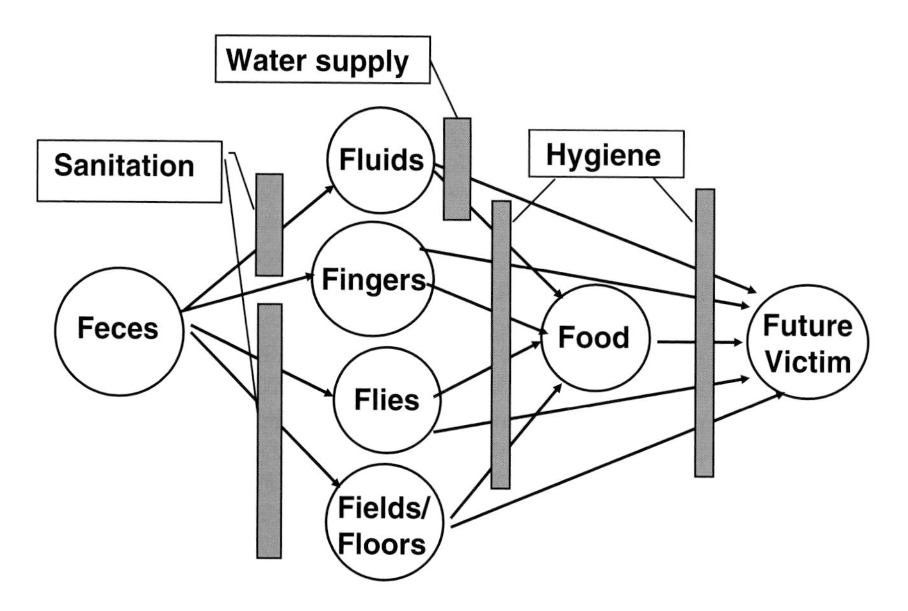

FIGURE WO-15 The "F" diagram.
SOURCE: Reprinted from Bostoen et al. (2007) and based on Wagner and Lanoix (1958) with permission from the World Health Organization.

cent of the world's illness is caused by fecal matter[27] (George, 2008). As Kolsky, of the World Bank, wryly observed, it is a frequently overlooked truth that drinking clean water does not entirely protect us from the "fecal peril." Fluids are but one route of invasion for fecal pathogens; as shown in Figure WO-15, sanitation and hygiene are necessary to close the others.

Yet health is not high on the list of reasons why people want sanitation, Kolsky reported. More often, they want privacy and dignity, and to avoid flies and foul odors. Moreover, he noted, even if public health improves following a sanitation intervention, its success or failure is likely to be judged on its reliability—how often breakdowns occur—compared to its cost. Be that as it may, any enticements to sanitation should be seized upon to "sell" interventions that ultimately benefit public health, Kolsky concluded.

[27]Approximately 80 percent of all diseases and more than a third of the deaths in developing countries are caused by the consumption of contaminated water, and on average up to a tenth of the productive lifetime of each person is taken up by water-related diseases (UNDESA, 1992).

Investment, Implementation, and Entrepreneurship

Investment and implementation: The World Bank perspective A deep financing gap separates investments needed for water and sanitation infrastructure and actual spending, according to speaker Vahid Alavian, of the World Bank. As Alavian observed, an estimated annual global investment of $25 to $30 billion in water and sanitation is necessary for meeting the MDG, but the world is spending only about half that amount.

As the largest global investor in water and sanitation, the World Bank follows a two-pronged approach, financing water and sanitation interventions that are intended to benefit either the entire economy of a nation or its poorest citizens. In either case, the bank focuses on the long term, supporting sustainable efforts to address endemic disease in developing countries, Alavian explained. This commitment requires not only financial resources, but also support for scaling up and expanding interventions, strengthening governance and institutions, and ensuring sustainability. In Senegal, for example, before financing the recovery of a water utility from bankruptcy, the World Bank established capacity-building and sector reform to create a viable environment for their later investment. As a result, the bank's relatively small contribution of about $500 million was leveraged into interventions that are likely to enable the country to meet the water and sanitation MDG, Alavian reported.

Since people invest in water and sanitation primarily for reasons unrelated to health, and therefore health improvement is not a priority for the design, construction, and operation of water supply and sanitation infrastructure, the World Bank must determine how to make investments in such projects as beneficial as possible from a public health standpoint, Kolsky said. This already difficult challenge is further complicated by the differing agendas of various sectors (e.g., health, urban development, utility, environment) involved in water and sanitation (see Boeston et al. in Chapter 4). Kolsky maintained, however, that when incorporated into water and sanitation projects from their inception, health interventions are relatively inexpensive to implement. Kolsky concluded, therefore, that "we need to build our public health perspective early on in project design, and we need to think about water and wastes and behavior."

Kolsky described one example of such a design process, which resulted from the discovery by an urban planner in India that plans to construct water infrastructure to serve a city center involved laying pipes through a slum area. Based on his determination that the cost of connecting the slum neighborhoods to the planned sewer and water network was negligible relative to the overall expense of the project, the planner was able to convince government officials to extend the benefits of the new network to the city's poorest residents.

Social entrepreneurship Much as distributed energy systems and microfinancing encourage the development of small-scale solutions for water and sanitation in developing countries, social entrepreneurship can provide the impetus for communities to use these tools to address their particular public health needs, explained speaker Sharon Hrynkow of the National Institute of Environmental Health Sciences (NIEHS).

Through a series of examples—ranging in scale from an NIEHS Superfund[28] basic research program on the remediation of arsenic-tainted wells in India to a community-run pay toilet organization in Indonesia—Hrynkow illustrated the tenets of social entrepreneurship as they apply to the development of water and sanitation interventions (see Chapter 4). Instead of viewing communities simply as beneficiaries of received services, a social entrepreneur recognizes a collection of experts who can judge the appropriateness of interventions and sustain successful ones, she said.

This perspective, which is not shared by medical researchers, offers a means to better understand why certain communities accept and develop new technologies, Hrynkow observed. To gain this perspective, she recommended that public health practitioners work more closely with social entrepreneurs ("we can put them on our boards, in our public slots; we can link them to our researchers on the ground in foreign countries") and use their ideas and enthusiasm to engage the next generation of public health researchers.

Needs and Opportunities

Central Themes

Over the course of the two days of this workshop, discussions returned to three main themes that underlie needs and opportunities for reducing the burden of water-related infectious diseases. The distillation of these discussion "themes" inform subsequent, specific considerations described below regarding the research agenda, opportunities for public heath intervention, infrastructure development and improvement, and strategic approaches to addressing issues in water, sanitation, and health. *They are not, nor should they be interpreted to be, conclusions or recommendations arrived at through a deliberative consensus study process.*

Global phenomena, local effects While the global water crisis may be viewed as a byproduct of interdependent global phenomena that includes population growth, industrialization, climate change, and urbanization, its impact on public health is locally variable, necessitating local solutions. Ecological factors contributing to infectious disease emergence are particularly influential in the case of water-related diseases.

[28]Superfund is the U.S. government's program to clean up the nation's uncontrolled hazardous waste sites. For more information, see http://www.epa.gov/superfund/.

Route and scale Each of the various water-related disease transmission processes can be interrupted at multiple points, permitting intervention on a range of geographic scales. Effective water and sanitation infrastructure is both appropriately scaled and sustainable for a given setting.

Human behavior From the implementation of individual interventions, to the resolution of border conflicts over water access, to international cooperation necessary to avert a global water crisis, success depends upon human actions and interactions motivated by diverse factors, of which a scientifically sound assessment of risk and benefit is but one.

Interventions to Address Water-Related Diseases

Interventions to improve health by increasing water quality, sanitation, and hygiene can be implemented at many points throughout the water distribution system, from source to household to consumer. As discussed by Clasen and Cairncross (2004), these interventions include the following:

- source water protection;
- removal of pathogens by physical methods (e.g., filtration, adsorption, and sedimentation), chemical treatment (e.g., assisted sedimentation, chemical disinfection, and ion exchange), or heat and UV radiation;
- maintaining the microbiological quality of safe drinking water through piped distribution, residual disinfection, and improved storage;
- steps to encourage proper disposal of human feces;
- increased access to and availability of safe water; and
- hygienic practices within domestic and community settings, such as handwashing.

Convenient access to "improved" water in quantity encourages better hygiene and limits the spread of diarrheal disease. Diarrheal disease morbidity may also be dramatically reduced by relatively simple interventions, as illustrated in Figure WO-16 (Cairncross and Valdmanis, 2006; WHO/UNICEF, 2005). Placing a water tap close to a home nearly doubles the odds of a mother cleaning her hands after contact with fecal material from a child. Poor women who spend hours per day collecting water usually view the time-saving aspect of an improved water supply as its greatest benefit. Similarly, access to even basic forms of improved sanitation, such as pit latrines, helps prevent exposures to diseases such as diarrhea, intestinal worm parasites, and trachoma. This outcome may result from improved hygiene practices that accompany better sanitation, which are also associated with social advantages such as higher status in the community, safety, convenience, and privacy.

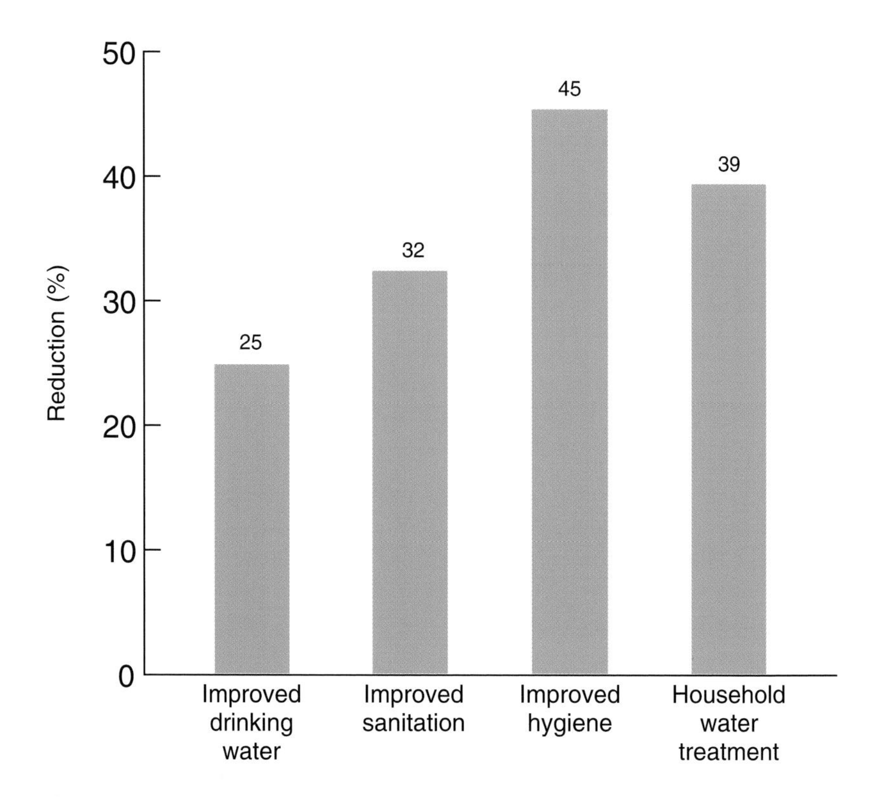

FIGURE WO-16 Reduction in diarrheal diseases morbidity resulting from improvements in drinking water and sanitation services.
SOURCE: Based on data in Fewtrell et al. (2005), and reprinted from WHO/UNICEF (2005) with permission from the World Health Organization.

In areas where the water supply is adequate, the adoption of simple and inexpensive methods to improve the microbiological quality of existing water supplies can significantly mitigate the disease burden due to diarrheal diseases. Many point-of-use interventions, such as filtration, disinfection with radiation, boiling or chlorine, or simply the provision of enclosed, protected containers have been shown to be effective (interestingly, the use of a combination of methods simultaneously was not shown to have any added benefit; Clasen et al., 2006; Fewtrell et al., 2005). However, any intervention or provision of clean water and sanitation will only be successful if it is used. Education of the population, as well as considerations of cost, accessibility, and acceptability of interventions, will be key issues in the design and implementation of these interventions.

According to the second edition of the report *Disease Control Priorities in Developing Countries* (Cairncross and Valdmanis, 2006), the main health benefit of water supply, sanitation, and hygiene is a reduction in diarrheal disease. However, the effect on the incidence and prevalence of other diseases, such as dracunculiasis, schistosomiasis, and trachoma, is substantial. Water, sanitation, and hygiene improvements could eliminate 3 to 4 percent of the global burden of disease (Cairncross and Valdmanis, 2006).

Sustained interventions to reduce vector-borne diseases fall into three main categories (Prüss-Üstün et al., 2008):

- Modifying the environment to permanently change the land, water or vegetation in ways that reduce vector habitats such as drainage, leveling land, contouring reservoirs, and altering river boundaries;
- Manipulating the environment to create temporary (and often repeated) unfavorable conditions for vector propagation by such means as removing aquatic plants that shelter mosquito larvae, alternately flooding and drying irrigated paddy fields, periodic flushing of natural and human-made waterways, and the introduction of predators, such as larvivorous fish; and
- Modifying or manipulating human habitation or behavior to reduce contact between humans and vectors with barriers such as window screens and nontreated mosquito nets, as well as by removing standing water in or near the home, which provide mosquito breeding sites.

The effectiveness of these and other interventions, such as sanitation and water quality management, to reduce the burden of water-related diseases will be strongly influenced by local conditions, which must be taken into account in order to make cost-effective choices to address water-related health risks in diverse contexts (Clasen and Cairncross, 2004).

Costs and Benefits of Interventions

The estimated global economic benefits of drinking water and sanitation improvements include (Prüss-Üstün et al., 2008)

- health-care savings of $7 billion per year for health agencies and $340 million for individuals;
- productivity gains of nearly $10 billion per year;
- time savings equivalent to $63 billion per year; and
- values of deaths averted (based on discounted future earnings) of more than $3 billion per year.

Taking these gains into account, an investment of $11.3 billion per year, as required to meet MDG7 sanitation and drinking water targets, would produce a return of approximately $84 billion.

Nevertheless, responding to recent shortfalls in progress toward achieving the water and sanitation targets of the MDG7, the editors of the *Lancet* observed that, despite such reports, most government donations to water and sanitation initiatives have not increased, nor are foundations or organizations "lining up to give money to build toilets or to fund education programmes to teach small children how to wash their hands" (Editorial, 2006). Moreover, the editorial continued, "the health-care community also seems to have lost sight of how fundamental clean water and sanitation are to health, preferring to get involved in more directly medical interventions, such as access to drugs and vaccines. It is dangerously short sighted to pour immense time and resources into vaccinating children only for them to die a few years later from diarrhoeal illnesses."

A Public Health Research Agenda

Workshop participants, most notably speakers Beach, Rose, Bradley, and Forum member Rima Khabbaz of the CDC, identified and discussed the following bulleted points as particularly important examples of water-health relationships that merit further exploration, including:

- What are the infectious disease risks associated with agricultural water uses?
- What are the direct and indirect connections between water and respiratory disease?
- What role does handwashing play in reducing the prevalence and incidence of respiratory diseases in the developing and developed world?
- What are the infectious disease risks associated with the unregulated components of the water distribution "system," including but not limited to private systems (private community and individual wells) and premise plumbing?

Workshop speakers Michael Beach, David Bradley, Kelly Reynolds, and Philip Singer also discussed the need for more information on the short- and long-term health and environmental consequences associated with individual exposures to water-associated diseases as well as the health and economic impacts associated with endemic water-related diseases. In particular, presenter David Bradley—who developed the Bradley Classification Scheme for water-related diseases in the 1970s—believed that additional factors could be incorporated into his original classification scheme to include behavioral and spatial aspects of changing or modifying water sources. These factors might add additional levels of complexity to our collective appreciation of the linkages between water, sanitation, and health. It was felt by presenter Reynolds that such knowledge could inform risk assessment methods development activities, and the design and priorities placed on different intervention strategies to reduce the health and economic impacts associated with these vector- and non-vector-borne diseases.

Conclusion

Safe water and sanitation pose universal challenges for public health. While global and regional phenomena such as climate change and geopolitical shifts threaten to intensify these challenges, effective solutions demand attention to local needs and opportunities. As *Running Dry* hopefully suggests, "most of the water crisis issues can be solved by a coordinated, global, environmentally sensitive, humanitarian effort."

REFERENCES

Arnold, B. F., and J. M. Colford, Jr. 2007. Treating water with chlorine at point-of-use to improve water quality and reduce child diarrhea in developing countries: a systematic review and meta-analysis. *American Journal of Tropical Medicine and Hygiene* 76(2):354-364.

Bartram, J., K. Lewis, R. Lenton, and A. Wright. 2005. Focusing on improved water and sanitation for health. *Lancet* 365(9461):810-812.

Black, R. E., S. S. Morris, and J. Bryce. 2003. Where and why are 10 million children dying every year? *Lancet* 361(9376):2226-2234.

Bostoen, K., P. Kolsky, and C. Hunt. 2007. Improving urban water and sanitation services: health, access and boundaries. In *Scaling urban environmental challenges: from local to global and back*, edited by P. M. Marcotullio and G. McGranahan. London: Earthscan Publications.

Bowen, A., H. Ma, J. Ou, W. Billhimer, T. Long, E. Mintz, R. M. Hoekstra, and S. Luby. 2007. A cluster-randomized controlled trial evaluating the effect of a handwashing-promotion program in Chinese primary schools. *American Journal of Tropical Medicine and Hygiene* 76(6):1166-1173.

Cairncross, S., and R. Feachem. 1993. *Environmental health engineering in the tropics (2nd edition).* Chichester, UK: John Wiley & Sons.

Cairncross, S., and V. Valdmanis. 2006. Water supply, sanitation, and hygiene promotion. In *Disease control priorities in developing countries (2nd edition)*, edited by D. T. Jamison, J. G. Breman, A. R. Measham, G. Alleyne, M. Claeson, D. B. Evans, P. Jha, A. Mills, and P. Musgrove. New York: Oxford University Press.

Calvin, L. 2007. Outbreak linked to spinach forces reassessment of food safety practices. *AmberWaves Magazine* (USDA Economic Research Service), http://www.ers.usda.gov/AmberWaves/June07/Features/Spinach.htm (accessed March 11, 2009).

CDC (Centers for Disease Control and Prevention). 2009. Outbreak of *Salmonella* serotype Saintpaul infections associated with eating alfalfa sprouts—United States, 2009. *Morbidity and Mortality Weekly Report* 58(18):500-503.

Clasen, T. F. 2008. *Scaling up household water treatment: looking back, seeing forward.* Geneva: World Health Organization.

Clasen, T. F., and S. Cairncross. 2004. Household water management: refining the dominant paradigm. *Tropical Medicine and International Health* 9(2):187-191.

Clasen, T. F., I. Roberts, T. Rabie, W. Schmidt, and S. Cairncross. 2006. Interventions to improve water quality for preventing diarrhoea. *Cochrane Database of Systematic Reviews* 3. Art. No. CD004794.

Clasen, T. F., L. Haller, D. Walker, J. Bartram, and S. Cairncross. 2007a. Cost-effectiveness of water quality interventions for preventing diarrhoeal disease in developing countries. *Journal of Water and Health* 5(4):599-608.

Clasen, T. F., W. P. Schmidt, T. Rabie, I. Roberts, and S. Cairncross. 2007b. Interventions to improve water quality for preventing diarrhoea: systematic review and meta-analysis. *British Medical Journal* 334(7597):782.

Clasen, T. F., C. McLaughlin, N. Nayaar, S. Boisson, R. Gupta, D. Desai, and N. Shah. 2008. Microbiological effectiveness and cost of disinfecting water by boiling in semi-urban India. *American Journal of Tropical Medicine and Hygiene* 79(3):407-413.

Cohen, M. J., and C. Henges-Jeck. 2001. *Missing water: the uses and flows of water in the Colorado River delta region.* Oakland, CA: Pacific Institute for Studies in Development, Environment, and Security.

Colwell, R. R. 2004. Infectious disease and environment: cholera as a paradigm for waterborne disease. *International Microbiology* 7(4):285-289.

Curriero, F. C., J. A. Patz, J. B. Rose, and S. Lele. 2001. The association between extreme precipitation and waterborne disease outbreaks in the United States, 1948-1994. *American Journal of Public Health* 91(8):1194-1199.

Dugger, C. W. 2006. Preventable disease blinds poor in third world. *New York Times*, March 31.

Editorial. 2006. Water and sanitation: the neglected health MDG. *Lancet* 368(9543):1212.

Emerson, P. M., M. Burton, A. W. Solomon, R. Bailey, and D. Mabey. 2006. The SAFE strategy for trachoma control: using operational research for policy, planning and implementation. *Bulletin of the World Health Organization* 84(8):613-619.

Feachem, R. G., D. J. Bradley, H. Garelick, and D. D. Mara. 1983. *Sanitation and disease: health aspects of excreta and wastewater management.* Chichester, UK: John Wiley.

Fewtrell, L., R. B. Kaufmann, D. Kay, W. Enanoria, L. Haller, and J. M. Colford, Jr. 2005. Water, sanitation, and hygiene interventions to reduce diarrhoea in less developed countries: a systematic review and meta-analysis. *Lancet Infectious Diseases* 5(1):42-52.

George, Rose. 2008. *The big necessity: the unmentionable world of human waste and why it matters.* New York: Henry Holt and Company.

Gil, A. I., V. R. Louis, I. N. Rivera, E. Lipp, A. Huq, C. F. Lanata, D. N. Taylor, E. Russek-Cohen, N. Choopun, R. B. Sack, and R. R. Colwell. 2004. Occurrence and distribution of *Vibrio cholerae* in the coastal environment of Peru. *Environmental Microbiology* 6(7):699-706.

Gleick, P. H. 2001. Making every drop count. *Scientific American* (February):28-33.

Hopkins, D. R. 2008. *Improving water, sanitation, and health at the grassroots.* Paper presented at the Global Issues in Water, Sanitation, and Health workshop, Washington, DC, September 23-24, 2008.

Hopkins, D. R., F. O. Richards, Jr., E. Ruiz-Tiben, P. Emerson, and P. C. Withers, Jr. 2008. Dracunculiasis, onchocerciasis, schistosomiasis, and trachoma. *Annals of the New York Academy of Sciences* 1136:45-52.

Hoyois, P., J.-M. Scheuren, R. Below, and D. Guha-Sapin. 2007. *Annual disaster statistical review: numbers and trends 2006.* Brussels: Center for Research on the Epidemiology of Disasters.

Hrudey, S. E., and E. J. Hrudey. 2007. Published case studies of waterborne disease outbreaks—evidence of a recurrent threat. *Water Environment Research* 79(3):233-245.

Hrudey, S. E., and R. Walker. 2005. Walkerton, 5 years later: tragedy could have been prevented. *Opflow* 31(6):1-5.

IOM (Institute of Medicine). 2006. *Addressing foodborne threats to health: policies, practices, and global coordination.* Washington, DC: The National Academies Press.

———. 2008. *Global climate change and extreme weather events: understanding the contributions to infectious disease emergence.* Washington DC: The National Academies Press.

———. 2009. *Microbial evolution and co-adaptation: a tribute to the life and scientific legacies of Joshua Lederberg.* Washington, DC: The National Academies Press.

IPCC (Intergovernmental Panel on Climate Change). 2008. Linking climate change and water resources: impacts and responses. In *Climate change and water*, edited by B. C. Bates, Z. W. Kundzewicz, S. Wu, and J. P. Palutikof. Technical Paper of the Intergovernmental Panel on Climate Change. Geneva: IPCC.
JMP (Joint Monitoring Programme for Water Supply and Sanitation). 2008. *Progress on drinking water and sanitation: special focus on sanitation*. New York and Geneva: UNICEF and WHO.
Liang, J. L., E. J. Dziuban, G. F. Craun, V. Hill, M. R. Moore, R. J. Gelting, R. L. Calderon, M. J. Beach, and S. L. Roy. 2006. Surveillance for waterborne disease and outbreaks associated with drinking water and water not intended for drinking—United States, 2003-2004. *Morbidity and Mortality Weekly Report* 55(SS12):31-58.
Lipp, E. K., A. Huq, and R. R. Colwell. 2002. Effects of global climate on infectious disease: the cholera model. *Clinical Microbiology Reviews* 15(4):757-770.
Lule, J. R., J. Mermin, J. P. Ekwaru, S. Malamba, R. Downing, R. Ransom, D. Nakanjako, W. Wafula, P. Hughes, R. Bunnell, F. Kaharuza, A. Coutinho, A. Kigozi, and R. Quick. 2005. Effect of home-based water chlorination and safe storage on diarrhea among persons with human immunodeficiency virus in Uganda. *American Journal of Tropical Medicine and Hygiene* 73(5):926-933.
Mac Kenzie, W. R., N. J. Hoxie, M. E. Proctor, M. S. Gradus, K. A. Blair, D. E. Peterson, J. J. Kazmierczak, D. G. Addiss, K. R. Fox, J. B. Rose, and J. P. Davis. 1994. A massive outbreak in Milwaukee of cryptosporidium infection transmitted through the public water supply. *New England Journal of Medicine* 331(3):161-167.
Miles, S. L., C. P. Gerba, I. L. Pepper, and K. A. Reynolds. 2008. Point-of-use drinking water devices for assessing microbial contamination in finished water and distribution systems. *Environmental Science and Technology* 43(5):1425-1429.
Mittelstaedt, M. 2009 (March 12). UN warns of widespread water shortages. *Globe and Mail*, http://www.theglobeandmail.com/servlet/story/RTGAM.20090311.wwater0312/BNStory/International/home (accessed April 22, 2009).
NICED (National Institute of Cholera and Enteric Diseases, Kolkata). 2005. Estimation of the burden of diarrhoeal diseases in India. In *National Commission on Macroeconomics and Health background papers—burden of disease in India*. New Delhi: Ministry of Health and Family Welfare.
Northern Territory Government, Department of Health and Families. 2009. *The Northern Territory Public Health Bush Book*. Darwin: Northern Territory Government, Department of Health and Families, http://www.nt.gov.au/health/healthdev/health_promotion/bushbook/volume2/chap2/intro.htm (accessed May 18, 2009).
NRC (National Research Council). 1993a. *Managing wastewater in coastal urban areas*. Washington, DC: National Academy Press.
———. 1993b. *Soil and water quality: an agenda for agriculture*. Washington, DC: National Academy Press.
———. 2002. *Countering bioterrorism: the role of science and technology*. Washington, DC: National Academy Press.
———. 2004a. *Review of the desalination and water purification technology roadmap*. Washington, DC: The National Academies Press.
———. 2004b. *Water and sustainable development: opportunities for the chemical sciences—a workshop report to the Chemical Sciences Roundtable*. Washington, DC: The National Academies Press.
———. 2004c. *Confronting the nation's water problems: the role of research*. Washington, DC: The National Academies Press.
———. 2005. *Water conservation, reuse, and recycling: proceedings of an Iranian-American workshop*. Washington, DC: The National Academies Press.

———. 2006. *Drinking water distribution systems: assessing and reducing risks*. Washington, DC: The National Academies Press.

———. 2007. *Colorado River basin water management: evaluating and adjusting to hydroclimatic variability*. Washington, DC: The National Academies Press.

———. 2008a. *Desalination: a national perspective*. Washington, DC: The National Academies Press.

———. 2008b. *Prospects for managed underground storage of recoverable water*. Washington, DC: The National Academies Press.

———. 2008c. *Urban stormwater management in the United States*. Washington, DC: The National Academies Press.

O'Reilly, C. E., M. C. Freeman, M. Ravani, J. Migele, A. Mwaki, M. Ayalo, S. Ombeki, R. M. Hoekstra, and R. Quick. 2008. The impact of a school-based safe water and hygiene programme on knowledge and practices of students and their parents: Nyanza Province, western Kenya, 2006. *Epidemiology and Infection* 136(1):80-91.

Osterholm, M. T. 2006. The food supply and biodefense: the next frontier of the food safety agenda. In *Addressing foodborne threats to health: policies, practices, and global coordination*. Washington, DC: National Academy Press.

Parker, A. A., R. Stephenson, P. L. Riley, S. Ombeki, C. Komolleh, L. Sibley, and R. Quick. 2006. Sustained high levels of stored drinking water treatment and retention of hand-washing knowledge in rural Kenyan households following a clinic-based intervention. *Epidemiology and Infection* 134(5):1029-1036.

Pond, K. 2005. *Water recreation and disease. plausibility of associated infections: acute effects, sequelae and mortality*. London: IWA Publishing on behalf of the World Health Organization.

Prüss-Üstün, A., R. Bos, F. Gore, and J. Bartram. 2008. Safer water, better health: costs, benefits and sustainability of interventions to protect and promote health. Geneva: World Health Organization.

Seas, C., J. Miranda, A. I. Gil, R. Leon-Barua, J. Patz, A. Huq, R. R. Colwell, and R. B. Sack. 2000. New insights on the emergence of cholera in Latin America during 1991: the Peruvian experience. *American Journal of Tropical Medicine and Hygiene* 62(4):513-517.

Thebaut, J. 2005. *Running dry*. Produced and directed by J. Thebaut. Redondo Beach, CA: The Chronicles Group. DVD.

UN (United Nations). 2009. *World water development report 3: water in a changing world*. New York: UNESCO and Earthscan.

UN Millennium Project. 2005. *Health, dignity, and development: what will it take?* UN Millennium Project Task Force on Water and Sanitation. London and Sterling, VA: Earthscan.

UNDESA (United Nations Department of Economic and Social Affairs). 1992. *Agenda 21 earth summit: United Nations program of action from Rio*. New York: United Nations.

UNDP (United Nations Development Programme). 2009. *About the MDGs: basics*, http://www.undp.org/mdg/basics.shtml (accessed March 17, 2009).

UNESCO (United Nations Educational, Scientific and Cultural Organization). 2006. *Water, a shared responsibility: the United Nations world water development report 2*. New York: UNESCO/Berghahn Books.

UNICEF (United Nations Children's Fund). 2006. *Progress for children: a report card on water and sanitation*. New York: UNICEF.

van der Werf, M. J., S. J. de Vlas, S. Brooker, C. Looman, N. Nagelkerke, J. Habbema, and D. Engels. 2003. Quantification of clinical morbidity associated with schistosome infection in sub-Saharan Africa. *Acta Tropica* 86(2-3):125-139.

Wagner, E. G., and J. N. Lanoix. 1958. Excreta disposal for rural areas and small communities. *WHO monograph series*. P. 39.

White, G. F., D. J. Bradley, and A. U. White. 1972. *Drawers of water: domestic water use in East Africa*. Chicago: University of Chicago Press.

WHO (World Health Organization). 2008a. *Dengue and dengue hemmorhagic fever fact sheet*, http://www.who.int/mediacentre/factsheets/fs117/en/ (accessed August 10, 2008).
———. 2008b. *Human african trypanosomiasis*, http://www.who.int/trypanosomiasis_african/disease/en/index.html (accessed August 10, 2008).
———. 2008c. *Initiative for vaccine research: schistosomiasis*, http://www.who.int/vaccine_research/diseases/soa_parasitic/en/index5.html (accessed August 10, 2008).
———. 2008d. *International network to promote household water treatment and safe storage*, http://www.who.int/household_water/network/en/ (accessed March 27, 2008).
———. 2008e. *Malaria fact sheet*, http://www.who.int/mediacentre/factsheets/fs094/en/index.html (accessed August 10, 2008).
———. 2008f. *Onchocerciasis disease information*, http://www.who.int/tdr/diseases/oncho/diseaseinfo.htm (accessed August 10, 2008).
WHO/UNICEF. 2005. *Water for life: making it happen*. Geneva: WHO/UNICEF.
———. 2006. *Meeting the MDG water and sanitation target: the urban and rural challenge of the decade*. New York and Geneva: UNICEF and WHO.
Wright, J., S. Gundry, and R. Conroy. 2004. Household drinking water in developing countries: a systematic review of microbiological contamination between source and point-of-use. *Tropical Medicine and International Health* 9(1):106-117.
www.Data360.org. 2009. *Average water use per person per day*, http://www.data360.org/dsg.aspx?Data_Set_Group_Id=757 (accessed June 3, 2009).
Zaidi, A. K. M., S. Awasthi, and H. J. deSilva. 2004. Burden of infectious diseases in South Asia. *British Medical Journal* 328(7443):811-815.

1

Global Problems, Local Solutions

OVERVIEW

In order to appreciate the complexity of the global water crisis, it must be viewed from multiple perspectives, and its effects considered on scales ranging from the individual to the planet. This chapter offers several opportunities for reflecting on the necessity of safe water and sanitation services, the consequences of their increasing scarcity, and the potential for addressing these consequences within communities, at the national level, and through international cooperation.

We begin, not with an essay or review, but with the 18-minute film *Running Dry*, on the DVD included with this volume and which opened the workshop. The film's writer, producer, and director, James Thebaut, said that his work was inspired by the book *Tapped Out* (Simon, 1998), by the late Senator Paul Simon. The film examines the growing global water crisis and its staggering toll of some 14,000 "quiet preventable deaths" per day. Focusing on China, the Middle East, Africa, India, and the United States, *Running Dry* presents compelling arguments for international cooperation on water issues and highlights some promising grassroots programs to improve access to safe water.

Grassroots efforts to improve water access, sanitation, and health are also featured in the chapter's second contribution, by Donald R. Hopkins of the Carter Center. In the course of recounting the histories of two programs supported by the Center and undertaken in Africa—the first to control trachoma in Ethiopia; the second to eliminate dracunculiasis (Guinea worm disease) in Ghana—Hopkins demonstrates the key role of behavioral change as a determinant of success for public health interventions.

A detailed understanding of waterborne pathogen transmission pathways can inform the development and implementation of effective interventions to prevent and control diseases. Such analyses, characterized in the chapter's third contribution by David Bradley of the London School of Hygiene and Tropical Medicine, highlight the importance of the household as a target for clean water interventions and the critical role of water access, as distinct from water quality, in preventing water-related diseases. In 1972, Bradley and coworkers published the first functional classification of water-related diseases according to their routes of transmission (White et al., 1972). This now widely used scheme structures and clarifies information critical to interdisciplinary efforts to address the health effects of water and sanitation (see also Workshop Overview). Bradley revisits and critically reviews his original taxonomy of water-related disease, and suggests several modifications to incorporate recent research findings, such as the involvement of the respiratory tract in certain water-related diseases (e.g., Legionnaire's disease; respiratory infections controlled by hand washing).

Taking the same approach to the study of sanitation and disease, Bradley and coworkers identified six sanitation-related transmission categories to inform the choice of preventive measures for a given disease (see Table WO-3 in the Workshop Overview). Bradley presents and explains this classification scheme and describes its possible integration with the functional taxonomy of water-related disease. To further advance the understanding of waterborne disease transmission processes provided by functional taxonomies, he also explores the systematization of hygiene behavior and the spatial structure of water and sanitation services. Finally, in order to assess the relevance of these concepts, Bradley applies them to a real-world system in southwest Uganda.

In much of the developing world, "water is often collected from sources of dubious quality, hauled over a distance, and stored in the home before it is consumed," observe workshop speaker Robert Tauxe, of the Centers for Disease Control and Prevention (CDC), and coworkers, who contributed this chapter's final essay. Water gathered in this way is vulnerable to contamination between its source and its point of use, and thus requires the most local of interventions in order to ensure its safety: household water treatment and storage interventions (also known as "point-of-use" strategies). Tauxe and colleagues discuss the concept and practice of point-of-use water treatment and review findings of recent implementation trials of this strategy in diverse settings that demonstrate its impact on public health; some of these trials featured the effective integration of point-of-use interventions with hand washing and other public health strategies. The authors also explore the critical connection between water- and foodborne disease through a series of case studies, all of which illustrate the global effects of local water quality.

IMPROVING WATER, SANITATION, AND HEALTH AT THE GRASSROOTS

Donald R. Hopkins, M.D., M.P.H.[1]
The Carter Center

While others are reaching the moon, we are trying to reach the villages.
Julius Nyerere

The Carter Center's motto is "Waging Peace, Fighting Disease, Building Hope." Our divisional motto in the Health Programs of The Carter Center is "Fighting Disease and Building Hope at the Grassroots." It is in the spirit of that motto that I shall review aspects of two programs we are assisting: illustrating the interaction of water and health in the fight against dracunculiasis (Guinea worm disease) in Ghana, and the interaction of sanitation and health in the fight to control trachoma in Ethiopia. The importance of behavioral change as well as biologic control measures is also greatly evident in both programs.

Trachoma Control in Ethiopia

Trachoma is a chronic bacterial infection of the eye that is caused by *Chlamydia trachomatis* (Figure 1-1). With an estimated 8 million blinded victims, comprising 16 percent of the global burden of blindness, trachoma is the leading cause of preventable blindness in the world, and an estimated three-quarters of its blind victims are women. About 500 million persons, or about 10 percent of the world's population, are at risk of the disease, and some 63 million are estimated to have active trachoma in 56 countries. In its early stages, the disease is characterized by inflamed inner surfaces of the eyelids, which yield pus and related secretions around the eye, to which certain species of flies (*Musca sorbens*) are attracted. The bacteria are spread from one person to another by the flies, and by contaminated fingers, cloths, towels, sheets, or other fomites. After repeated infections, the inflamed inner surfaces of the eyelids become scarred and contract, turning the eyelashes inward, causing them to scrape the eyeball, which is extremely painful. The subsequent scarring of the cornea of the eye itself causes blindness. Women are affected more than men because young children are the reservoirs of infection in the remote, poor, dusty, and unhygienic villages where this infection flourishes, and women suffer repeated reinfections because of their close association with their children (Mabey et al., 2003).

---

[1]Vice President, Health Programs.

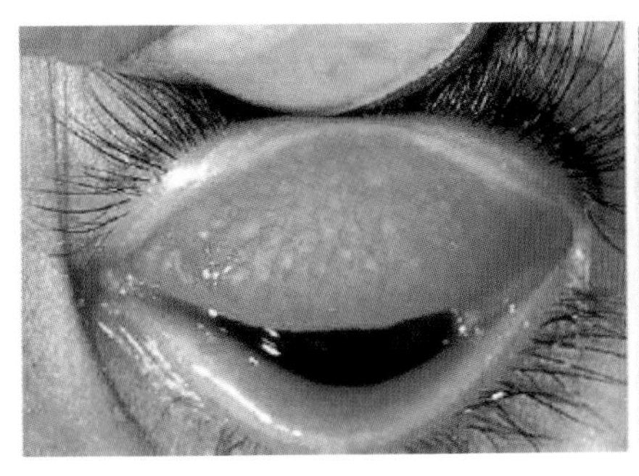
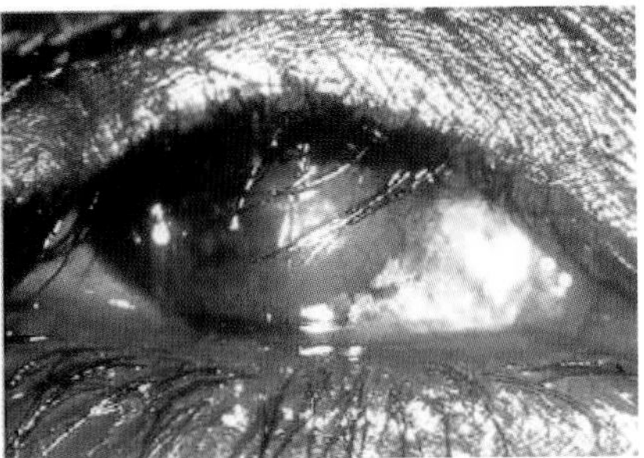

FIGURE 1-1 Patients with early (left) and late (right) trachomatous infections.
SOURCE: Reprinted from WHO (2009) with permission from the World Health Organization.

The World Health Organization (WHO) has developed a four-pronged approach, called the "SAFE strategy," which combines curative and preventive control measures for achieving sustained reductions in prevalence of the disease (Emerson et al., 2006). The S stands for *surgery* to correct inward-turning eyelashes and thus prevent progression to blindness if the surgery is conducted early enough. The A is for mass distribution of *antibiotics* to cure active phases of the infection, although people do not develop immunity and are still subject to reinfection. The F stands for *face* (and hand) washing, which decreases attraction of the disease-bearing flies by removing secretions and dirt. The E denotes *environmental* improvement, especially measures to prevent humans from defecating on the ground, where such deposits are preferred breeding sites for *M. sorbens* (Emerson et al., 2001).

Worldwide, Ethiopia[2] is most affected by trachoma, and the Amhara Region (population ~19 million) of Ethiopia contains 31 percent of the active trachoma in the country (Emerson et al., 2006). In partnership with national and regional health authorities and local Lions Club members, and with support of the Lions Club International Foundation, Pfizer Inc. (which donates Zithromax® antibiotic via the International Trachoma Initiative), and other donors, in October 2000, The Carter Center began assisting efforts to eliminate blinding trachoma from the Amhara Region of Ethiopia by 2012. This elimination campaign emphasizes all four components of the SAFE strategy, but the E component is of particular relevance here.

Since latrines are the main intervention to prevent people from defecating on the open ground and thus denying suitable breeding places for *M. sorbens*, the program in Amhara Region began encouraging villagers to construct and use latrines as part of the effort to prevent trachoma. With the support of local

[2]Population estimated at 85.2 million as of June 2009 (CIA, 2009a).

health and administrative leaders and Ethiopian Lions Club members, villagers in an initial area comprising parts of four districts built 1,333 simple latrines in 2002 at negligible cost beyond the labor for digging the individual pits, by taking advantage of favorable local geology and a plentiful supply of available wood, mud, and thatch (Figure 1-2). After other villagers in the area built 2,151 latrines in 2003, the program set an ambitious goal to build 10,000 latrines in 2004. What happened next astonished all who were concerned.

FIGURE 1-2 Example of a latrine in Amhara Region of Ethiopia.
SOURCE: Courtesy of The Carter Center/L. Rotondo.

As villagers were mobilized to build latrines to prevent trachoma, women and girls began to take special interest for their own reasons, since local tradition allowed men to defecate in the open during the day, but forced women to wait until dark so no one could see them. One woman commented, "I am a prisoner of the daylight." Women's groups began pushing their husbands and other male relatives to dig latrines, which soon became status symbols and a focus for competition among families, villages, and officials. Villagers built not just 10,000, but more than 89,000 latrines in 2004 (O'Loughlin et al., 2006), and over 144,000 more latrines in 2005. Another woman was quoted as declaring, "Now we are equal to the men!" and another vowed, "We'll never go back [to defecating in the field]!" After temporarily diminished output due to insecurity related to national elections, and program emphasis on distributing mosquito nets, over a quarter million latrines were constructed between January and August 2008 (Figure 1-3). This surprising feminist-led explosion in latrine building will not only help prevent trachoma but will undoubtedly reduce the spread of several intestinal parasites and diarrheal diseases as well. This experience also shows how quickly people will change their behavior when they perceive a rational reason for doing so.

Dracunculiasis Elimination in Ghana

Dracunculiasis (Guinea worm disease) is a parasitic infection caused by the nematode *Dracunculus medinensis* (Figure 1-4). The infection is manifest by one-meter-long thin white worms that emerge directly and slowly through the skin on any part of the body. The resulting pain incapacitates victims for periods averaging two to three months and severely constrains agricultural productivity and school attendance. People are infected by drinking water that contains tiny water fleas that have ingested immature forms of the parasite that have been spewed into stagnant ponds from emerging adult worms. This infection is only transmitted by contaminated drinking water, and there is a one-year lag between infection and emergence of the adult worm. Dracunculiasis also has no animal reservoir other than humans, no vaccine or cure, and infection confers no immunity to reinfection. It can be prevented, however, by teaching villagers to filter their drinking water through a fine cloth and to not enter sources of drinking water when a worm is emerging; by treating unsafe sources monthly with a mild larvicide, ABATE®, which is safe for humans; and by providing safe sources of drinking water such as from borehole wells (Hopkins and Ruiz-Tiben, 1991).

In 1986, dracunculiasis infected an estimated 3.5 million persons and some 120 million persons were at risk of the disease in impoverished rural areas of India, Pakistan, Yemen, and 17 affected African countries (Watts, 1987). Ghana[3] was the second-most highly endemic country for dracunculiasis, enumerating

[3]Population estimated at 23.8 million as of June 2009 (CIA, 2009b).

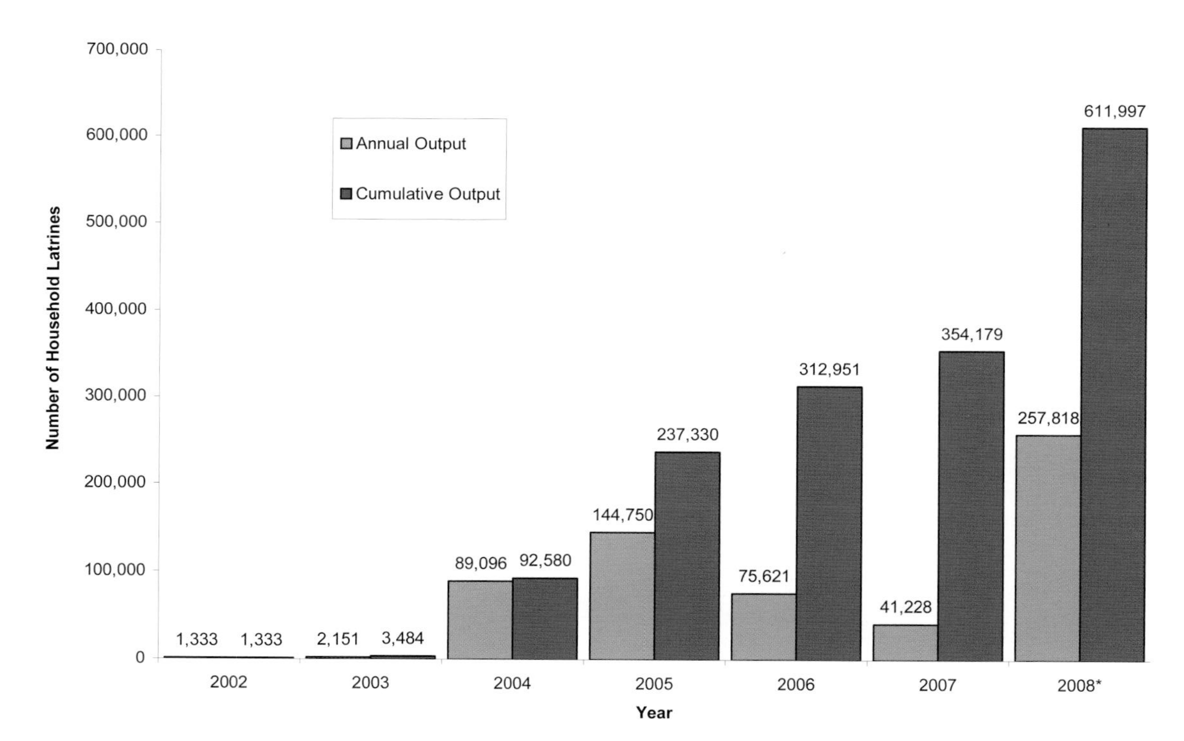

FIGURE 1-3 Carter Center-supported household latrine construction in Ethiopia. *2008 data are provisional, January-August. SOURCE: Courtesy of The Carter Center.

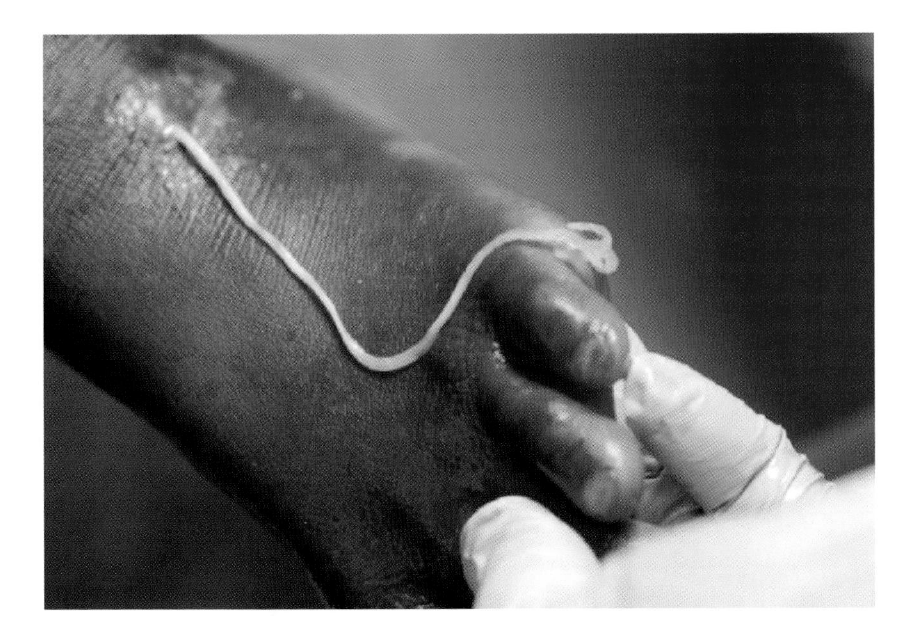

FIGURE 1-4 Guinea worm emerging.
SOURCE: Courtesy of The Carter Center/L. Gubb.

nearly 180,000 cases during its first nationwide village-by-village search for cases of the disease in 1989-1990. Ghana's poor, arid, and long-neglected Northern Region, which includes one-third of the land mass of the country and the lowest rates of literacy, school attendance, and access to safe drinking water and medical care, was found to contain 56 percent of all cases.

With early external support from The Carter Center, the U.S. Centers for Disease Control and Prevention (CDC), the U.S. Agency for International Development (USAID), the Japan International Cooperation Agency (JICA), the United Nations Children's Fund (UNICEF), the Danish Bilharziasis Laboratory, and enthusiastic support from then head of state Flight Lieutenant (retired) Jerry John Rawlings, Ghana's Guinea Worm Eradication Program made impressive progress initially, reducing cases by 95 percent between 1989 and 1994, from 179,670 to 8,432. As a result of JICA's targeted provision of 159 borehole wells in the Northern Region's agriculturally fertile Nanumba District in 1988-1989, for example, cases of dracunculiasis in the district plummeted by 77 percent in one year, from ~14,000 to ~3,000 between 1989 and 1990, with substantial associated increases in local yam production.

A disastrous and ill-timed outbreak of ethnic fighting in the most highly endemic areas of the Northern Region in 1994-1995 inaugurated a 12-year-long period of programmatic stagnation that has been overcome only in the past two years (Figure 1-5). Some of the long-standing historical enmities at play in the volatile region include memories of slave raiding during the Atlantic Slave Trade, and earlier twentieth century battles that preceded the "Guinea Fowl War" of 1994-1995 (Dawson, 2000; Skalník, 1987). These and other factors such as strongly held traditional beliefs about the disease were manifest over the past five years by the predominance of cases in one or another of the different ethnic groups. Addressing the latter challenges required targeting training, mobilization, and health education efforts to members of specific ethnic groups and their leaders, once their relatively low participation was belatedly recognized.

Without a vaccine or curative drug, in Ghana and elsewhere the global Guinea Worm Eradication Program has had to rely on persuading large numbers of conservative villagers to change their behavior. The obvious horror and easy diagnosis of clinical dracunculiasis are advantages in that regard, but the one-year-long incubation period between infection by drinking contaminated water and appearance of the disease, and the strong traditional beliefs associated with the disease (Bierlich, 1995), are distinct disadvantages. Filtering water from the local source of drinking water, backwashing the filtered material into a clear jar or glass, and letting villagers see the numerous water fleas (and other microscopic organisms) swimming around in the water they were drinking is a very effective tool for convincing villagers to filter their water before drinking it. It also helps them understand how and why the infection comes from their drinking water.

In Ghana and other dracunculiasis endemic countries, unfavorable geologic (Hunter, 1997) and sociologic factors, expense, political pressures, indifference, corruption, and managerial incompetence have often impeded attempts to bring the water supply sector to bear in endemic areas. The dramatic impact realized by the JICA project cited above has been much rarer than it ought to be in the Guinea worm eradication campaign. But advocacy by health workers in this eradication campaign for consideration to be given to distribution of *disease* in establishing priorities for providing and rehabilitating sources of drinking water has at least helped affirm that important principle for future decision makers.

Meanwhile, the global Dracunculiasis Eradication Program has reduced cases of the disease from the estimated 3.5 million cases in 20 countries in 1986 to a projected total of less than 5,000 cases in 6 countries in 2008, with the overwhelming majority (98 percent) of remaining cases in Sudan, Ghana, and Mali, and falling fast (Hopkins et al., 2008a,b; Figure 1-6). The goal is to try to stop all transmission of dracunculiasis by the end of 2009.

Acknowledgments

I am grateful to Dr. Ernesto Ruiz-Tiben, director of The Carter Center's Guinea Worm Eradication Program, and his staff; to Dr. Paul Emerson, director of The Carter Center's Trachoma Control Program, and his staff; to the many

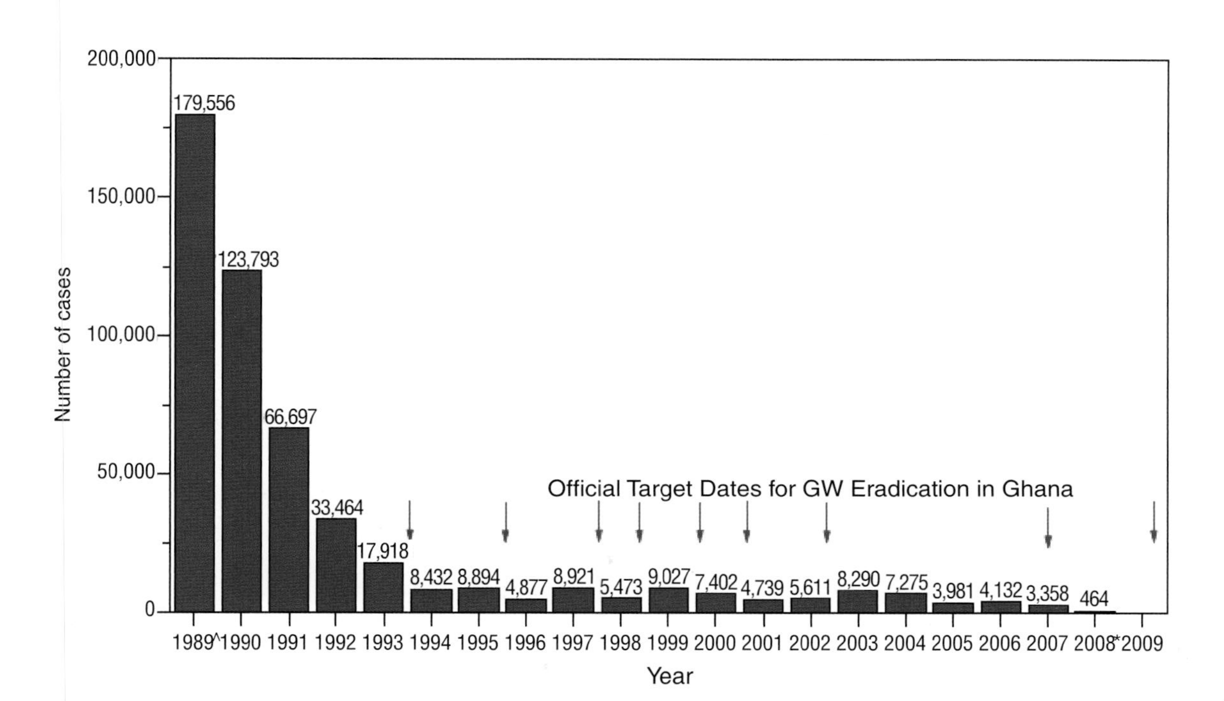

FIGURE 1-5 Ghana Guinea Worm Eradication Program. Number of cases of dracunculiasis reported by year, 1989-2008 (1989^ national case search; 2008* number is provisional).
SOURCE: Courtesy of The Carter Center.

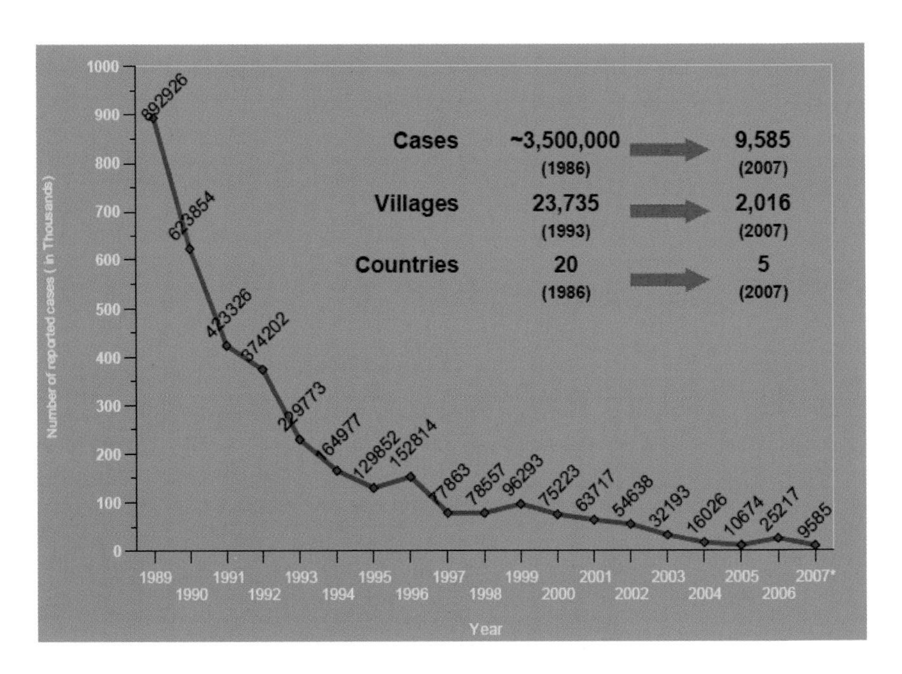

FIGURE 1-6 Number of reported cases of dracunculiasis by year, 1989-2007.
SOURCE: Courtesy of The Carter Center.

village volunteers, health workers, local Lions Club members in Ethiopia; and to The Carter Center's numerous donors for their support and participation in the work reported here. I also thank my executive assistant, Ms. Shandal Sullivan, for assisting me in preparing this manuscript.

THE SPECTRUM OF WATER-RELATED DISEASE TRANSMISSION PROCESSES

David J. Bradley[4]
London School of Hygiene and Tropical Medicine

Introduction

This paper seeks to do three things, all related to functional classification, or taxonomy, of water-related diseases. The first is to revisit the taxonomies of diseases related to water and to sanitation that I put forward more than 30 years

[4]Ross Professor of Tropical Hygiene Emeritus.

ago, review them critically, and suggest modifications. The second is to venture into some unsystematized areas that need attention in the epidemiology of water-related diseases in space and time. Third, I shall use a small area of southwest Uganda to show the complexities, related to change, opportunities, and research questions, being raised by water-related disease. Overall, this paper will tend to take a "water's eye view" of disease transmission.

In Defense of Functional Taxonomies

A classification is a qualitative model using distinctions that matter. Useful classifications of living beings or other phenomena enable simplification of the natural world or of human experience and are a form of conceptual model which, by making generalizations possible, can make communication across different disciplines easier. In public health, where analysis is often performed by physicians or by epidemiologists, while successful interventions often depend on engineers or on policy makers and planners, classifications can be very helpful, especially functional classifications that point toward common interventions and are formulated in the technical language of those who are expected to implement control measures.

In the case of water-related diseases, progress in explaining the issues to engineers and others who manage water had been impeded in two ways: a limited perception and an inappropriate classification. Because water-related pathogens fall into several different biological groupings (i.e., criteria, viruses, protozoa, and helminths), they have been conventionally discussed in those terms, although the modes of water-related transmission cut across these biological categories. Hence, a functional classification of modes of transmission is more useful to those in charge of interventions to control disease. It also avoids what they might view as biological jargon since they would lack the useful connotations that would be apparent to a microbiologist. The limited perceptions of water-related disease in the world as a whole resulted from using the simplified list of infections seen in richer temperate countries and addressed in the conventional texts of those countries. Since piped water to all dwellings was taken as a given, the problem focus was upon pathogens in the water itself, contracted by drinking the water. The broad spectrum of water-related diseases was therefore reduced to a subset and the term *waterborne diseases* was inappropriately used for the whole lot, with the consequent narrow focus on microbiological water quality that has limited the range of interventions as well as implying a single transmission route, which is far from true.

Our introduction of a broader approach—of dividing water-related communicable disease transmission into four functional categories—seemed to meet the need for a way of communicating across disciplines and has become widely accepted (White et al., 1972). It had quite large policy consequences for water supply in particular and has been used chiefly for work in tropical poor countries, where the populations are exposed to the full range of water-related pathogens.

However, it has been rarely critically reviewed, and this is done in the next section.

Since our group turned its attention from water to sanitation (here used in the narrow sense of excreta management), it was logical to attempt a functional categorization of sanitation-related infections (Feachem et al., 1983). This proved to be somewhat more complex, and so less elegant, and, because sanitation is less often discussed than water, the classification is less well known and therefore is set out and discussed later in the section on functional classification of sanitation-related diseases.

Both water- and sanitation-related functional classifications of disease transmission were addressed, particularly to public heath engineers, planners, and geographers—chiefly the first of these professions. Can functional classifications have a wider relevance? A major focus of water- and sanitation-related health research in the past two decades has been on human behavior. The spatial element of sanitary activity has also become more tractable with the development of geographical information systems. Moreover, the dynamic nature of many poor communities has become apparent recently—even rural areas may be undergoing rapid environmental and population changes that are highly relevant to water-related disease patterns. The section on new areas for systematic attention gives a tentative exploration of types of functional taxonomy that might prove relevant to these dimensions of the problem. Lest these endeavors appear too abstract, the final section applies these concepts to a very rapidly changing area of Uganda and explores their relevance in the real world.

The Functional Classification of Water-Related Disease Transmission Revisited

The original classification of water-related disease transmission was developed for a study of domestic water use in East Africa. It aimed to be more comprehensive than the previous narrow "waterborne disease" textbook focus and to separate out the epidemiological processes involved. One consequence was that some diseases fell into two categories—but this was not a disadvantage if it showed that two distinct interventions were required to control transmission (White et al., 1972). There were four categories, as set out in Table 1-1, which also links interventions to categories. From a policy viewpoint, the clear separation of the first two categories, the "strictly waterborne" and the water-washed aspects of transmission, had the most significance, in particular because it drew attention to problems of access to enough water for personal and household hygienic purposes as well as water quality. Since there were many opportunities for improving one aspect at modest cost, even when the other might be financially unaffordable, this spurred operational improvement of rural water supplies and gave emphasis to opportunities that had previously been neglected in favor of urban activities. The formulation of a "waterwashed" modality, whereby disease

TABLE 1-1 Revised Classification of Water-Related Disease Transmission

1. WATER-BORNE
Classical
Other
2. WATER-WASHED
Diarrhea
Skin
Eyes
Water-washed respiratory infection transmission*
3. WATER-AEROSOL*
4. WATER-BASED
Percutaneous
Oral
5. WATER-RELATED INSECT VECTORS
Breeding
Biting
6. WATER-SOURCE CROWDING TRANSMISSION*

*These new categories are discussed in the text.

burden might be reduced by making water more available even if its quality were not fully ideal, also pointed to research questions that had been neglected, the answers to which also led on to the issues of hygiene behavior which were of great importance from the 1990s onward.

Because both water access (quantity used) and microbiological quality affected transmission of many diarrheal diseases, those primarily wanting a simple "pigeon-hole" for each infective disease sometimes lumped them together as fecal-oral diseases, which simplified grouping but lost the analytical and interventional distinctions between quantity and quality, which is of particular importance in highly resource-constrained environments and communities. The third and fourth categories—of water-based diseases (helminths with aquatic intermediate hosts) and water-related insect vectors—both drew attention to further hazards of undeveloped or poorly managed surface water sources and also brought together two previously disparate literatures and areas of study: domestic water supply and diseases of water resource development (such as reservoir construction and irrigation projects), to their mutual benefit (Bradley, 1977a,b; Ensink et al., 2002; Moriarty et al., 2004).

Overall, the water classification has been little criticized and rather widely accepted. However, detailed epidemiological and pathophysiological study of pathogens, and in particular the discovery of many previously unknown pathogenic viruses and bacteria in recent years, has made reassessment overdue.

I believe there is a need for a fifth primary category of transmission, and development of the secondary categories is also needed to incorporate recent research.

It is now clear that bacteria of the genus *Legionella* are spread from infected persons in water but normally produce secondary infections by inhalation and consequent infection of the respiratory tract, and usually not through the intestinal tract (Szewzyk et al., 2000). This makes for important epidemiological differences in terms of patterns of transmission and preventative measures. *Legionella* has become the most common single cause of water-related infection in the United States and therefore is no longer a rarity. Its transmission is often related to air-conditioning systems and not only domestic water use in the conventional sense, giving an added reason for its separate categorization. For all those reasons a fifth primary category of water-aerosol transmission has been added to Table 1-1.

The respiratory tract is also involved in an addition to the secondary categories. Currently the water-washed diseases, primary category 2, are subdivided into the diarrheal diseases and the body surface infections that are affected by water availability for hygiene purposes: this latter group may be viewed as a single category or, more usefully, as two: skin infections and trachoma, a potentially blinding infection of the conjunctiva of the eye.

Recent work has demonstrated that provision of water for hygiene purposes, coupled with soap and education in their proper use, is in some poor communities associated with a reduction in respiratory tract infections. The number of such studies is limited (Rabie and Curtis, 2006), but there does appear to be a need for a category of water-washed respiratory infection transmission, which has also been added to Table 1-1.

There is a further way in which water affects disease transmission. In semi-arid areas, sources of surface water may be both few and small in size, and in great demand not only by people and their livestock but also by wildlife in the area. The resulting multispecies crowding around waterholes may greatly facilitate both intraspecific and interspecific transmission of communicable diseases. If our interest extends to new pathogens and emerging diseases this is highly relevant, particularly because the majority of human emerging diseases in recent years have been zoonoses. One could, therefore, add a sixth category to Table 1-1—water-crowding diseases; this is placed tentatively with the other more directly water-related transmission.

Functional Classification of Sanitation-Related Diseases

The apparent utility of the functional classification of water-related disease naturally led to exploration of a similar approach to sanitation and disease. This was presented in a short chapter of a rather large volume (Feachem et al., 1983) and seems to have largely escaped critical attention. It is therefore outlined here as a preliminary to exploring how the two classifications may be integrated. There is a substantial overlap between the diseases related to water and those related to sanitation, both because they are different phases of a single transmission cycle in many cases—sanitation deals with the pathway out of the infected host while water may be route of entry into a new host by a pathogen—and also because the public health engineer is heavily involved in both aspects.

Three characteristics of excreted pathogens strongly affect transmission: latency, persistence, and multiplication. Some pathogens are immediately infectious on emerging from the host in the feces but others have a latent period lasting from a few hours to several weeks before they are infectious. The latter are more likely to be transmitted some way away from where the feces were passed. Excreted pathogens vary greatly in their ability to survive outside the human body: eggs of the helminth *Enterobius* survive only briefly and are readily transmitted from the anus (where they adaptively give rise to itching) to the mouth in young children. By contrast, the eggs of the large roundworm *Ascaris* are very resistant and persist for up to several years in the ground, so that there is much more time for them to be carried to other places. Often latency and persistence are combined in the same helminth species so that these have a very different epidemiology than do short-lived species without latency. The former are much affected by the long-term fate of excreta and their management, while the latter are more influenced by personal hygiene behavior and the immediate condition of the latrine or toilet. The third bioenvironmental variable is whether the pathogen multiplies in the excreta, or in some component of the biological environment, or not at all outside the human host.

An empirical classification on the basis of these three variables gives rise to five categories (Table 1-2). The first two lack latency, with the second having a longer persistence and sometimes multiplication in the excreta. Categories III-V show latency, marked persistence in the case of category III, while category IV undergoes development in another mammal before becoming infective to people at the mature sexual stage, and category V develops in aquatic intermediate hosts. A sixth category is for infections transmitted by insects that are sanitation-related, such as the *Culicine* mosquitoes that thrive in heavily organically polluted waters, cockroaches, and some flies.

The categories are related to the relative efficacy of various sanitary interventions. Human hygienic behavior (and particularly hand washing with soap) is most effective in the first two categories of disease transmission, the sanitary infrastructure affects categories II-IV the most, and the last two categories are most dependent on rather specific measures unless the general sanitary standard is relatively high.

Thus, the classification gives a sense of the most important means of control for particular diseases: whether personal hygiene behavior, domestic water supply, provision of toilets and how they are maintained, eventual treatment of the excreta, or very specific methods of attack are most cost-effective. These in turn indicate what level of intervention is likely to be most effective, whether at the individual (e.g., hand washing), the household (toilet provision), or the community level (sewage treatment).

TABLE 1-2 Excreta-Related Transmission

Type	Latency	Persistence	Multiplication	Biology, predominantly:	Examples	Routes of entry and egress*
I	No	Short	No	Viruses, Protozoa, Helminths	*Enterobius vermicularis,*	or, pa
					Rotavirus	or, fe
II	No	Longer	Yes	Bacteria	*Salmonella typhi,*	or, ur/fe
					Leptospira interrogans	pc, ur
III	Yes	Long	No	Helminths	*Ascaris lumbricoides,*	or, fe
					Necator americanus	pc, fe
IV	Yes	[Long]	In cow/pig	Helminths	*Taenia solium, T*	or, fe
					aenia saginata	or, fe
V	Yes	[Long]	In aquatic organisms	Helminths	*Schistosoma haematobium,*	pc, ur
					Schistosoma mansoni	pc, fe
VI	Spread by excreta-related insects			Some mosquitoes		

*For entry: or = oral; pc = percutaneous or through mucosa; for egress: fe = in feces; pa = perianal region; ur = in urine.
SOURCE: David Bradley.

The utility of the separate water-related disease and sanitation-related disease classifications is, for public health engineers, particularly in relation to choosing appropriate water and sanitation improvements, respectively, in conditions of limited resources. It is possible, as shown in Figure 1-7, to link the two systems, though this tends to make things look more complicated than is necessary. If this is to be done, the third component of the complex system—the importance of hygiene behavior—needs to be imported, as in the next section.

Three New Areas for Systematic Attention

The two preceding classification systems have facilitated communication between epidemiologists and public health engineers by linking transmission ecology to specific modes of intervention. But there are other aspects that they ignore or address indirectly. Two such areas are hygiene behavior and the spatial structure of water-related disease transmission.

Behavior

Over the past 40 years, there has been a progression in our focus from water in the 1960s and 1970s, through sanitation in the 1980s, to sanitation behavior in the 1990s (Cairncross and Kochar, 1994). The key stimulus for the focus to extend from structures to behavior was the demonstration that hand washing with soap reduced the transmission of cholera and other sanitation-related diseases (Curtis and Cairncross, 2003). Other behavioral issues became apparent also: whether children's feces were perceived as a health risk or not, the way that distance to water (typically a standpipe) outside the home did not affect the amount of water carried between the source and household over a considerable range of distances, and the recent observation that handwashing with soap also appears to reduce transmission of upper respiratory infections.

Is there a way to classify this growing body of behavioral information relevant to control of water- and sanitation-related diseases to guide the parent, teacher, or health educator? This needs exploration by social scientists. It may be that a simple six- or eight-cell categorization into four levels (person, household, community, and state) and two types (positive actions that are to be encouraged, and things to avoid doing and which need regulation either formally or by cultural pressure, Table 1-3) may be of value in thinking about behavioral-change agendas. Such an approach would be a classification of behaviors, to which the various diseases can be linked. Anything that served to group healthy behaviors in a usable way would be of public health value. There is a sense in which the water-related transmission categories point to appropriate behavior: boiling or filtering water for waterborne diseases, hand washing with soap for the water-washed transmission, avoiding water bodies with water-based disease transmission, and so on. But could there be a wider and more informative approach?

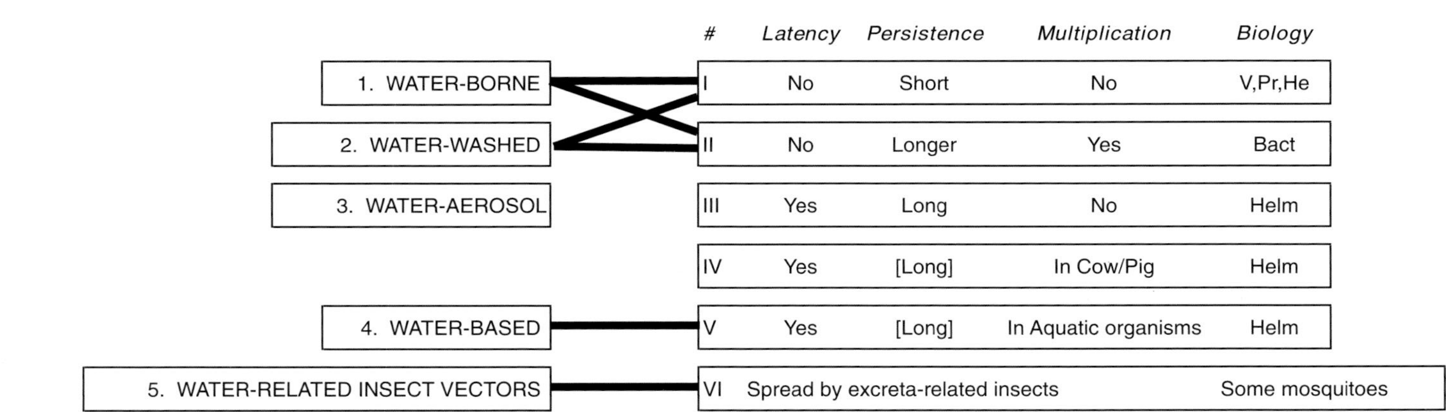

FIGURE 1-7 Relation of water- and excreta-related transmission categories. Some waterborne and some water-washed diseases are also related to excreta, in both categories I and II; the water-aerosol diseases are not excreta-related, nor are the excreta-related disease categories III and IV primarily water-related diseases. However, water-based diseases also mostly fall into category V of excreta-related diseases, and some few mosquito vectors of disease prefer to breed in highly excreta-polluted waters.

TABLE 1-3 A Possible Way to Group Behavior Change for Water and Sanitation Interventions

	Action Type	
Level	Positive Actions	Things to Avoid Doing
Person		
Household		
Local Community		
State		

NOTE: The scheme set out in this table is emphatically not proposed to be of definitive value as it stands. It rather serves to provoke attention from the appropriate specialists in the hope of providing better guidance to those whose activities will, whether by chance or design, affect the public health through changing the transmission of water-related diseases.

Space

The other aspect of epidemiology that has developed dramatically over recent years concerns spatial processes. In rural and peri-urban sites, especially under the conditions of rapid environmental change that prevail in many developing countries, the factors related to disease transmission are spatially variable and complex. In the periphery of expanding African cities, where a mixture of urban and agricultural activities persists, the location of on-site wells and pit latrines affects disease risk. The management, distribution, and allocation of space between the individual and the community; the dichotomy between public space and private space emphasized by Cairncross; and the successive circles of spaces discussed by Kolsky, may be guides to a possible classification, and matters of scale need attention. Such a spatial analysis could primarily relate to geographical space and distance, as for the sanitation diseases that are best controlled by personal cleanliness after excreting, or by toilet provision, or by the ultimate processing of the excreta. Alternatively, it may relate to political subdivisions of space—communal or public versus personal and private. There is also the publicly funded versus privatized, leading onto allocation of construction and of maintenance costs appropriately: consideration of this led to the challenge of received wisdom by the suggestion that the role of the community was more suitably in building water supplies than in their maintenance.

The types of spatial processes affecting these diseases are diverse, and it is unclear how best to reach a simple but useful way to perceive matters. It is clear that, for water and sanitation, there are some situations where widely used interventions, often assumed to be universally applicable, are inadequate. For example, nomadic populations require special arrangements. As for many health interventions, "one size does not fit all." If one has a very small number of people using a water source that is polluted only by themselves, is the risk of disease much reduced compared to larger common sources? What are the extra hazards when surface water sources are shared among livestock and people? What special opportunities for disease control result from the many large environmental changes affecting water resources? It is far from clear whether convenient generalizations or spatial classifications can clarify this complexity and aid the public health worker, though there is certainly a need to plan more often in spatial terms than has been usual. A strongly spatial theory of disease transmission is needed for communication with physical planners, especially at the urban periphery.

Spatial issues are of particular relevance in considering emerging diseases. The opportunities for influenza viruses to evolve rapidly in parts of East Asia where ducks, pigs, and wild birds are brought together at water bodies are well known.

Many of the water-related problems of vectors are particularly linked to what goes on in "untidy" pieces of water. In most irrigation schemes or other water developments, there are areas where the ecotone between water and land is relatively unmanaged (as around the junctions of drains, rather than canals, in irrigation schemes) and there are small areas where quite complex, highly biodiverse areas, involving a variety of domestic animals, as well as people, come into contact. Similarly, small irrigation dams are found in large numbers throughout the tropics and are usually in uncontrolled, disorderly contact with people and livestock. The majority of emerging diseases in recent years have been zoonoses. Other examples are given in the last section.

Finally, one can perhaps usefully classify "agendas" (McGranahan et al., 2001) for change in water distribution and abundance, since in our crowded world, different objectives for change are both coming into conflict and overlapping. The earlier perception of perhaps two agendas (economic development and public health) that could sometimes conflict and at other times be synergetic is now too simple. A basic table of some of these agendas may serve to remind those about to intervene in the name of one of these aims to also consider the impact of their proposed action upon the other agendas. Tables 1-3 and 1-4 show how one might categorize behaviors, although they are not proposed to be of definitive value as they currently stand.

Water, Health, and Disease in a Changing African Community

The earlier sections of this paper have sought, with variable levels of confidence, simple generalizations to help understand and control water-related disease. This final section is an account of changing patterns of water in an area of Uganda and is used to illustrate the rich complexity and rapidity of change, and

TABLE 1-4 List of Some "Agendas" to Be Considered When Changing Aspects of Water or Sanitation

Blue Agenda	Water availability for domestic and agricultural needs
Brown Agenda	Sanitation to control communicable diseases
Green Agenda	Environmental sustainability, biodiversity
Red Agenda	Less vector-borne disease
Purple Agenda	Less non-biological pollution

NOTE: The scheme set out in this table is emphatically not proposed to be of definitive value as it stands. It rather serves to provoke attention from the appropriate specialists in the hope of providing better guidance to those whose activities will, whether by chance or design, affect the public health through changing the transmission of water-related diseases.

the implications for health. A larger scale picture of change is given by the two East Africa-wide surveys with comparable methodology and the same sites but 30 years apart (Thompson et al., 2001; White et al., 1972).

As one travels from Kampala, the capital of Uganda, toward the southwest, the landscape gradually changes from matoke (plantain, cooking banana) plantations to more open and hilly country at a greater altitude with scrub and thornbush in the lower areas and rolling grasslands above. Rainfall is unreliable, and for many decades the land has primarily been used for grazing by the beautiful long-horned cattle of the Hima nomadic pastoralists. Under great pressure from the President of Uganda (a Hima himself), they have become sedentarized, leading to great and complex changes in environment, livelihoods, and economy. The large herds of cattle, kept and viewed as capital assets by the traditional Hima, were to be reduced, leading to livestock production primarily for sale as beef, and with hybridization with exotic breeds to increase milk yield in those kept for dairy herds.

Settlement took place in two stages. The first was the creation of huge ranches and later their subdivision under popular political pressure. The second phase was implemented with disregard for water access so that the traditional routes to water now lie across privatized land and the sources themselves are on private land. This precipitated both the universal creation of farm ponds by settlers and the excavation of communal small "dams" to provide dry-season water. Increasing human populations tended to pump boreholes dry in the arid seasons, cattle spraying against ticks has polluted ponds with insecticide residues, and new matoke plantations have increased transpiration and reduced stream flow. New nucleated settlements have increased point-source pollution; charcoal burning has removed tree cover from large areas. The water source of last resort in drought, Lake Mburo, is in a National Park, and cattle, buffalo, and zebra can be seen drinking its water beside each other. Some communal dams are well looked after and provide "stepping stones" for the fauna to spread out from Lake Mburo and increase biodiversity. Road construction and repair has created borrow pits where malaria vectors breed. Temporary "beehive" huts of grass have been replaced by long-term mud-walled structures with thatched roofs, and those by metal-roofed concrete-block permanent dwellings.

Effects upon water- and sanitation-related diseases have been diverse and complex; some aspects are not well understood and causal chains may be inaccurately inferred or misunderstood.

The Hima traditionally live on a largely milk diet (Jelliffe and Blackman, 1962), so that they are not used to drinking water on a large scale. With the move to a monetary economy and sale of beef and milk, their diet has become more diverse and the degree to which water is used for drinking is changing and has been little studied. Settled populations need their water sources to be more reliable than do households accustomed to a nomadic life moving from one source to another.

Although the policy has been to develop boreholes for domestic water supply, while large parish dams have been made for watering livestock, this distinction is often not followed through. Children minding livestock also bring domestic water back for the household, often from the cattle ponds. The epidemiology of shared polluted water supplies, common to both cattle and people, has been poorly studied. Still less has this situation been recognized in water policy formulation.

Malaria has increased markedly during the past two decades, and this increase was perceived by the Hima as a consequence of sedentarization. It was also hypothesized that the mechanism might be reduced deviation to cattle for blood meals of the putative vector, *Anopheles arabiensis*, as herd sizes decreased and livestock were housed separately from their owners. However, it was found that malaria had increased across the whole population, including those still nomadic; indeed the burden was heaviest among the recently settled. Currently, the more likely hypothesis is that land allocation is the causal factor. Disruption of routes to water had led to a proliferation of local farm ponds that bred the efficient anthropophilic vector *Anopheles gambiae* near households. New large parish dams have added to the malaria. Anophelines breed to a limited degree in the dam itself as predators abound in the perennial water, but the water is pumped to cattle troughs surrounded by an inadequate concrete apron, beyond which the spilled water forms puddles that are ideal breeding sites for *Anopheles gambiae.*

Sanitation for nomadic households in a semi-arid environment is very different from settled farming circumstances, where the same area may be repeatedly exposed to excreta. Moreover, the matoke plantations, where latrines may be dug, provide the shade and moisture suitable for hookworm development and infection of the people.

The combination of National Park and water source of last resort for cattle that comprises Lake Mburo provides an obvious site for the exchange of ticks and other ectoparasites between livestock and wildlife, and it is likely that other pathogens may occasionally be transferred between domestic and wild animals or birds there. Indeed, because water is a need for most mammals, limited surface water sources may lead to multiple species crowding for access and may facilitate transmission of diseases that are not waterborne in any usual sense of the word.

All these specific situations occur in addition to the water-related diseases that are associated with increasing population density, incipient urbanization, poverty, and other widespread processes in tropical rural areas. The detailed long-term interdisciplinary study of populations and environments under complex rapid change is likely to provide much helpful input to water and sanitation policies and their implementation.

Acknowledgments

Many of the ideas summarized in these classifications owe much to my collaborators over the years, particularly Sandy Cairncross, John Thompson, Richard Feachem, and above all, Gilbert and Anne White. Central to the work in southwestern Uganda have been Joseph Okello-Onen and Charles Muchunguzi. To all of them, I am most grateful.

SAFER WATER, CLEANER HANDS, AND SAFER FOODS: DISEASE PREVENTION STRATEGIES THAT START WITH CLEAN WATER AT THE POINT OF USE[5]

Robert V. Tauxe, M.D., M.P.H.[6]
Centers for Disease Control and Prevention

Robert E. Quick, M.D., M.P.H.[7]
Centers for Disease Control and Prevention

Eric D. Mintz, M.D., M.P.H.[8]
Centers for Disease Control and Prevention

Introduction

The developing world is hazardous for young children. An estimated 10 million young children died there in 2006 (Anon, 2008). Of these deaths, the World Health Organization (WHO) estimates that 16.5 percent, or at least 1.65 million,

[5]The findings and conclusions in this publication are those of the authors and do not necessarily represent the views of the Centers for Disease Control and Prevention.

[6]Corresponding author. Deputy Director, Division of Foodborne, Bacterial and Mycotic Diseases (DFBMD), National Center for Zoonotic, Vector-borne and Enteric Diseases (NCZVED), 1600 Clifton Road, Mailstop C-09, Atlanta, Georgia 30333; E-mail: rvt1@cdc.gov; Tel: 404-639-3818; Fax: 404-639-2577.

[7]Enteric Disease Epidemiology Branch, DFBMD, NCZVED, CDC.

[8]Leader, Diarrheal Diseases Team, Enteric Disease Epidemiology Branch, DFBMD, NCZVED, CDC.

were due to diarrheal diseases, many of which were caused by contaminated water (WHO, 2008). In addition to deaths related to diarrheal illnesses, deaths caused by nondiarrheal infections like typhoid fever are also related to contaminated water (Crump et al., 2004).

These high mortality rates resemble those of the United States in the late nineteenth century, before the "sanitary revolution" improved urban water and sewer systems. City-by-city investments in those systems dramatically lowered illness and death rates (Cutler and Miller, 2005). For example, in Pittsburgh, Pennsylvania, the overall death rate from typhoid fever dropped from 130 per 100,000 in 1907, just before the water treatment plant opened, to just 24 in 1909, when 75 percent of the population was supplied with treated water (Rosenau, 1928). Municipal waterworks became objects of civic pride, celebrated architecturally as a Greek temple in Philadelphia, and as a Georgian palace in Cambridge, Massachusetts. Now municipalities are challenged to maintain and upgrade those systems as they enter their second century of service.

Similar improvements have been seen recently in parts of Latin American and other emerging economies. These also require substantial financial investment. In 1991, the Pan American Health Organization (PAHO) estimated that $200 billion would be required to complete and modernize the entire water and sanitary infrastructure of Latin America (de Macedo, 1991). Some of this investment has now occurred, resulting in decreases in disease that in some countries are as dramatic as that observed a century ago in the United States and Europe. In 1991, following the arrival of epidemic cholera, a long-planned sewage treatment system was built in Santiago, Chile, that interrupted the use of raw sewage to irrigate food crops (Alcayaga et al., 1993). Within a year, the incidence of typhoid fever in that city decreased by 85 percent, hepatitis A by 58 percent, and no further cases of cholera were reported. In Mexico, the infant death rate due to diarrheal illness dropped from 11.6 per 1,000 live births in 1980 to 0.71 in 2005, a decrease of 93 percent (Sepulveda et al., 2006). The fastest drop occurred from 1990 to 1993, when an interdepartmental Clean Water Program increased the proportion of the population with access to potable water from 58 to 95 percent. While the mortality rates in Latin American and Caribbean children still vary substantially from country to country, the general mortality of children less than 5 years old per 1,000 live births dropped from 55 in 1990 to 26 in 2006 (Anon, 2008), and in 2006, only 5 percent of those deaths were due to acute diarrheal illness (PAHO, 2008).

Although progress has been made, this effort remains incomplete. In the last decade, epidemic cholera has disappeared from Latin America, though it remains rampant in sub-Saharan Africa and South Asia (Gaffga et al., 2007). In South Asia, the estimated death rates for children under age 5 were 83 per 1,000 live births, and in sub-Saharan Africa, 158 per 1,000 live births (Anon, 2008). In many countries in the developing world, particularly those in South Asia and sub-Saharan Africa, large water systems serving the entire population remain a distant hope, because these countries lack the technological and financial tools to support and maintain such systems. Those countries critically need interim solutions that can improve the safety of water now, without waiting for major municipal investments. To succeed, these solutions need to be simple, inexpensive, reliable and sustainable, and implementable at smaller scales, down to the household level.

We review here the general concept of point-of-use water treatment, and some recent implementation trials that have shown what this strategy can achieve in a variety of settings, with demonstrable impact on health outcomes. These interventions grew from our field experience with epidemic and endemic diarrheal diseases around the world. Interventions were evaluated for acceptability and health impact in trials in a variety of settings. If expanded, these and similar interventions can contribute to reaching the Millennium Development Goals by decreasing childhood morbidity and mortality and by increasing access to safe drinking water at the point of use.

These interim solutions also represent a down payment on long-term solutions. They engage the community and demonstrate immediately the value of safe water and hand washing to individuals, schools, clinics, and other community groups, which can ultimately translate into increased demand for adequate municipal services.

These interventions are centered on simple strategies to make water safe at the point of use, to keep it safe until it is consumed, and to promote hand washing with soap. Although building latrines is also a public and private good, good data on health outcomes from intervention trials with latrines are largely lacking, and are not covered in this paper.

Interventions to Make Water Safe at the Point of Use

Traditionally, water and sanitation programs have been the domain of engineers, while preventing the specific diseases that result from unsafe water, poor sanitation, or hygiene has occupied the public health professions, including physicians, epidemiologists, microbiologists, and other public health professionals. In many countries, the two camps have been housed in different agencies or ministries, have had different funding sources, and have been relatively independent of each other. The evolution of decentralized technologies for treatment of drinking water and sewage now allows the public health professions to play a more active role in implementing "engineered" solutions, while engineers are thinking beyond the traditional solutions of piping treated drinking water to consider more broadly how to serve populations without access to clean water or the means to pay for it. In the mid-1990s, we suggested that both groups should consider that point-of-use water treatment by consumers might be a practical and effective means of reducing the risks of waterborne disease (Mintz et al., 1995). Since then a substantial and growing body of research has borne this out (Clasen et al., 2006, 2007a; Wright et al., 2004).

All human communities have some access to water, but the safety of the water they consume is less often assured. At least a billion people fetch their drinking water from surface sources that are inherently unsafe, such as rivers or water holes, and many hundreds of millions more collect their drinking water from "improved sources," such as wells and municipal standpipes that are more convenient, but which may not protect water from contamination. Both groups then store their drinking water in the home, where further contamination can occur. Indeed, in many parts of the world, municipal systems provide "economic water," usable for laundry and flushing latrines or toilets and other household purposes, but that requires further treatment in order to make it potable (Moe and Rheingans, 2006).

Many a Slip 'Twixt Source and Lip (Adapted from the Proverb)

Some "improved" water systems may start with water that is microbiologically safe. However, once the water has flowed past cracked well heads and casings, through poorly maintained pipes laid adjacent to sewer pipes, and has been subject to low and sometimes negative water pressure and other flaws, it is not surprising that the water is often contaminated by the time it reaches the point of collection. As the water is carried home and stored, it can be further compromised by hands and utensils that are dipped into the bucket and by other intrusions. The end result is that the water may be heavily contaminated at the moment that it is consumed, even if it started out as potable. During the cholera epidemic of 1991, in the town of Trujillo, Peru, water tested at the municipal well head was found to have a geometric mean total coliform count of 1 per 100 ml (range 0-1), rising to 6 (range 0-1,100) by the time it reached the standpipes used by neighborhoods with cholera, and to 794 by the time it was stored in the patients' homes (Swerdlow et al., 1992). When water is hauled home in open containers, this increase can occur even when the source water is free of contamination at the point of collection. This was well illustrated during an outbreak of cholera on the Pacific island of Ebeye in 2000 (Beatty et al., 2004). Many of the 9,000 residents of Ebeye drank water that was fetched from the U.S. military installation on the neighboring island of Kwajalein, where piped, chlorinated, and safe water was available at a dockside tap. Cholera was strongly associated with drinking that piped water after it was hauled back to Ebeye, though water from the Kwajalein tap was free of contamination, and no cases occurred on Kwajalein. A principle co-factor was the use of wide-mouthed containers to transport and store the water. Contamination of water after it is collected may be the rule, rather than the exception. A recent systematic review of data from 57 field studies concluded that microbiological contamination of water between source and point-of-use is widespread, and often significant (Wright et al., 2004). This highlights the need for point-of-use disinfection and safe storage as part of the strategy to improve the safety of the water that is consumed in the developing world, and the need for surveys of water at the point of use rather than at the source to accurately assess the quality of water that is consumed (Mintz et al., 2001; Wright et al., 2004).

Four different approaches to point-of-use water treatment have demonstrated effectiveness at improving water quality and reducing diarrheal diseases: chlorination, combined chlorination-flocculation, filtration, and solar disinfection. These four approaches and programs used to implement them have been well described recently (Lantagne et al., 2007). All have the following in common: (1) a physical or chemical process that removes or inactivates pathogens in the water and (2) a properly designed water vessel that protects the treated water from recontamination during storage. The processes vary in ease of use, in their effects on the taste and appearance of the finished water, and in the cost and "shelf-life" of the materials they use. For example, locally manufactured dilute sodium hypochlorite packaged and promoted for water treatment is easily added to water and will quickly inactivate bacterial pathogens without affecting the appearance of the water, but with a noticeable effect on its taste. A bottle of hypochlorite solution costs on average about 50 cents, and will last a family one month. A silver-impregnated ceramic filter costs approximately $15, will remove pathogens from water, possibly improving its appearance and without affecting its taste, and will last many months if properly maintained. Examples of safe storage containers include: (1) a simple 1-liter polyethylene terephthalate (PET) plastic soda bottle, painted black on one side and used for solar disinfection of water on the roof of a home; (2) a 20-liter jerry can or a modified clay vessel with a narrow mouth, a lid, and a tap that allows users to remove water by pouring while limiting the potential to contaminate the stored water by dipping in cups, utensils, or hands; or (3) the bottom of two interconnected buckets separated by a filter and equipped with a tap for water removal.

Protective Efficacy and Cost-Effectiveness

Intervention trials of point-of-use treatment and safe storage can provide several different types of information. The outcomes can be measures of the microbiological quality of the water at the point of consumption, the presence of disinfectant in the water as an objective indicator of use, disease incidence, and the knowledge and attitudes of the participants. Most point-of-use water treatment practices depend on using specific products, so the uptake, dissemination, and sustainability of the practices can be readily assessed. Where an intervention depends on the marketing of goods, such as a storage vessel or water disinfectant, marketing strategies and business models can be compared.

Trial results show the efficacy of the intervention strategy, as well as a measure of the fraction of diarrheal illness that is related to water. For example, a trial of two methods of point-of-use disinfection in 49 villages in western Kenya showed that diarrheal illness was decreased by between 19 and 26 percent in families that treated their water, and between 17 and 25 percent among children less than 2 years old; all-cause mortality itself was reduced by 42 percent (Crump et al., 2005). This study illustrates both the large health burden of unsafe water on this population and the means to reduce it. An intervention that is based in clinic settings can directly engage the health community in an area that has traditionally been left to the "water ministry." For example, in Uganda, a trial providing home water treatment and storage to persons with HIV infection consulting an AIDS clinic showed they had a 25 percent reduction in the incidence of diarrheal illnesses, and 33 percent fewer days with diarrhea, compared to others not using the intervention (Lule et al., 2005). As a result of this trial, point-of-use safe water was added to the bundle of routine preventive measures provided to persons with HIV infections though the Global Aids Program clinics in Uganda and four other countries.

It had been thought in the 1980s that improvements in water quality (as measured at the source) yielded less health benefit than did improvements in sanitation, hygiene, and water "quantity." Perhaps the unnoticed deterioration of water quality from source to the point of use contributed to that perception. Now that paradigm appears to have been incomplete, and it is changing (Clasen and Cairncross, 2004). A growing number of randomized controlled intervention trials conducted in a variety of developing world settings demonstrate that point-of-use water treatment and safe storage strategies are generally effective in reducing the incidence of diarrheal disease and are more effective than interventions focused on improving the quality of water at the source. A Cochrane library meta-analysis of 38 published studies found that point-of-use interventions had an aggregated effectiveness of 44 percent reduction in diarrheal illnesses (range 26 to 58 percent), while interventions focused on the water source reduced diarrheal illness by only 13 percent (range 2 to 26) (Clasen et al., 2006, 2007a). A second meta-analysis of 21 studies, focusing on the impact of home chlorination strategies on children less than 5 years old, found an aggregated reduction of diarrheal disease of 29 percent (range 13 to 42 percent) in that age group, as well as an 80 percent reduction in the frequency of fecal contamination in stored household water samples (Arnold and Colford, 2007).

Proof of health impact and microbiologic effectiveness in research studies provides a strong scientific base, but, to make a difference in the real world, point-of-use treatment methods have to be economically "scalable" and sustainable. An analysis of the cost-effectiveness of water quality interventions found that for sub-Saharan Africa and Southeast Asia, both source and point-of-use interventions were cost-effective, and that a household-based chlorination strategy was the most cost-effective intervention (Clasen et al., 2007b). Larger scale, sustainable implementation models for these point-of-use treatment technologies are the subject of ongoing operational research. Social marketing through the commercial sector by the non-profit nongovernmental organization (NGO), Population Services International has successfully brought locally manufactured hypochlorite-based water treatment products to millions of people across the world (Lantagne et al., 2007). Subsidized or free distribution has been utilized to reach populations affected by natural disasters, and vulnerable populations including persons living with HIV/AIDS, pregnant women, and the poorest poor. Education and distribution through clinics, schools, and other community institutions also boost awareness and adoption.

Integrating Safe Water with Hand Washing Promotion and Other Interventions

Hand washing has been a fundamental public health intervention since Ignaz Semmelweiss (1818-1865) demonstrated that the risk of puerperal fever could be lowered by implementing hand washing with soap among labor and delivery personnel (Semmelweis, 1983). The health impact of hand washing is still readily demonstrable, particularly in children. For example, a large randomized trial that included weekly home visits to promote hand washing was conducted in the slums of Karachi, Pakistan, with approximately 300 households allocated to each of two intervention groups receiving free soap and one control group (Luby et al., 2004). In this trial, in children less than 15 years old in the intervention group, diarrheal disease was reduced by 53 percent, in children less than 12 months old by 39 percent, and in severely malnourished children by 42 percent compared to the children in the control group. The health benefits went beyond diarrheal illness. Compared with controls, children in the intervention group who were less than 5 years old had a 50 percent lower incidence of pneumonia and children less than 15 years old had a 34 percent lower incidence of impetigo (Luby et al., 2005). A meta-analysis of 7 intervention trials and 10 observational studies produced similar results, suggesting that hand washing reduces the risk of diarrhea by 43 percent, severe enteric infections by 49 percent, and shigellosis by 58 percent (Curtis and Cairncross, 2003). Because the implementation approaches used in many of the aforementioned studies were expensive and intensive, it would not be possible to scale up and sustain them by themselves. However, hand washing can be effectively co-promoted with safe water programs and integrated with other public health interventions.

Integrating Safe Water and Hand Washing Promotion in the Clinical Setting

Integration of combined safe water and hand hygiene interventions into health facilities has proven to be feasible and effective. In 2004, the CDC collaborated with the Kenyan Ministry of Health and CARE Kenya to co-promote safe water and hand washing with soap at a district hospital (Parker et al., 2006). Project implementers installed special clay water pots in all outpatient departments and hospital wards and provided sodium hypochlorite solution (with the brand name WaterGuard®) for water treatment and soap for hand washing (Figure 1-8). Hospital nurses were trained in water treatment and hand washing and instructed

FIGURE 1-8 Supervising nurse at a drinking water station at a community clinic in Nyanza Province, Kenya. The water is disinfected with a chlorine product, which the clinic also sells, and the clay pot is equipped with a tap to allow easy dispensing of the water without contaminating it.
SOURCE: Figure courtesy of Amy Parker, CDC.

to provide this teaching to their clients. Exit interviews with 220 clinic clients found that 85 percent of mothers had received education about water treatment, 80 percent on hand washing with soap, and 76 percent on both topics. Two weeks after the exit interviews, home visits were conducted with a random sample of 98 clinic clients and confirmed that 68 percent were using WaterGuard® (by measuring chlorine residuals in stored water), 44 percent of respondents could demonstrate the six steps of hand washing taught by the nurses, and 81 percent could demonstrate at least four hand washing steps. One year later, 71 percent of evaluation households were found to still have detectable chlorine residuals in treated water, 34 percent of respondents could demonstrate all six steps of hand washing, and 98 percent could demonstrate four steps. Five years later, the intervention is still in place at the district hospital (CDC, unpublished data). In the interim, CARE Kenya continued to implement this intervention in hospitals, health centers, and dispensaries. To date, an additional 108 health facilities have installed hand washing and drinking water stations. The curriculum used to train nurses in these programs has been adapted for use in at least five other countries in sub-Saharan Africa and Asia.

In 2007, a project was launched in antenatal clinics in Malawi to provide free hygiene kits consisting of safe water storage containers, WaterGuard® solution, soap, two sachets of oral rehydration salts, and education to pregnant mothers attending the clinics (Figure 1-9; Sheth et al., 2008). Pregnant mothers also received free refills of WaterGuard® and soap when they returned for follow-up visits. An evaluation of 400 clinic clients showed that, from baseline to follow-up, there was an increase in confirmed use of WaterGuard® from 2 to 61 percent, confirmed purchase and use of WaterGuard® from 2 to 32 percent (suggesting the potential sustainability of water treatment behavior), and the ability to demonstrate correct hand washing procedure from 22 to 68 percent. A concurrent study of friends and relatives of clinic clients showed an increase of WaterGuard® use from 2 to 25 percent and of the ability to demonstrate correct hand washing procedure from 18 to 60 percent, suggesting diffusion of knowledge and practices (Dr. Anandi Sheth, CDC, personal communication, January 2009). A program to

FIGURE 1-9 Water and hygiene kit currently being distributed to expectant women at antenatal clinics in Malawi, including WaterGuard® chlorine disinfectant, Safe Water storage vessel with spigot and lid, and packet with soap and oral rehydration solution.
SOURCE: Figure courtesy of Dr. Anandi Sheth, CDC.

offer hygiene kits of WaterGuard® solution and soap as incentives to motivate mothers to bring their infants to clinic is currently underway in western Kenya.

Integrating Safe Water and Hand Washing Promotion into Schools

In 2003, a pilot program at a single private school in Kenya suggested that combined programs of safe water treatment and hand washing with soap were feasible (Migele et al., 2007). In this program, drinking water stations and hand washing stations were provided to the school; teachers were trained about water treatment and hand washing and taught their pupils about these two topics; and safe water clubs were formed so pupils could engage in water and hygiene learning projects. Over the course of the project, trips to the doctor decreased by more than half and the school saved more than $5.00 per pupil. Following this pilot, similar programs were implemented in 60 public primary schools in Kenya (Figures 1-10A and B). An evaluation in 2005-2006 showed that, from baseline to follow-up, pupils' knowledge of correct water treatment procedure increased from 21 to 65 percent, and of the need to wash hands after using the toilet increased from 73 to 90 percent (O'Reilly et al., 2008). Among parents, increases were observed from baseline to follow-up in reported current use of WaterGuard® (6 percent increasing to 14 percent, $p < 0.01$), confirmed use of WaterGuard® (5 to 8 percent, $p = 0.20$), and having soap in the home (74 to 90 percent, $p < 0.01$). From baseline to follow-up, pupil absentee rates decreased by 35 percent, compared to a 5 percent increase in neighboring control schools.

In China, a randomized trial in primary schools where safe water was already available compared schools with a standard program of hand washing education; an expanded program of education, soap, and peer hygiene monitors; and controls schools with no hand washing promotion (Figure 1-11; Bowen et al., 2007). The trial showed that students in the expanded program had significantly fewer episodes and days of absence than did students in control schools; students in schools with the standard program had intermediate rates of both indicators that did not exhibit a statistically significant difference from either of the other two groups. Similar assessments are underway in the Philippines and Pakistan, and an assessment of the long-term impact of hand washing programs on growth and cognitive development is in the planning stages.

Bundling Safe Water and Hand Washing Promotion in the Community Setting

Another strategy that holds promise is to integrate hand washing with soap with other interventions in community programs. The Nyando Integrated Child Health and Education (NICHE) program promotes several proven products for child health: WaterGuard®solution, PuR® flocculent disinfectant (Procter and Gamble, Cincinnati, Ohio), Aquatabs® (Medentech, Ltd., Wexford, Ireland),

A

B

FIGURE 1-10 Two students in the Safe Water Club at Sino SDA Primary School in Nyanza Province collect and treat the water for the school each morning to (A) fill the handwashing station reservoir and (B) the clay pot with treated water for drinking.
SOURCE: Figure courtesy of Dr. Ciara O'Reilly, CDC.

FIGURE 1-11 Students at a primary school in Fujian Province, China, washing their hands at the start of the lunch hour.
SOURCE: Figure courtesy of Dr. Anna Bowen, CDC.

safe water storage containers, soap, insecticide-treated bed nets, micronutrient Sprinkles (Heinz Co., Toronto, Canada), and de-worming with albendazole (CDC, 2007). These products are part of a basket of goods sold door-to-door to community members at a low price by HIV self-help groups as an income-generating activity. The self-help groups also provide health education and work with schools, local clinics, churches, and local political leaders to expand the promotional messages as widely as possible. Assessment of the success of this program is in progress.

Health Benefits Related to Safer Foods

Safer Weaning Foods

Water is used to prepare many foods in the home, including weaning food. Weaning is a particularly hazardous time for growing children and has been associated with an increase in diarrheal illness and an associated delay in growth (Martorell et al., 1975). Wet foods and gruels used as weaning foods may often be contaminated with fecal coliforms (Henry et al., 1990). In Peru, cereals and purees prepared especially for infants were more contaminated than were the foods prepared for the rest of the family (Black et al., 1989). Household drinking water used in the preparation of weaning foods is one of several important sources of contamination (Motarjemi et al., 1993). It is likely that using safe water and washed hands to prepare these foods would reduce their contamination. It is possible that some of the general reduction in diarrheal illness observed when household water disinfection is practiced is the result of safer foods and safer drinking water. An intervention study remains to be done that evaluates the microbial quality of weaning foods prepared in homes with and without point-of-use water disinfection.

Safer Foods on the Street

Street vendors provide fast, inexpensive, and convenient food and drink in much of the developing world, filling the same social niche that fast food restaurants do in the developed world. Street-vended foods can be safe if they are cooked hot and served fast (Abdussalam and Kaferstein, 1993). However, many are not, and unsafe water often appears to play a supporting role. High levels of contamination with fecal indicator bacteria have been demonstrated in surveys of street-vended food and beverages (Garin et al., 2002), and disease resulting from consuming them has been well documented. In the early 1980s, eating street-vended flavored ices was a risk factor for typhoid fever in children in Santiago, Chile (Black et al., 1985). In 1989, in Manila, Philippines, a cholera epidemic in an area not served by piped water was linked to consuming street-vended foods, particularly to a rice noodle dish, and to mussel soup (Lim-Quizon et al., 1994). In 1991, consuming street-vended foods and beverages was a risk factor for epidemic cholera in Piura, Peru, where the municipal water supply was not chlorinated, and the ice used in beverages may have been particularly risky (Ries et al., 1992). In 1993, a cholera epidemic in Guatemala City, where the municipal water supply was chlorinated, was linked to eating street-vended foods and flavored ices (Koo et al., 1996). From 2001 to 2003, endemic paratyphoid fever was linked to eating foods prepared outside the home, largely as street-vended foods, while typhoid fever was associated with lack of clean water and sanitation inside the home in Jakarta, Indonesia (Vollaard et al., 2004b). The street vendors of Jakarta often had fecally-contaminated hands, ice, and drinking water, and poor hand hygiene (Vollaard et al., 2004a).

Even if street vendors understand the principles of safe food preparation, there is little that they can do without access to safe water for preparing food, washing their hands, and cleaning their utensils and dishes (Mahon et al., 1999). In 1994, we conducted a survey of the knowledge and practices of street vendors in two Guatemalan towns. Most vendors and their customers were aware of the importance of using clean water and hand washing, but none of the vendors did so, perhaps because none had treated water available. In 1996, we provided a cohort of street vendors in Guatemala City with Safe Water® system containers and chorine-based disinfectant (Sobel et al., 1998). Vendors were given soap and encouraged to wash their hands with the clean water, as well as use it to prepare the drinks they sold, made from fruits and cereals. We tested the stored water, beverages, and hands of the 41 intervention vendors and 42 similar but unequipped control vendors for evidence of fecal contamination before and during the six-week intervention period. The intervention was associated with a significant decrease in fecal coliforms in the stored water, and in the beverages that they were selling, though there was little change in the contamination of their hands. The intervention was well accepted. The vendors thought it increased their sales by increasing customer confidence. In a spontaneous innovation, some vendors offered the hand washing platform to their customers, so they could wash their hands before eating. Five months later, the original intervention group of vendors was still using the water disinfection system. Though no assessment of health impact was attempted, the intervention trial showed that making clean water and soap available at the street vendors' point of use resulted in a less contaminated product. This intervention can be easily adopted as a public health policy and incorporated into ongoing efforts to educate and license street vendors.

Safer Food Processing

Water is used in many stages of food production and processing. For fresh produce that is eaten without further cooking, the quality of the water that is used to spray, wash, and chill the food is linked directly to its safety. This means that the microbiological quality of the water in the developing world is of immediate consequence to consumers in the industrialized world, as well as affecting the health of consumers in the country where the produce is grown. In 2001, 17 percent of fresh and frozen vegetables and 23 percent of the fresh and frozen fruit consumed in the United States was imported, largely from Latin America (Jerardo, 2003). In Europe, a growing fraction of fresh produce is imported from Africa. Outbreaks of foodborne diseases have been associated with important lapses in maintaining the safety of water used to process foods before they were imported. Investigating these outbreaks and tracing them back to their sources is difficult and requires effort, luck, and the cooperation of many parties, so it is likely that the identified outbreaks represent only the tip of the metaphorical iceberg (Tauxe et al., 2008).

Shigellosis and Parsley

In 1998 in Minnesota, two outbreaks occurred of a febrile bloody diarrheal illness caused by the fecal bacterium *Shigella sonnei.* In one outbreak, an epidemiological investigation linked illness with eating foods made with parsley, while in the other, parsley was strongly suspected though not proven; the parsley source for both events was the same (Naimi et al., 2003). The state health department used molecular methods to fingerprint the *Shigella* bacteria and showed that the *Shigella* strains from the two outbreaks were indistinguishable. Public health authorities around the continent were alerted and quickly reported six more outbreaks caused by *Shigella sonnei* with the same DNA fingerprint as the first two. In each outbreak, parsley was implicated or was strongly suspected. In addition, two other outbreaks of gastroenteritis in Minnesota caused by a second microbe, enterotoxigenic *Escherichia coli*, also appeared to be linked to parsley. In all, 486 persons were ill with shigellosis and 77 with enterotoxigenic *E. coli* diarrhea. The combination of two different enteric diseases linked to the same food item across multiple states and countries suggested that contamination by raw sewage had somehow occurred near the point of production. The parsley in these outbreaks was traced to the likely source, a farm in Mexico. Investigators from the Food and Drug Administration (FDA), Mexico, and CDC visited that farm. The farm used local municipal water to rinse and chill the parsley, though the chlorination of the municipal system was inadequate and intermittent. The farm also placed ice made from apparently unchlorinated water on the parsley to ship it chilled to the United States. It is likely that the local water and/or ice contaminated the parsley. Preventing such contamination would require a guaranteed potable water supply for washing and chilling the produce. It seems self-evident that water used to process fresh foods that are eaten without further cooking must be safe to drink.

Jaundice and Green Onions

In 2003, large outbreaks of hepatitis A infections affected customers of restaurants in Tennessee, Georgia, North Carolina, and Pennsylvania (Amon et al., 2005; Wheeler et al., 2005). A total of 1,023 cases were reported. In Pennsylvania, at least 124 of the 601 identified patients were hospitalized, and 3 died. Symptoms were jaundice, abdominal pain, fever, and the prolonged malaise characteristic of this infection in adults. In each location, illness was linked to eating green onions, which were traced back through the supply chain to likely source farms in northern Mexico. In rural Mexico, hepatitis A is a common infection in young children, acquired as a result of poor sanitation and unsafe drinking water, and usually causes relatively mild diarrhea. Investigation of the farms by the FDA, Mexican authorities, and CDC revealed dubious quality of water used in packing sheds and ice machines, poor sanitation and hand washing facilities, and the possibility that young children in diapers were sitting on harvested produce (FDA, 2003). Prevention strategies include ensuring that water used for washing and icing the produce is potable, improving local sanitation and health conditions of farm worker families, and separating young children from harvested food.

Salmonella *Newport Infections and Mangoes*

One outbreak is a cautionary tale showing how phytosanitary food regulations intended to prevent one problem created another. In 1999, a large outbreak of *Salmonella* Newport infections sickened at least 78 persons in 13 states and was linked to imported mangoes (Sivapalasingam et al., 2003). The mangoes came from one large orchard in Brazil that also exported mangoes to Europe, where no cases were identified by public health authorities. Investigation of the orchard by Brazilian authorities, FDA, and CDC revealed a difference in how the mangoes were handled that depended on their destination. To prevent the introduction of Mediterranean fruitflies, U.S. regulations mandated that the mangoes be disinfested. However, mangoes destined for Europe were not disinfested, because Mediterranean fruit flies are native to Europe. In the past, the mangoes were fumigated with ethylene dioxide gas, which has toxic effects on workers and the ozone layer, as well as on fruit flies. In the 1990s, the United States began to require dipping mangoes in a hot water bath instead. At the Brazilian orchard, the hot water dip was followed by a cold water dip to cool the mangoes. The sudden transfer of a hot mango into a cold water bath makes the interior of the fruit contract slightly, pulling water inside along with any bacteria that are present (Penteado et al., 2004). Although the quality of the water is critical to the safety of the fruit, treatment specifications did not mandate reliably safe water. The hot water was not chlorinated, and the cold water was only chlorinated once a week, though the tank was open to the tropical environment and was used daily. A point-of-use system that continuously disinfects and protects the water is needed to prevent the contamination of the fruit. One hopes that this is required wherever disinfestation via a hot water dip is mandated.

As these outbreaks demonstrate, the safety of the water used to process foods is related to the health of the consumer. These scenarios offer a window into the safety of food production, whether the food is then consumed locally or exported. Using potable water for processing and chilling is a fundamental good food manufacturing practice. The health of the workers is also likely to be important, so providing clean drinking water, hand washing, and sanitation for workers and their families is not only humanitarian but also a prudent business practice.

Summary and Conclusions

The experiences outlined above show that substantial progress is possible with low-cost, sustainable, and effective interventions. They provide broad health benefits including fewer diarrheal illnesses and respiratory and skin infections, in both children and adults, and they decrease school absenteeism in students and teachers. Hand washing promotion can be easily added to a program that provides safe water at the point of use. Clinics and schools are good venues for introducing these interventions, to educate the public in their use, to target persons at high risk, and to improve institutional hygiene. Social entrepreneurs can distribute and sell safe water treatment products and soap along with other point-of-use health products. Safe water and hand washing is likely to make weaning foods and street-vended foods safer. In addition to local health benefits, safer water for drinking and food processing in the developing world makes fresh produce safer when exported to the developed world.

These interventions are also likely to stimulate social development. Local production and sale of containers and disinfectants can create micro-economies. Successful interventions help people recognize the value of making their drinking water safe and keeping it safe, building a constituency for reliably safer water. Treating household drinking water can be a daily reminder of the desirability of a long-term solution. Rather than slowing demand for treated piped water, the practice of point-of-use disinfection may be an impetus for small community water systems and micro-utilities of increasing social and technical complexity.

However, implementation to date has been limited, and few programs have been scaled up to the national level. Considering all the point-of-use interventions together, only about 1 percent of the billion persons lacking access to safe water have been reached to date (Clasen, 2008). Progress will accelerate if the health-care providers can engage and partner with the water department authorities, bridging the two bureaucracies that are often separate realms. One important step is to introduce water interventions in health clinics, which would likely reduce the risk of disease transmission, model healthy behavior for patients and their families, and bring the medical community into immediate contact with water issues. Progress will also accelerate if household water treatment can be coupled with other major health intervention programs, and promoted as an investment for community health. It can be a low-cost addition to programs for childhood vaccination, malaria control, and for programs to improve the quality of life in persons with AIDS. It can be integrated with maternal and child health programs, preventing illness in the youngest and most vulnerable children. Beyond the clinic, water treatment strategies can be promoted by pharmacists, traditional healers, and birth attendants.

Progress in reaching the benchmarks of the Millennium Development Goals will also be faster if safe drinking water at point of use and handwashing are recognized and promoted as useful strategies. These interventions can reduce a substantial fraction of the childhood morbidity and probably can reduce the mortality that is targeted by Goal #4.[9] Goal #7[10] aims to increase access to safe drinking water, but even if the water comes from "improved" sources there is a

[9]Reduce by two-thirds the mortality rate among children under five (UNDP, 2009a).

[10]Target 7a: Integrate the principles of sustainable development into country policies and programmes; reverse loss of environmental resources; Target 7b: Reduce biodiversity loss, achieving, by 2010, a significant reduction in the rate of loss; Target 7c: Reduce by half the proportion of people without sustainable access to safe drinking water and basic sanitation; Target 7d: Achieve significant improvement in lives of at least 100 million slum dwellers, by 2020 (UNDP, 2009b).

demonstrated risk for contamination as it is collected, transported, and stored in the home. Therefore, this goal would have even more health impact if it were amended to focus on increasing access to water that is safe at the moment that it is consumed.

Finally, progress will be faster if decision makers find that interim point-of-use water disinfection is a logical step along the path towards safe piped water in every home.

OVERVIEW REFERENCES

Simon, P. 1998. *Tapped out: the coming world crisis in water and what we can do about it.* New York: Welcome Rain Publishers.

White, G. F., D. J. Bradley, and A. U. White. 1972. *Drawers of water: domestic water use in East Africa.* Chicago: University of Chicago Press.

HOPKINS REFEFERENCES

Bierlich, B. 1995. Notions and treatment of Guinea worm in northern Ghana. *Social Science and Medicine* 41(4):501-509.

CIA (Central Intelligence Agency). 2009a. *The World Factbook: Ethiopia*, https://www.cia.gov/library/publications/the-world-factbook/geos/et.html (accessed July 14, 2009).

———. 2009b. *The World Factbook: Ghana*, https://www.cia.gov/library/publications/the-world-factbook/geos/gh.html (accessed July 14, 2009).

Dawson, C. A. 2000. Becoming Konkomba: recent transformations in a Gur Society of Northern Ghana. Thesis submitted to the Department of Anthropology, University of Calgary, Alberta Canada.

Emerson, P. M., R. L. Bailey, G. E. Walraven, and S. W. Lindsay. 2001. Human and other feces as breeding media of the trachoma vector *Musca sorbens*. *Medical and Veterinary Entomology* 15(3):314-320.

Emerson, P. M., M. Burton, A. W. Solomon, R. Bailey, and D. Mabey. 2006. The SAFE strategy for trachoma control: using operational research for policy, planning and implementation. *Bulletin of the World Health Organization* 84(8):613-619.

Hopkins, D. R., and E. Ruiz-Tiben. 1991. Strategies for dracunculiasis eradication. *Bulletin of the World Health Organization* 69(5):533-540.

Hopkins, D. R., E. Ruiz-Tiben, P. Downs, P. C. Withers, Jr., and S. Roy. 2008a. Dracunculiasis eradication: neglected no longer. *American Journal of Tropical Medicine and Hygiene* 79(4):474-479.

Hopkins, D. R., E. Ruiz-Tiben, M. L. Eberhard, and S. Roy. 2008b. Update: progress toward global eradication of dracunculiasis, January 2007-June 2008. *Morbidity and Mortality Weekly Report* 57(43):1173-1176.

Hunter, J. M. 1997. Geographical patterns of Guinea worm infestation in Ghana: an historical contribution. *Social Science and Medicine* 44(1):103-122.

Mabey, D. C., A. W. Solomon, and A. Foster. 2003. Trachoma. *Lancet* 362(9379):223-229.

O'Loughlin, R., G. Fentie, B. Flannery, and P. M. Emerson. 2006. Follow-up of a low cost latrine promotion programme in one district of Amhara, Ethiopia: characteristics of early adopters and non-adopters. *Tropical Medicine and International Health* 11(9):1406-1415.

Skalník, P. 1987. On the inadequacy of the concept of the traditional state-illustrated with ethnographic material on Nanun, Ghana. *Journal of Legal Pluralism* 25 and 26:301-325.

Watts, S. J. 1987. Dracunculiasis in Africa in 1986: its geographical extent, incidence, and at-risk population. *American Journal of Tropical Medicine and Hygiene* 37(1):119-125.

WHO (World Health Organization). 2009. *Trachoma*, http://www.who.int/blindness/causes/priority/en/index2.html (accessed April 21, 2009).

BRADLEY REFERENCES

Bradley, D. J. 1977a. Health aspects of water supplies in tropical countries. In *Water, wastes and health in hot climates*, edited by R. G. Feachem, M. G. McGarry, and D. Mara. New York: Wiley. Pp. 3-17.

———. 1977b. The health implications of irrigation schemes and man-made lakes in tropical environments. In *Water, wastes and health in hot climates*, edited by R. G. Feachem, M. G. McGarry, and D. Mara. New York: Wiley. Pp. 18-29.

Cairncross, S., and V. Kochar, eds. 1994. *Studying hygiene behaviour: methods, issues, and experiences*. Thousand Oaks, CA: Sage Publications.

Curtis, V., and S. Cairncross. 2003. Effect of washing hands with soap on diarrhoea risk in the community: a systematic review. *Lancet Infectious Diseases* 3(5):275-281.

Ensink, J. H. J., M. R. Aslam, F. Konradsen, P. K. Jensen, and W. van der Hoek. 2002. *Linkages between irrigation and drinking water in Pakistan*. Working Paper 46. Colombo, Sri Lanka: International Water Management Institute.

Feachem, R. G., D. J. Bradley, H. Garelick, and D. D. Mara. 1983. *Health aspects of excreta and wastewater management*. Chichester, UK: John Wiley.

Jelliffe, D. B., and V. Blackman. 1962. Bahima disease. Possible "milk anemia" in late childhood. *Journal of Pediatrics* 61:774-779.

McGranahan, G., P. Jacobi, J. Songsore, C. Surjadi, and M. Kjellen. 2001. *The citizens at risk: from urban sanitation to sustainable cities*. London: Earthscan.

Moriarty, P., J. Butterworth, and B. van Koppen. 2004. *Beyond domestic: case studies on poverty and productive uses of water at the household level*. Technical Paper Series, no. 41. Delft, the Netherlands: IRC International Water and Sanitation Centre.

Rabie, T., and V. Curtis. 2006. Handwashing and risk of respiratory infections: a quantitative systematic review. *Tropical Medicine and International Health* 11(3):258-267.

Szewzyk, U., R. Szewzyk, W. Manz, and K.-H. Schleifer. 2000. Microbiological safety of drinking-water. *Annual Review of Microbiology* 54:81-127.

Thompson, J., I. T. Porras, J. K. Tumwine, M. R. Mujwahuzi, M. Katui-Katua, N. Johnstone, and L. Wood. 2001. *Drawers of water II. 30 years of change in domestic water use and environmental health in East Africa*. London: International Institute for Environment and Development

White, G. F., D. J. Bradley, and A. U. White. 1972. *Drawers of water: domestic water use in East Africa*. Chicago: University of Chicago Press.

TAUXE ET AL. REFERENCES

Abdussalam, M., and F. K. Kaferstein. 1993. Safety of street foods. *World Health Forum* 14(2):191-194.

Alcayaga, S., J. Alcayaga, and P. Gassibe. 1993. Cambios del perfil de morbilidad en algunas patologias de transmisión enterica con posterioridad a un brote de colera. Servicio de Salud Metropolitano Sur. Chile. *Revista Chileana Infectiologia* 1:5-10.

Amon, J. J., R. Devasia, G. Xia, O. V. Nainan, S. Hall, B. Lawson, J. S. Wolthuis, P. D. Macdonald, C. W. Shepard, I. T. Williams, G. L. Armstrong, J. A. Gabel, P. Erwin, L. Sheeler, W. Kuhnert, P. Patel, G. Vaughan, A. Weltman, A. S. Craig, B. P. Bell, and A. Fiore. 2005. Molecular epidemiology of foodborne hepatitis A outbreaks in the United States, 2003. *Journal of Infectious Diseases* 192(8):1323-1330.

Anon. 2008. *Goal 4: Reduce child mortality,* http://go.worldbank.org/ZNUWVAHQD0 (accessed December 14, 2008).

Arnold, B. F., and J. M. Colford, Jr. 2007. Treating water with chlorine at point-of-use to improve water quality and reduce child diarrhea in developing countries: a systematic review and meta-analysis. *American Journal of Tropical Medicine and Hygiene* 76(2):354-364.

Beatty, M. E., T. Jack, S. Sivapalasingam, S. S. Yao, I. Paul, B. Bibb, K. D. Greene, K. Kubota, E. D. Mintz, and J. T. Brooks. 2004. An outbreak of *Vibrio cholerae* O1 infections on Ebeye Island, Republic of the Marshall Islands, associated with use of an adequately chlorinated water source. *Clinical Infectious Diseases* 38(1):1-9.

Black, R. E., L. Cisneros, M. M. Levine, A. Banfi, H. Lobos, and H. Rodriguez. 1985. Case-control study to identify risk factors for paediatric endemic typhoid fever in Santiago, Chile. *Bulletin of the World Health Organization* 63(5):899-904.

Black, R. E., G. Lopez de Romana, K. H. Brown, N. Bravo, O. G. Bazalar, and H. C. Kanashiro. 1989. Incidence and etiology of infantile diarrhea and major routes of transmission in Huascar, Peru. *American Journal of Epidemiology* 129(4):785-799.

Bowen, A., H. Ma, J. Ou, W. Billhimer, T. Long, E. Mintz, R. M. Hoekstra, and S. Luby. 2007. A cluster-randomized controlled trial evaluating the effect of a handwashing-promotion program in Chinese primary schools. *American Journal of Tropical Medicine and Hygiene* 76(6):1166-1173.

CDC (Centers for Disease Control and Prevention). 2007. Baseline data from the Nyando Integrated Child Health and Education Project–Kenya, 2007. *Morbidity and Mortality Weekly Report* 56(42):1109-1113.

Clasen, T. F. 2008. *Scaling up household water treatment: looking back, seeing forward.* Geneva: World Health Organization.

Clasen, T. F., and S. Cairncross. 2004. Household water management: refining the dominant paradigm. *Tropical Medicine and International Health* 9(2):187-191.

Clasen, T., I. Roberts, T. Rabie, W. Schmidt, and S. Cairncross. 2006. Interventions to improve water quality for preventing diarrhoea. *Cochrane Database of Systematic Reviews* 3:CD004794.

Clasen, T., W.-P. Schmidt, T. Rabie, I. Roberts, and S. Cairncross. 2007a. Interventions to improve water quality for preventing diarrhoea: systematic review and meta-analysis. *British Medical Journal* 334(7597):782

Clasen, T., L. Haller, D. Walker, J. Bartram, and S. Cairncross. 2007b. Cost-effectiveness of water quality interventions for preventing diarrhoeal disease in developing countries. *Journal of Water and Health* 5(4):599-608.

Crump, J. A., S. P. Luby, and E. D. Mintz. 2004. The global burden of typhoid fever. *Bulletin of the World Health Organization* 82(5):346-353.

Crump, J. A., P. O. Otieno, L. Slutsker, B. H. Keswick, D. H. Rosen, R. M. Hoekstra, J. M. Vulule, and S. P. Luby. 2005. Household based treatment of drinking water with flocculant-disinfectant for preventing diarrhoea in areas with turbid source water in rural western Kenya: cluster randomised controlled trial. *British Medical Journal* 331(7515):478.

Curtis, V., and S. Cairncross. 2003. Effect of washing hands with soap on diarrhoea risk in the community: a systematic review. *Lancet Infectious Diseases* 3(5):275-281.

Cutler, D., and G. Miller. 2005. The role of public health improvements in health advances: the twentieth-century United States. *Demography* 42(1):1-22.

de Macedo, C. G. 1991. Presentation of the PAHO regional plan. *Proceedings of the conference: confronting cholera, the development of a hemispheric response to the epidemic*, North-South Center, University of Miami, Miami, Florida. Pp. 39-44.

FDA (Food and Drug Administration). 2003. *FDA update on recent hepatitis A outbreaks associated with green onions from Mexico*, www.fda.gov/bbs/topics/NEWS/2003/NEW00993.html (accessed February 4, 2007).

Gaffga, N. H., R. V. Tauxe, and E. D. Mintz. 2007. Cholera: a new homeland in Africa? *American Journal of Tropical Medicine and Hygiene* 77(4):705-713.

Garin, B., A. Aidara, A. Spiegel, P. Arrive, A. Bastaraud, J. L. Cartel, R. B. Aissa, P. Duval, M. Gay, C. Gherardi, M. Gouali, T. G. Karou, S. L. Kruy, J. L. Soares, F. Mouffok, N. Ravaonindrina, N. Rasolofonirina, M. T. Pham, M. Wouafo, M. Catteau, C. Mathiot, P. Mauclere, and J. Rocourt. 2002. Multicenter study of street foods in 13 towns on four continents by the food and environmental hygiene study group of the international network of Pasteur and associated institutes. *Journal of Food Protection* 65(1):146-152.

Henry, F. J., Y. Patwary, S. R. Huttly, and K. M. Aziz. 1990. Bacterial contamination of weaning foods and drinking water in rural Bangladesh. *Epidemiology and Infection* 104(1):79-85.

Jerardo, A. 2003. *Import share of U.S. food consumption stable at 11 percent*, www.ers.usda.gov/Publications/FAU/July02/FAU6601 (accessed February 1, 2007).

Koo, D., A. Aragon, V. Moscoso, M. Gudiel, L. Bietti, N. Carrillo, J. Chojoj, B. Gordillo, F. Cano, D. N. Cameron, J. G. Wells, N. H. Bean, and R. V. Tauxe. 1996. Epidemic cholera in Guatemala, 1993: transmission of a newly introduced epidemic strain by street vendors. *Epidemiology and Infection* 116(2):121-126.

Lantagne, D., R. Quick, and E. D. Mintz. 2007. Household water treatment and safe storage options in developing countries: a review of current implementation practices. In *Water stories: expanding opportunities in small-scale water and sanitation projects*, edited by M. Parker, A. Williams, and C. Youngblood. Washington, DC: Environmental Change and Security Program, Woodrow Wilson International Center.

Lim-Quizon, M. C., R. M. Benabaye, F. M. White, M. M. Dayrit, and M. E. White. 1994. Cholera in metropolitan Manila: foodborne transmission via street vendors. *Bulletin of the World Health Organization* 72(5):745-749.

Luby, S. P., M. Agboatwalla, J. Painter, A. Altaf, W. L. Billhimer, and R. M. Hoekstra. 2004. Effect of intensive handwashing promotion on childhood diarrhea in high-risk communities in Pakistan: a randomized controlled trial. *Journal of the American Medical Association* 291(21):2547-2554.

Luby, S. P., M. Agboatwalla, D. P. Feikin, W. Billheimer, A. Altaf, and R. M. Hoekstra. 2005. Effect of handwashing on child health: a randomized controlled trial. *Lancet* 366(9481):225-233.

Lule, J. R., J. Mermin, J. Ekwaru, S. Malamba, R. Downing, R. Ransom, D. Nakanjako, W. Wafula, P. Hughes, R. Bunnell, F. Kaharuza, A. Coutinho, A. Kigozi, and R. Quick. 2005. Effect of home based water chlorination and safe storage on diarrhea among persons with human immunodeficiency virus in Uganda. *American Journal of Tropical Medicine and Hygiene* 73(5):926-933.

Mahon, B. E., J. Sobel, J. M. Townes, C. Mendoza, M. Gudiel Lemus, F. Cano, and R. V. Tauxe. 1999. Surveying vendors of street-vended food: a new methodology applied in two Guatemalan cities. *Epidemiology and Infection* 122(3):409-416.

Martorell, R., J. P. Habicht, C. Yarbrough, A. Lechtig, R. E. Klein, and K. A. Western. 1975. Acute morbidity and physical growth in rural Guatemalan children. *American Journal of Diseases of Children* 129(11):1296-1301.

Migele, J., S. Ombeki, M. Ayalo, M. Biggerstaff, and R. Quick. 2007. Diarrhea prevention in a Kenyan school through the use of a simple safe water and hygiene intervention. *American Journal of Tropical Medicine and Hygiene* 76 (2):351-353.

Mintz, E. D., F. M. Reiff, and R.V. Tauxe. 1995. Safe water treatment and storage in the home: a practical new strategy to prevent waterborne disease. *Journal of the American Medical Association* 273(12):948-953.

Mintz, E. D., J. Bartram, P. Lochery, and M. Wegelin. 2001. Not just a drop in the bucket: expanding access to point-of-use water treatment systems. *American Journal of Public Health* 91(10):1565-1570.

Moe, C. L., and R. Rheingans. 2006. Global challenges in water, sanitation and health. *Journal of Water and Health* 4(Suppl 1):41-57.

Motarjemi, Y., F. Kaferstein, G. Moy, and F. Quevedo. 1993. Contaminated weaning food: a major risk factor for diarrhoea and associated malnutrition. *Bulletin of the World Health Organization* 71(1):79-92.

Naimi, T. S., J. H. Wicklund, S. J. Olsen, G. Krause, J. G. Wells, J. M. Bartkus, D. J. Boxrus, M. Sullivan, H. Kassenborg, J. M. Besser, E. D. Mintz, M. T. Osterholm, and C. W. Hedberg. 2003. Concurrent outbreaks of *Shigella sonnei* and enterotoxigenic *Escherichia coli* associated with parsley: implications for surveillance and control of foodborne illness. *Journal of Food Protection* 66(4):535-541.

O'Reilly, C. E., M. C. Freeman, M. Ravani, J. Migele, A. Mwaki, M. Ayalo, S. Ombeki, R. M. Hoekstra, and R. Quick. 2008. The impact of a school-based safe water and hygiene programme on knowledge and practices of students and their parents: Nyanza Province, western Kenya, 2006. *Epidemiology and Infection* 136(1):80-91.

PAHO (Pan American Health Organization). 2008. *Health situation in the americas: basic indicators 2008*, http://www.who.int/pmnch/topics/paho2008healthstats/en/index.html (accessed February 15, 2009).

Parker, A. A., R. Stephenson, P. L. Riley, S. Ombeki, C. Komolleh, L. Sibley, and R. Quick. 2006. Sustained high levels of stored drinking water treatment and retention of hand-washing knowledge in rural Kenyan households following a clinic-based intervention. *Epidemiology and Infection* 134(5):1029-1036.

Penteado, A. L., B. S. Eblen, and A. J. Miller. 2004. Evidence of *Salmonella* internalization into fresh mangos during simulated postharvest insect disinfestation procedures. *Journal of Food Protection* 67(1):181-184.

Ries, A. A., D. J. Vugia, L. Beingolea, A. M. Palacios, E. Vasquez, J. G. Wells, N. Garcia Baca, D. L. Swerdlow, M. Pollack, and N. H. Bean. 1992. Cholera in Piura, Peru: a modern urban epidemic. *Journal of Infectious Diseases* 166(6):1429-1433.

Rosenau, M. J. 1928. *Preventative medicine and hygiene*. New York: D. Appleton and Company. P. 111.

Semmelweis, I. 1983. *Etiology, concept and prophylaxis of childbed fever (1861)*. Madison, WI: University of Wisconsin Press.

Sepulveda, J., F. Bustreao, R. Tapia, J. Rivera, R. Lozano, G. Olaiz, V. Partida, L. Garcia-Garcia, and J. L. Valdespino. 2006. Improvement of child survival in Mexico: the diagonal approach. *Lancet* 368(9551):2017-2027.

Sheth, A. N., E. Russo, M. Menon, A. C. Kudzala, J. D. Kelly, M. Weinger, K. Sebunya, H. Masuku, K. Wannemuehler, and R. Quick. 2008. Successful promotion of water treatment and hand hygiene through a pilot clinic-based intervention for pregnant women seeking antenatal care-Malawi, May 2007-March 2008, Abstract #16. 57th Annual Conference, American Society for Tropical Medicine and Hygiene, New Orleans, December 7-11, 2008.

Sivapalasingam, S., E. Barrett, A. Kimura, M. S. Van Duyne, W. De Witt, M. Ying, A. Frisch, Q. Phan, E. Gould, P. Shillam, V. Reddy, T. Cooper, M. Hoekstra, C. Higgins, J. P. Sanders, R. V. Tauxe, and L. Slutsker. 2003. A multistate outbreak of *Salmonella enterica* serotype Newport infections linked to mango consumption: impact of a water-dip disinfestation technology. *Clinical Infectious Diseases* 37(12):1585-1590.

Sobel, J., B. Mahon, C. E. Mendoza, D. Passaro, F. Cano, K. Baier, F. Racioppi, L. Hutwagner, and E. Mintz. 1998. Reduction of fecal contamination of street-vended beverages in Guatemala by a simple system for water purification and storage, handwashing, and beverage storage. *American Journal of Tropical Medicine and Hygiene* 59(3):380-387.

Swerdlow, D. L., E. D. Mintz, M. Rodriguez, E. Tejada, C. Ocampo, L. Espejo, K. D. Greene, W. Saldana, L. Seminario, R. V. Tauxe, J. G. Wells, N. H. Bean, A. A. Ries, M. Pollack, B. Vertiz, and P. A. Blake. 1992. Waterborne transmission of epidemic cholera in Trujillo, Peru: lessons for a continent at risk. *Lancet* 340(8810):28-32.

Tauxe, R. V., S. J. O'Brien, and M. Kirk. 2008. Outbreaks of food-borne diseases related to the international food trade. In *Imported foods: microbial issues and challenges*, edited by M. Doyle and M. Erickson. Washington, DC: American Society for Microbiology Press.

UNDP (United Nations Development Programme). 2009a. *Goal 4: reduce child mortality*, http://www.undp.org/mdg/goal4.shtml (accessed February 24, 2009).

———. 2009b. *Goal 7: ensure environmental sustainability*, http://www.undp.org/mdg/goal7.shtml (accessed February 24, 2009).

Vollaard, A. M., S. Ali, H. A. van Asten, I. S. Ismid, S. Widjaja, L. G. Visser, Ch. Surjadi, and J. T. van Dissel. 2004a. Risk factors for transmission of foodborne illness in restaurants and street vendors in Jakarta, Indonesia. *Epidemiology and Infection* 132(5):863-872.

Vollaard, A. M., S. Ali, H. A. van Asten, S. Widjaja, L. G. Visser, C. Surjadi, and J. T. van Dissel. 2004b. Risk factors for typhoid and paratyphoid fever in Jakarta, Indonesia. *Journal of the American Medical Association* 291(21):2607-2615.

Wheeler, C., T. M. Vogt, G. L. Armstrong, G. Vaughan, A. Weltman, O. Nainan, V. Dato, G. Xia, K. Waller, J. Amon, T. M. Lee, A. Highbaugh-Battle, C. Hembree, S. Evenson, M. A. Ruta, I. T. Williams, A. E. Fiore, and B. P. Bell. 2005. An outbreak of hepatitis A associated with green onions. *New England Journal of Medicine* 353(9):890-897.

WHO (World Health Organization). 2008. *World health statistics 2008*, http://www.who.int/whosis/whostat/EN_WHS08_Table1_Mort.pdf (accessed February 15, 2009).

Wright, J., S. Gundry, and R. Conroy. 2004. Household drinking water in developing countries: a systematic review of microbiological contamination between source and point-of-use. *Tropical Medicine and International Health* 9(1):106-117.

2

Lessons from Waterborne Disease Outbreaks

OVERVIEW

This chapter is comprised of three case studies of waterborne disease outbreaks that occurred in the Americas. Each contribution features an outbreak chronology, an analysis of contributing factors, and a consideration of lessons learned. Together, they illustrate how an intricate web of factors—including climate and weather, human demographics, land use, and infrastructure—contribute to outbreaks of waterborne infectious disease.

The chapter begins with an account of the massive cholera epidemic that began in urban areas of Peru in 1991 and swept across South America by Carlos Seas and workshop presenter and Forum member Eduardo Gotuzzo, of Universidad Peruana Cayetano Heredia and Hospital Nacional Cayetano Heredia in Lima, Peru. The authors describe current understanding of the role of *Vibrio cholerae* in marine ecosystems, and consider how climatic and environmental factors, as well as international trade, may have influenced the reintroduction of this pathogen to the continent after nearly a century's absence. The epidemic persisted for five years, then reappeared, with diminshed intensity, in 1998. While attempts to control the epidemic through educational campaigns aimed at improving sanitation were unsuccessful in the short term, Seas and Gotuzzo report that, following a significant investment in sanitation in the wake of this public health disaster, transmission rates of other waterborne infectious diseases, including typhoid fever, declined in Peru. They note that, by understanding the ecology of *V. cholerae*, researchers may be able to predict relative risk for pathogen transmission from marine environments and thereby aid efforts at preventing epidemics.

In 1993, two years after cholera struck Peru, an epidemic of cryptosporidiosis in Miluwaukee, Wisconsin, sickened hundreds of thousands of people and caused at least 50 deaths, demonstrating that even "modern" water treatment and distribution facilities are vulnerable to contamination by infectious pathogens. In their contribution to this chapter, workshop presenter Jeffrey Davis and coauthors recount their investigation of this outbreak, which resulted from the confluence of multiple and diverse environmental and human factors. Based on lessons learned from their discoveries, the authors made—and authorities undertook—recommendations to prevent further outbreaks in the Milwaukee water system, resulting in significant improvements in water quality. Their findings have proven applicable to other water treatment facilities that share Lake Michigan and have received attention from water authorities worldwide.

The final paper in this chapter, by workshop presenter Steve Hrudey and Elizabeth Hrudey of the University of Alberta, Canada, discusses an episode of bacteria contamination of the water in Walkerton, Ontario, in 2000. The outbreak sickened nearly half of the town's 5,000 residents and caused 7 deaths, as well as 27 cases of hemolytic uremic syndrome, a severe kidney disease. Several incidents of human error and duplicity figure prominently among the causes of this entirely preventable outbreak, the authors explain. "Because outbreaks of disease caused by drinking water remain comparatively rare in North America," they conclude, "complacency about the dangers of waterborne pathogens can easily occur." Based on their findings, they present a framework for water system oversight intended to save other communities from Walkerton's fate.

THE CHOLERA EPIDEMIC IN PERU AND LATIN AMERICA IN 1991: THE ROLE OF WATER IN THE ORIGIN AND SPREAD OF THE EPIDEMIC

Carlos Seas, M.D.[1]
Universidad Peruana Cayetano Heredia

Eduardo Gotuzzo, M.D., FACP[1,2]
Universidad Peruana Cayetano Heredia

At Athens a man was seized with cholera. He vomited, and was purged and was in pain, and neither the vomiting nor the purging could be stopped; and his voice failed him, and he could not be moved from his

[1]Insitituto de Medicina Tropical Alexander von Humboldt. Universidad Peruana Cayetano Heredia, Lima, Peru, and Departamento de Enfermedades Infecciosas, Tropicales y Dermatológicas. Hospital Nacional Cayetano Heredia. Lima, Peru.

[2]Corresponding author. Av. Honorio Delgado 430, Lima 31, Peru. Phone: 51-1-4823910, Fax: 51-1-4823404, E-mail: egh@upch.edu.pe.

bed, and his eyes were dark and hollow, and spasms from the stomach held him, and hiccup from the bowels. He was forced to drink, and the two (vomiting and purging) were stopped, but he became cold.

Hippocrates

After an absence of almost one century, cholera reappeared in South America in Peru during the summer of 1991. This event was totally unexpected by the scientific community, which had anticipated the spread of cholera to the continent from Africa and had hypothesized its introduction by Brazil following well-recognized routes of dissemination of the disease that involve trade and commerce. The further spread of the epidemic was very rapid; all Peruvian departments had reported cholera cases in less than six months; almost all Latin American countries, with the exception of Uruguay, had reported cases within one year of the beginning of the epidemic. The chains of events that triggered and disseminated the epidemic into the continent have not been fully elucidated, but evidence is being gathered on the possible role of marine ecosystems, climate and environmental factors, and the pivotal role of water. We discuss here the evidence in support of water's role in cholera dynamics.

The Environmental Life Cycle of *Vibrio cholerae*

The natural reservoirs of *V. cholerae* are aquatic environments, where O1 and non-O1 serogroups coexist. *V. cholerae* survives by attaching to and forming symbiotic associations with algae or crustacean shells (Figure 2-1). In these environments, *V. cholerae* multiplies and can persist for years in a free-living cycle without human intervention, as it has been elegantly described by Dr. Colwell and her associates at the International Centre for Diarrheal Diseases Research in Dhaka, Bangladesh (Colwell et al., 1990).

A number of environmental factors modulate the abundance of *Vibrio*, including, but not limited to, temperature, pH, salinity, and nutrient availability. Under adverse conditions, *V. cholerae* survives in a dormant state with all metabolic pathways shut down, which can be reactivated again when suitable conditions return. Additionally, *V. cholerae* can produce biofilms—surface-associated communities of bacteria with enhanced survival under negative conditions—which can switch to active bacteria and induce epidemics.

The ability of *V. cholerae* to regulate its metabolism based on the environmental conditions of its natural reservoir may explain the endemicity of cholera in many parts of the world. During the cholera epidemic in Peru, *V. cholerae* was isolated from many aquatic environments, including not only marine ecosystems, but riverine and lake environments. Even one of the highest commercially navigable freshwater lakes in the world, Lake Titicaca, located 3,827 meters above sea level on the border of Peru and Bolivia, was impacted by the cholera epidemic of 1991.

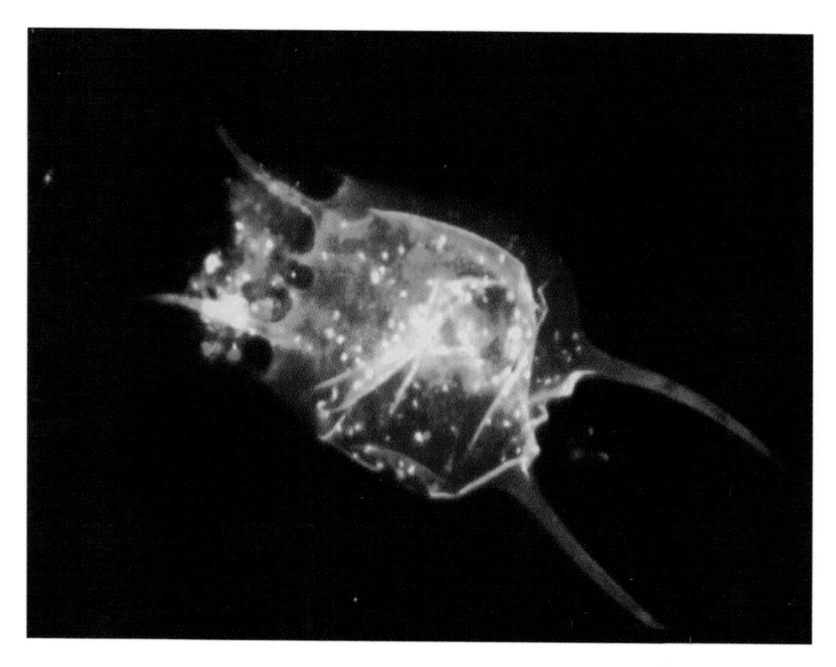

FIGURE 2-1 *Vibrio cholerae* O1 attached to a copepod.
SOURCE: Courtesy of Rita Colwell, Ph.D., University of Maryland.

Humans are only temporary reservoirs of *V. cholerae*. Interestingly, lytic phages modulate the abundance of *V. cholerae* in the human intestine, but on the other hand, *V. cholerae* are able to up-regulate certain genes in the intestine of humans resulting in a short-time hyperinfectious state. As illustrated in Figure 2-2, *V. cholerae* is introduced to humans from its aquatic environment through contamination of food and water sources.

The Origins of the Latin American Epidemic

The Latin American cholera epidemic was officially declared in Peru during the third week of January 1991, almost simultaneously in three cities along the north coastal area of the country. By the end of that year, almost 320,000 cases had been officially reported to the Pan American Health Organization by the Peruvian Ministry of Health. Nearly 45,000 cases occurred every week, in what was considered the worst cholera epidemic of the century in Peru (Gotuzzo et al., 1994). There were several distinctive features of this epidemic:

- Very high attack rates were reported soon after the epidemic started.

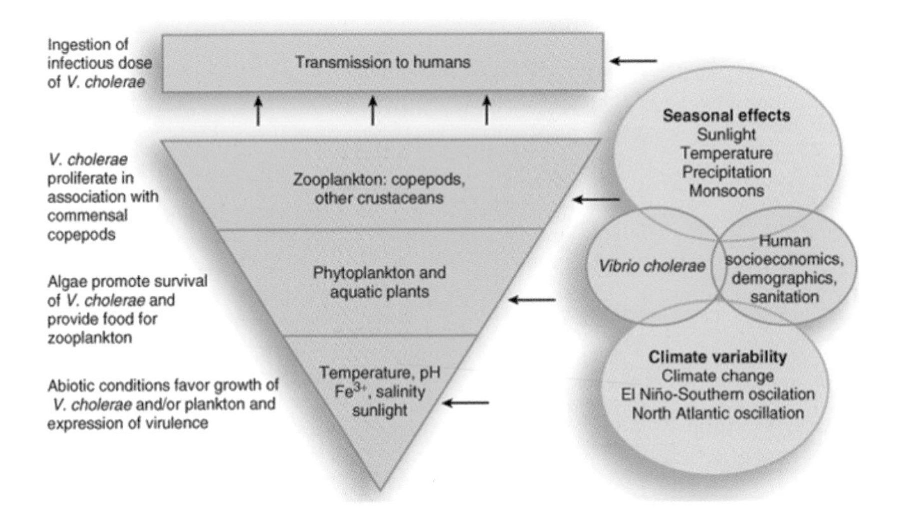

FIGURE 2-2 A hierarchical model for cholera transmission.
SOURCE: Reprinted from Lipp et al. (2002) with permission from the American Society for Microbiology.

- Cholera accounted for almost 80 percent of all acute diarrhea cases in the country irrespective of the degree of dehydration and age group.
- The epidemic was initially concentrated in urban areas, where it spread very rapidly, suggesting a common source of dissemination.
- Transmission was halted in very few areas, where treatment and chlorination of municipal water was possible, suggesting a critical role of water in the transmission of the disease.
- Very low case-fatality rates were reported from urban areas where patients had access to treatment by well-trained health personnel, but higher figures were reported from isolated communities where patients did not have access to health centers, a situation similar to those reported from Africa in refugee settings under political instability.

Although the epidemic spread to neighboring countries, it never reached the magnitude seen in Peru, which suffered that year from serious economic constraints and reported the lowest level of sanitary coverage and sanitary investment in the region. During 1991, approximately 50 percent of the population in urban cities of Peru received treated municipal water; intermittent supply and clandestine connections were common in many cities of the country (Figure 2-3). Additionally, less than 10 percent of sewerage water was treated properly. These conditions prevailed before the beginning of the epidemic and were responsible for its very rapid spread. The epidemic lasted for five years until 1995, only to reappear again in 1998 with much less intensity, as shown in Figure 2-4 (WHO, 2008). The message conveyed to the population at the beginning of the epidemic to curtail transmission focused on avoiding eating raw fish and shellfish and to boil water for drinking purposes.

Massive investment in sanitation followed the epidemic, which was responsible for a reduction in transmission not only of cholera but also of other enteric infections, such as numerous parasitic infections and typhoid fever. The case of typhoid fever deserves special mention. Many experienced doctors in Lima saw a marked reduction in the incidence of typhoid fever in their practices as a consequence of improvements in sanitation and hygiene, a situation that was also seen at our Institute (Figure 2-5).

FIGURE 2-3 A shantytown in Peru during 1991.
SOURCE: Instituo de Medicina Tropical Alexander von Humboldt, Lima, Peru.

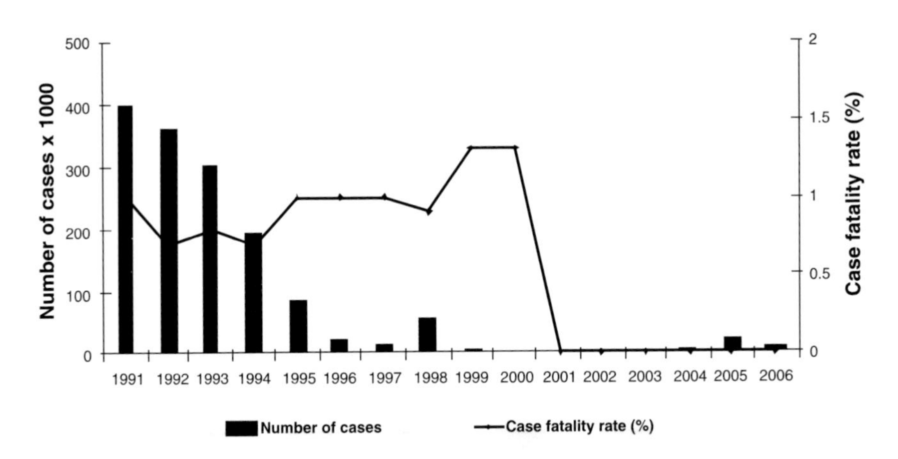

FIGURE 2-4 Cholera in the Americas, 1991-2006.
SOURCE: Based upon data compiled from and reported in the WHO's Weekly Epidemiological Record.

Before 1990, typhoid fever was responsible for the majority of episodes of undifferentiated fever lasting at least five days in Lima. Approximately 70 to 100 patients with complicated typhoid fever were hospitalized yearly in our institution the decade before the cholera epidemic; these figures were reduced tenfold after 1991. The reduction in typhoid fever incidence was so dramatic that the disease is almost unknown by the generation of physicians trained after 1991, with the subsequent delay in diagnosis and development of complications, an unthinkable situation the decade before 1990.

Still, a question remains unanswered: From where did this huge cholera epidemic originate? Although both nontoxigenic O1 and non-O1 *V. cholerae* strains had been isolated from environmental sources and from patients in Peru and other countries in the region, the hypothesis that suggested that these *Vibrio* became residents in aquatic environments of coastal Peru with further acquisition of virulence genes that mediated for toxigenic expression through phage infection seems unlikely. Additionally, genetic comparison of the *Vibrio* responsible for the epidemic; *V. cholerae* O1 serotype Inaba and biotype El Tor, with endemic agents in Asia, disclosed very similar patterns, suggesting common ancestors or spread from one place to another. The latter option seems more reasonable. Another hypothesis suggests that *V. cholerae* was seeded into the marine ecosystems of northern Peru a few months before the epidemic started, which seems more likely in light of what was discussed earlier—that *Vibrio* was imported from Asia transported by crew ships, or emptied from vessels discharging bilge water contaminated with the bacterium.

From its aquatic environment *Vibrio* was first amplified along the north coast

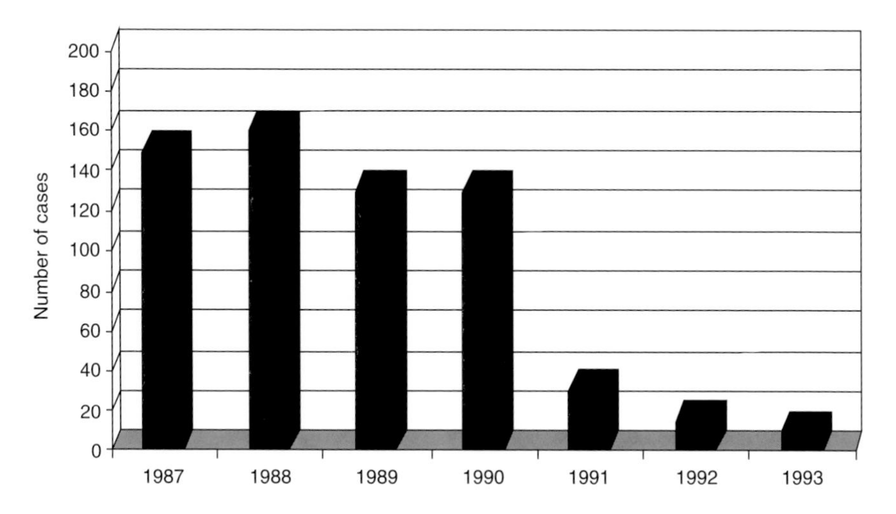

FIGURE 2-5 Typhoid fever cases seen at the Alexander von Humboldt Tropical Medicine Institute in Lima, Peru, 1987-1993.

and then introduced almost simultaneously into several cities of the country (Figure 2-6). This hypothesis was proposed after analyzing data generated by a retrospective study that reviewed charts of patients who had attended several hospitals along the Peruvian North Coast in 1989 and 1991, and disclosed that seven patients fulfilled the clinical definition of cholera proposed by the World Health Organization three months before the epidemic had started (Seas et al., 2000). These adult patients attended with severe dehydration and watery diarrhea, clinical presentation that had not been at these health centers the year before the epidemic. Although no convincing evidence proves definitively that these cases were due to cholera (clinical laboratory cultures had not been obtained for these cases), the clinical presentation is similar to that described in other epidemic areas for cholera, and also similar to that which many Peruvian doctors subsequently saw (Figure 2-7).

Which Forces Drove the Spread of *Vibrio* into the Pacific Coastal Areas of Peru?

Dr. Rita Colwell's theory on the environmental niche for *V. cholerae* in aquatic ecosystems is crucial for understanding cholera dynamics. Factors that modify the survival of *Vibrio* in the environment may dramatically influence cholera transmission. Climate change and climate variability are among these

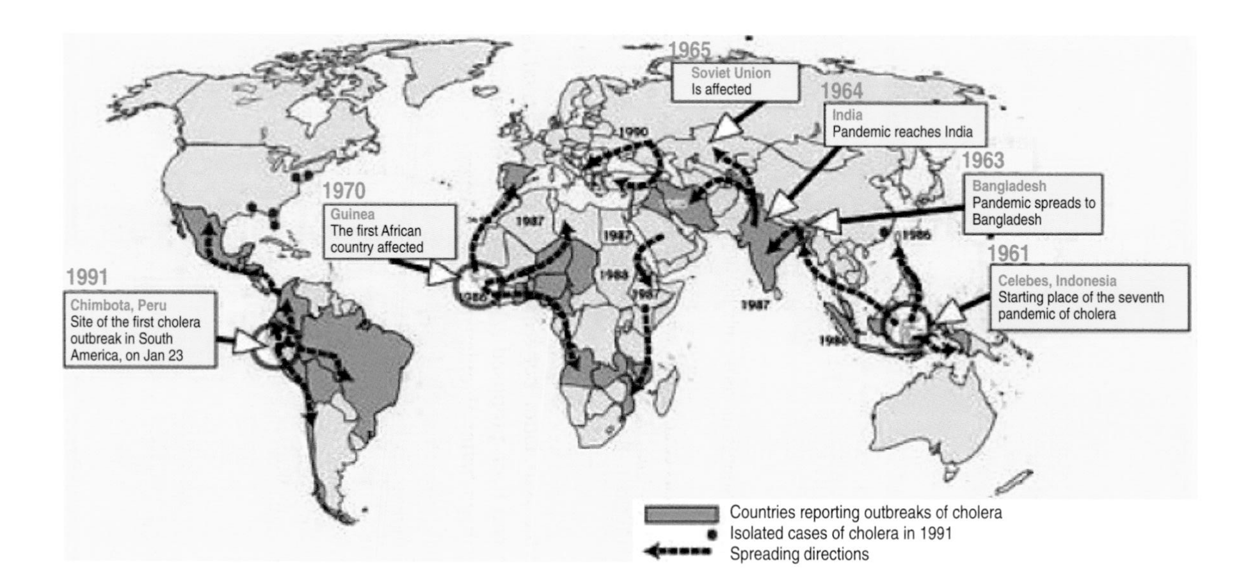

FIGURE 2-6 The seventh cholera pandemic.
SOURCE: Carlos Seas. Cólera. Medicina Tropical. CD-ROM. Version 2002. Instituto de Medicina Tropical, Príncipe Leopoldo. Amberes, Bélgica. Instituto de Medicina Tropical Alexander von Humboldt; Lima, Peru. Universidad Mayor de San Simón, Cochababmba, Bolivia.

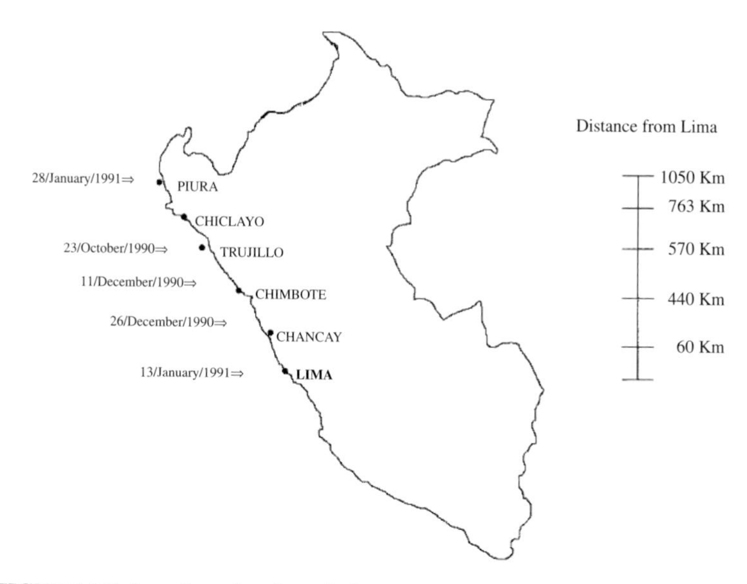

FIGURE 2-7 Location of patients in Peru with presumed cholera, identified before the epidemic of 1991.
SOURCE: Reprinted from Seas et al. (2000) with permission from the *American Journal of Tropical Medicine and Hygiene*.

critical forces. While the association between climate, environmental factors, and cholera transmission had been proposed a long time ago, the role of climate in cholera dynamics has been better elucidated in recent years.

Time series analysis has demonstrated a relationship between the appearance of cholera cases in Bangladesh and occurrence of El Niño-Southern Oscillation (ENSO). Observations have linked the interannual variability of ENSO with the proportion of cholera cases in Dhaka, Bangladesh. Additionally, climate variability due to ENSO and temporary immunity explained the interannual cycles of cholera in rural Matlab, Bangladesh, for a period of almost 30 years (Pascual et al., 2000). The net effect of ENSO—rise in both sea water temperature and planktonic mass—modifies the abundance of *V. cholerae* in the environment by affecting the concentration of plankton to which *V. cholerae* is attached and affects the concentration of nutrients and salinity.

Water temperature affects cholera transmission, as has been observed in the Bay of Bengal, Bangladesh. All these data support the role of ENSO in the interannual variability of endemic cholera. An unproven hypothesis suggests that El Niño triggered the epidemic of cholera in Peru in 1991 by amplifying the planktonic mass and dispersing existing *Vibrio* along the north coast of the country. Then *Vibrio* was introduced into the continent by contaminated food and subsequently by contamination of the water supply system (Seas et al., 2000).

Several studies conducted in Peru after 1991 have shown an association between warmer air temperatures and cholera cases in children and adults. Additionally, toxigenic *V. cholerae* O1 has been isolated from aquatic environments in the coastal waters of Peru, suggesting that it has been successful in adapting to these environments, as has been described in Bangladesh and India. These findings support the theory of an environmental niche for *V. cholerae* O1 in Latin America and temporal associations between ENSO and cholera outbreaks from 1991 onward (Salazar-Lindo et al., 2008).

The Spread of the Cholera Epidemic in Peru as a Model to Understand Transmission in the Region

As illustrated earlier in Figure 2-3, the cholera transmission cycle involves infection of humans by the consumption of contaminated food and water and further shedding of the bacteria into the environment via contaminated stools. Incredibly high attack rates accompany human infection under favorable conditions, especially in previously nonexposed populations. Very high household transmission rates also occur.

Transmission via contaminated water and food has been long recognized. During the Latin American epidemic, acquisition of the disease by drinking contaminated water from rivers, ponds, lakes, and even tube well sources were documented. Contamination of municipal water was the main route of cholera transmission in Trujillo, Peru, during the epidemic in 1991. Drinking unboiled water, introducing contaminated hands into containers used to store drinking water, drinking beverages from street vendors, drinking beverages when contaminated ice had been added, and drinking water outside the home are recognized exposure risk factors for cholera. In addition to the crucial role of water in the transmission of cholera, poor hygienic conditions also contribute to the spread of cholera by exposing susceptible persons to the pathogen. Educational campaigns were implemented throughout the country with little effect in the short term.

Certain host factors may have played a role in the transmission of cholera. Infection by *Helicobacter pylori*, the effect of the O blood group, and the protective effect of breast milk deserve to be mentioned. Studies from Bangladesh and Peru show that people infected by *H. pylori* are at higher risk of acquiring cholera than people not infected by *H. pylori* (León-Barúa et al., 2006). Additionally, the risk of acquiring severe cholera among people coinfected with *H. pylori* is higher in patients without previous contact with *V. cholerae*, as measured by the absence of vibriocidal antibodies in the serum (Clemens et al., 1995).

H. pylori is highly endemic in developing countries, particularly among low-income status individuals. Infection causes a chronic gastritis that induces hypochlorhydria, which in turn reduces the ability of the stomach to limit the

Vibrio invasion. Patients carrying the O blood group, which is widespread in Latin America, have a higher risk of developing severe cholera. Higher affinity of the cholera toxin to the ganglioside receptor in patients with O blood group and lower affinity in patients of A, B, and AB blood groups may explain this association. Finally, the protective effect of breast milk, possibly mediated by a high concentration of secretory IgA anti-cholera toxin, has been proposed.

Preventing Future Epidemics

The scarce number of autochthonous cases reported from developed countries, such as the United States and Australia where *Vibrio cholerae* O1 is a resident of aquatic environments, provides additional support for the well-known concept that hygiene and sanitation can control cholera transmission. These relatively simple measures are very difficult to implement in the developing world (Zuckerman et al., 2007).

Alternative ways to prevent cholera transmission have been explored, including but not limited to the boiling and/or chlorination of water, exposing water to sunlight, filtering water using Sari cloth, and educating the population at risk on appropriate hygienic practices (Colwell et al., 2003). Using new information generated from the studies that delineated the ecological niche of *Vibrio* may help in predicting the onset of an epidemic, which may have a tremendous impact on prevention. Searching for *V. cholerae* O1 in municipal sewage and environmental samples in endemic areas could be used as a warning signal of future epidemics (Franco et al., 1997), and monitoring the movement and abundance of plankton by satellite seems attractive, but more studies are needed to support the implementation of these methods.

Conclusions

The cholera epidemic in Latin America was characterized by an explosive beginning with rapid spread in urban areas of Peru and other poor neighboring countries. The available information suggests that environmental factors amplified the existing *Vibrio* population and induced an epidemic, which was further amplified by contamination of municipal water and food. Water played a key role not only in maintaining *Vibrio* in its natural reservoir but also in disseminating the epidemic.

Acknowledgments

We would like to express our most sincere gratitude to Dr. Rita Colwell and Dr. Bradley Sack for sharing with us valuable information and images that were reproduced in this manuscript.

LESSONS FROM THE MASSIVE WATERBORNE OUTBREAK OF *CRYPTOSPORIDIUM* INFECTIONS, MILWAUKEE, 1993

Jeffrey P. Davis, M.D.[3]
Wisconsin Division of Public Health

William R. Mac Kenzie, M.D.[4]
Centers for Disease Control and Prevention

David G. Addiss, M.D., M.P.H.[5]
Fetzer Institute

The Investigation Begins[6]

On Monday, April 5, 1993, the City of Milwaukee Health Department (MHD) received reports of increased school and workplace absenteeism due to diarrheal illness in Milwaukee County, Wisconsin. This appeared to be quite widespread, particularly on the south side of the city. In one particular hospital, during the previous weekend, more than 200 individuals had been cultured for bacterial enteric pathogens. The hospital ran out of bacterial culture media, yet none of the patients tested positive for bacterial enteric pathogens. Routine tests for ova and parasites done on many stools did not reveal pathogens. Pharmacies were experiencing widespread shortages of antidiarrheal medications. Because of the clinical profile of illness and the apparent magnitude of the outbreak, we considered this outbreak to be due to a product with wide local distribution with drinking water being the most likely vehicle. The Wisconsin Division of Health (now the Wisconsin Division of Public Health [DPH]) offered onsite assistance to investigate and control this outbreak. The offer was accepted; lead staff arrived on April 6 and additional team members arrived on April 7.

The Director of the Bureau of Laboratories, MHD, requested and received water quality and treatment data from the Milwaukee Water Works (MWW). While preliminarily reviewing these data, he noted striking spikes in turbidity of water treated in one of the two MWW treatment plants (the southern plant), and these spikes in finished water turbidity occurred on multiple days in late March and early April. This was reminiscent of a large waterborne outbreak of *Cryptosporidium* infections in Carrollton, Georgia, that occurred among customers of a municipal water supply (Hayes et al., 1989). On April 6, following discussion with DPH staff, the laboratory director selected some representative stool speci-

---

[3]Madison, Wisconsin.

[4]Division of Tuberculosis Elimination, Atlanta, Georgia.

[5]Kalamazoo, Michigan.

[6]See Mac Kenzie et al. (1994b).

mens among those that had tested negative for enteric pathogens and tested them for protozoan parasites including *Cryptosporidium*, which initially had not been done in the clinical laboratories. Early on April 7, DPH, MHD, and Wisconsin Department of Natural Resources staff met with MWW officials. By late afternoon on April 7, positive results for *Cryptosporidium* were reported found in stool specimens from three adults conducted by the MDH laboratory, and stools from five adults tested at other Milwaukee laboratories. These adults resided at widespread locations in southern Milwaukee and one neighboring municipality within Milwaukee County. Following a meeting with city and state public health and water treatment officials, Milwaukee's mayor, John Norquist, issued a boil water advisory on the evening of April 7. The outbreak received considerable media attention for more than two weeks. Inordinate numbers of people were inconvenienced. Pharmacies continued to sell a lot of antidiarrhea medications. Many industries with processes dependent on treated water were challenged.

The city of Milwaukee occupies most of Milwaukee County. Three rivers—the Milwaukee, the Menomonee, and the Kinnickinnic—flow through the county and converge in the city, where they empty into Lake Michigan within a breakfront; the ambient flow of the river water entering the lake is southerly (Figure 2-8).

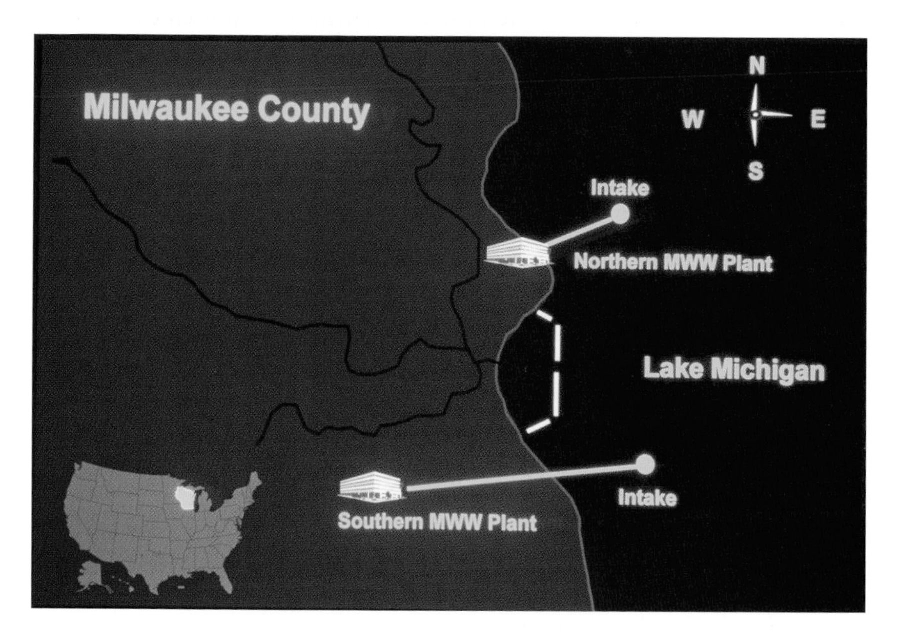

FIGURE 2-8 Location of the three rivers that flow through Milwaukee County, Wisconsin, the breakfront protecting the city of Milwaukee harbor, and the northern and southern Milwaukee Water Works water treatment plants and their intake grids.
SOURCE: Addiss et al. (1995). Reproduced by permission of The Royal Society of Chemistry.

Figure 2-8 also shows the location of the two MWW water treatment plants: the northern plant received water by gravity flow through an intake 1.2 miles offshore, and the southern plant received water by gravity flow through an intake 1.4 miles offshore.

Figure 2-9, which depicts daily turbidity values for treated water at both plants during March and April 1993, demonstrates several spikes in treated water turbidity at the southern plant (Mac Kenzie et al., 1994b). The first peak, which occurred on or about March 23, was the largest recorded at the southern plant in more than 10 years. It was followed by a considerably larger, sustained peak with maximum turbidity measurements on March 28 and 30, and another peak on April 5. The southern plant was closed on April 7, but the water there was sampled on April 8 (Mac Kenzie et al., 1994b).

At the time of the outbreak, not much was known about *Cryptosporidium* in water. The pathogen had first been documented in humans in 1976 (Meisel et al., 1976; Nime et al., 1976), and by the early 1980s, was recognized as an AIDS-defining illness (Current et al., 1983). There had been several water-associated outbreaks in the United States and in the United Kingdom prior to the Milwaukee event, although most were associated with surface water contamina-

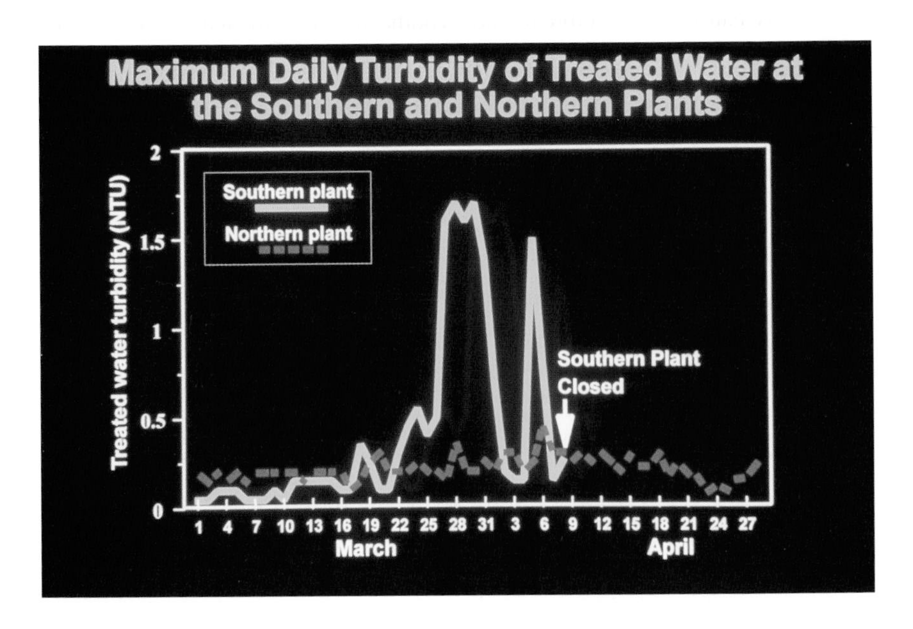

FIGURE 2-9 Maximal turbidity of treated water in the northern and southern water treatment plants of the Milwaukee Water Works from March 1 through April 28, 1993. NTU denotes nephelometric turbidity units.
SOURCE: Reprinted from Mac Kenzie et al. (1994b).

tion (D'Antonio et al., 1985; Gallagher et al., 1989; Joseph et al., 1991; Leland et al., 1993; Richardson et al., 1991). The 1987 outbreak in Carrollton, Georgia, affected an estimated 13,000 customers of a municipal water supply, but it was not associated with high turbidity of treated water (Hayes et al., 1989).

To evaluate for other microbiologic etiologies for the Milwaukee outbreak we reviewed the results of laboratory examinations of stool samples conducted in 14 different local laboratories between March 1 and April 16, 1993 (Mac Kenzie et al., 1994b). No increase in bacterial enteric pathogens was found. Prior to recognition of the outbreak, between March 1 and April 6, only 42 *Cryptosporidium* tests (nearly all of them on samples from patients with HIV/AIDS) were conducted, but after the outbreak was recognized more than 1,000 *Cryptosporidium* tests were conducted in a seven-day period. During both intervals, nearly one-third of these samples tested positive for *Cryptosporidium* (Mac Kenzie et al., 1994b). The percentage of positive *Cryptosporidium* tests, although not as high as one might expect in an outbreak, were similar to the rates of positive tests noted during the Carrollton event (39 percent). We believe that these results reflect the limits of standard microbiologic testing for *Cryptosporidium* available at that time.

Thus, early in our investigation we established that *Cryptosporidium* was the most likely cause of the outbreak and hypothesized that treated water from the southern water treatment plant was the vehicle for the majority of human infections associated with this outbreak. In addition to the primary task of testing this hypothesis, there were many tasks and questions we sought to address, which included determining

- the magnitude and timing of cases associated with the outbreak,
- the spectrum of clinical symptoms experienced in a large population of persons infected with *Cryptosporidium*,
- the incubation period of cryptosporidiosis following exposure,
- the timing of contamination of Milwaukee water,
- the secondary attack rate of cryptosporidiosis among family members not exposed to Milwaukee water,
- the frequency of recurrence of the symptoms of cryptosporidiosis after initial recovery,
- the presence of *Cryptosporidium* oocysts in Milwaukee water in water archived during the time of putative exposure of Milwaukee residents,
- factors at the MWW southern water treatment plant that allowed *Cryptosporidium* oocysts to pass through in treated water to infect the public,
- mortality associated with the outbreak,
- the frequency of asymptomatic infection among exposed Milwaukee residents, and
- the ultimate source of these *Cryptosporidium* oocysts: animals or humans.

To develop an epidemiologic case definition of cryptosporidiosis, we compared people with laboratory-confirmed infections with those who had clinical diagnoses (Mac Kenzie et al., 1994b). The age and gender profiles of these two patient classes were similar, although laboratory-confirmed cases were skewed, as one might expect, toward more serious illness. There was a uniform occurrence of diarrhea/watery diarrhea in all cases. Cramps, fatigue, muscle aches, vomiting, and fever occurred more frequently in the laboratory-tested individuals. Temporal distribution of the two patient classes was virtually identical (Mac Kenzie et al., 1994b).

Rapid Hypothesis Testing—Nursing Home Study

To rapidly test the hypothesis that the southern water treatment plant was the likely source of the outbreak, we examined rates of diarrhea among geographically fixed populations—residents of nursing homes—in different parts of Milwaukee (Mac Kenzie et al., 1994b). Due to their relative geographic location, nine nursing homes received drinking water primarily from the north plant; seven received water primarily from the south plant. Information on diarrhea was collected routinely at these nursing homes, so we were able to review their logs to establish the rate of diarrhea (defined as three or more loose stools per 24-hour period).

We found a spike in diarrheal illness peaking between April 1 and 6 among nursing home residents served by the south water plant. High rates of diarrhea continued into the following week and returned to baseline by April 19 (Mac Kenzie et al., 1994b). By contrast, diarrhea rates at nursing homes served by the north water plant remained at baseline throughout March and April. Importantly, the one nursing home in the south that obtained its water from a well had no increase in diarrhea rates. We tested stools from 69 nursing home residents with diarrhea from the south, and 12 from the north, for *Cryptosporidium*. Thirty-five (51 percent) of the southern samples were positive, but every northern sample was negative (Mac Kenzie et al., 1994b).

Magnitude and Impact of the Milwaukee Outbreak—Random Telephone Survey

To assess the magnitude of this outbreak, we conducted a random telephone survey of 840 households in Milwaukee and in the four surrounding counties, asking about the number of cases of watery diarrhea experienced between March 1 and April 28 (Mac Kenzie et al., 1994b). The response rate was 73 percent, and included 1,663 household members whose demographic features closely tracked 1990 census data. Among this sample, 436 (26 percent) were reported to have had watery diarrhea during the survey period, with the peak number of cases occurring during April 3 through April 5. As may be seen in Figure 2-10, the attack rate

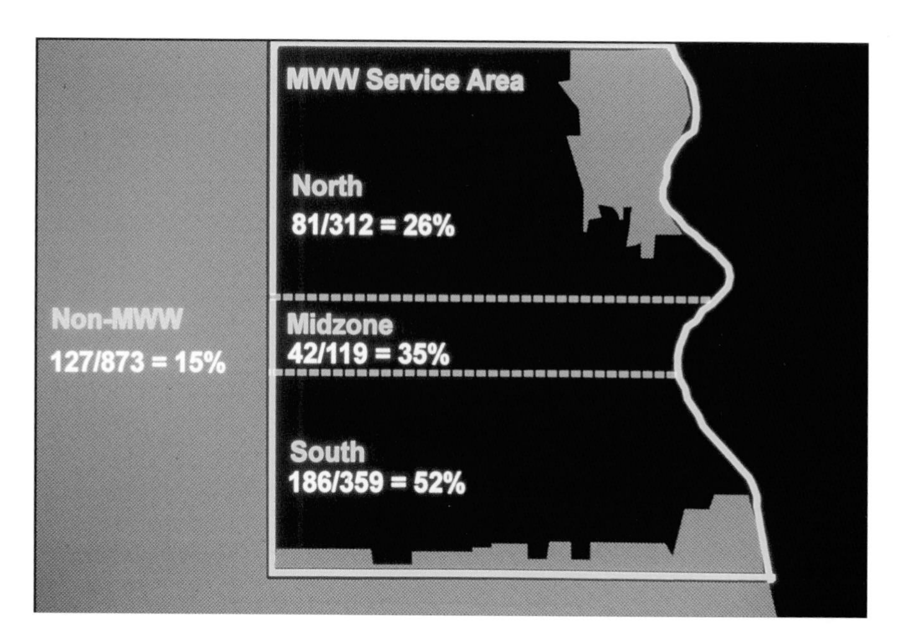

FIGURE 2-10 Rate of watery diarrhea from March 1 through April 28, 1993, among respondents in a random-digit telephone survey of households in the five county Greater Milwaukee area, by Milwaukee Water Works region.
SOURCE: Mac Kenzie et al. (1994b).

among MWW customers whose homes were served principally by the northern plant was 26 percent, compared with 52 percent of those from homes served principally by the southern plant. Residents receiving a mixture of water from both plants had an intermediate attack rate of 35 percent (Mac Kenzie et al., 1994b).

In our survey of Milwaukee and the four surrounding counties there was an overall attack rate of watery diarrhea of 26 percent. Applying this to the population of the five-county area, and subtracting a background rate of 0.5 percent, we estimated that 403,000 residents had watery diarrhea associated with this outbreak (certainly other people outside the survey area had also become ill, but we could not estimate their numbers; Mac Kenzie et al., 1994b). In our survey, 11 percent of people with watery diarrhea visited health-care providers (44,000 estimated visits) and 1.1 percent were hospitalized (4,400 estimated hospitalizations).

Approximately 1.8 days of productivity (school or work) were lost per case patient, which projects to 725,000 person-days lost, including about 479,000 person-days among those in the workforce (ages 18 to 64 years; Mac Kenzie et al., 1994b). Based mainly on review of death certificate data, we attributed 69 deaths to this outbreak; most of these were premature deaths among people with AIDS/HIV infection (Hoxie et al., 1997).

Study of Short-Term Visitors—Determining the Timing of Exposure, Incubation Period, and Frequency of Secondary Transmission

We studied short-term visitors to the Milwaukee area to answer the following questions:

- When was *Cryptosporidium* present in the water system?
- How long was the incubation period?
- How frequent was secondary (person-to-person) transmission?

Specifically, we studied people who visited the five-county Milwaukee area one time between March 15 and April 15, unaccompanied by other members of their households (Mac Kenzie et al., 1995b). We identified 94 such individuals who had stayed in the area less than 48 hours and had either laboratory-confirmed cryptosporidiosis (n = 54) or clinical cryptosporidiosis (n = 40) following their visit. Two-thirds of these visitors had stayed for less than 24 hours, and all had drunk beverages that contained unboiled tap water (the median amount consumed was 16 ounces while in Milwaukee; 32 percent of these ill visitors drank less than eight ounces) (Mac Kenzie et al., 1995b).

We examined the dates of arrival in Milwaukee and the dates of illness onset among the 94 brief interval visitors, as shown in Figure 2-11. Using dates of arrival as data of initial exposure, oocysts were presumed present in the treated water during 13 consecutive days (March 24 through April 5). By subtracting the date of arrival from the date of illness onset for each ill visitor, incubation periods could be calculated and the median incubation period was 7 days (range: 1-14 days). Diarrhea abated only to recur in 39 percent of visitors with laboratory-confirmed infection as it had in Milwaukee residents, suggesting that recurrence of diarrhea was not due to reinfection (Mac Kenzie et al., 1995b).

To determine the rate of secondary household transmission, we looked at nonvisiting members of the 94 visitors' households. We surveyed 74 people who fit this description, of whom 5 percent experienced watery diarrhea; thus, we concluded that the rate of secondary household transmission was quite low (Mac Kenzie et al., 1995b).

To evaluate for the presence of *Cryptosporidium* oocysts in the public water supply earlier in the outbreak, we needed to identify a large quantity of archived water. We obtained large blocks of ice for sculpture made by one southern Milwaukee ice manufacturer. Because of visible impurities the ice blocks frozen on specific days could not be used as intended, but fortunately they had been saved. We sampled melted water from these ice blocks made with water coming from the southern treatment plant. These blocks were available for two different days of manufacture around the time of the outbreak (March 25 and April 9; Addiss et al., 1995; Mac Kenzie et al., 1994b). To gauge *Cryptosporidium* oocyst levels in the water, we melted the blocks from each production day and sepa-

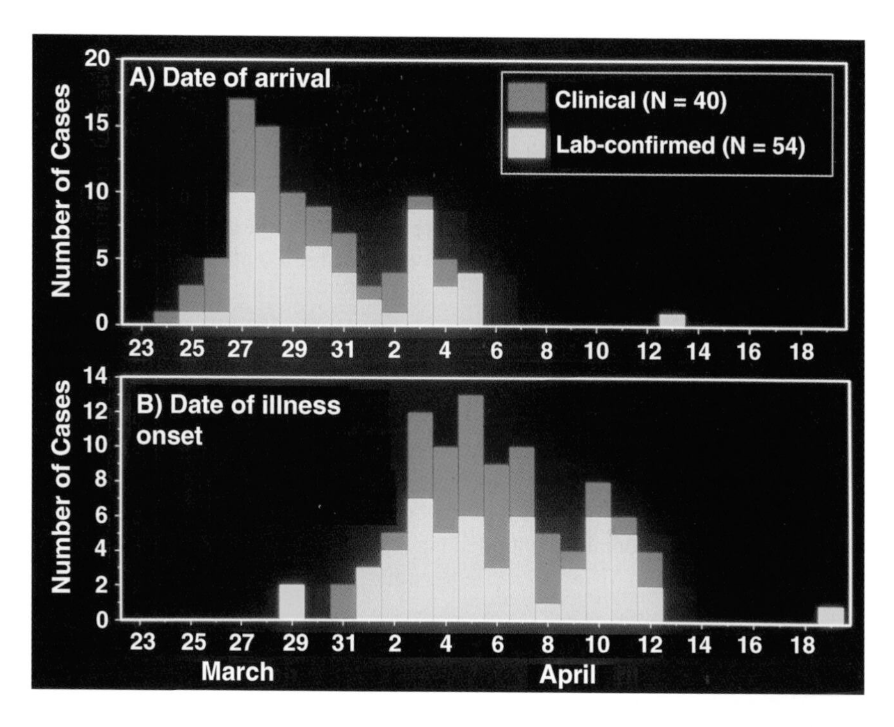

FIGURE 2-11 (A) Dates of arrival and (B) dates of onset of illness for 54 persons with laboratory-confirmed *Cryptosporidium* infection (black bars) and 40 persons with clinically defined *Cryptosporidium* infection (gray bars) among patients who had brief (<48 hours) exposure to the Milwaukee Water Works service area.
SOURCE: Adapted from Mac Kenzie et al. (1995b).

rated each of the respective samples into aliquots, which were filtered to recover oocysts using a peristaltic pump and two different kinds of filters: a 0.45-micron (absolute pore size) membrane filter (one aliquot for each production day), and a 1.0-micron (nominal pore size) spun polypropylene cartridge (the other aliquot for each production day). At the time, polypropylene cartridges were the standard filtration technique to determine the number of oocysts per liter in raw or finished water; membrane filters were "cutting edge," but these proved to be a much more sensitive means of detecting oocysts. Using the membrane filter, we detected 13.2 oocysts per 100 liters of melted ice from March 25 (before the peak in turbidity), and 6.7 per 100 liters on April 9 (after the boil-water advisory was invoked; Mac Kenzie et al., 1994b). While quite elevated, these likely underestimate concentrations originally in the water because freezing disrupts *Cryptosporidium* oocysts. The median infectious dose of *C. parvum* among healthy adult volunteers with no serologic evidence of past infection is 132 oocysts (Du Pont et al., 1995). More recently, the 50 percent infectious dose (ID50) of *C. hominis* among healthy adult volunteers with no serologic evidence of past infection was estimated to be 10 oocysts using a clinical definition of infection and 83 oocysts using a microbiologic definition (Chappell et al., 2006).

A Confluence of Events and Contributing Factors Leading to This Outbreak and the Investigation of the Milwaukee Water Treatment Plants

We collected considerable data regarding the operation of the two MWW plants and operating conditions during March and April 1993. Figure 2-12 depicts the water treatment process used by the MWW in 1993 in both treatment plants.

Raw water was introduced by gravity and rapidly mixed with chlorine for disinfection and a coagulant for mechanical flocculation. Following sedimentation, the water was rapidly filtered through sand-filled filtration beds (16 in the north plant and 8 in the south plant) and then stored in a large clear well prior to entry into water distribution pipes (Addiss et al., 1995; Mac Kenzie et al., 1994b). The capacity for producing treated water in each plant was substantial; if one plant was shut down, the full catchment area would still be fully served by the other plant remaining in operation.

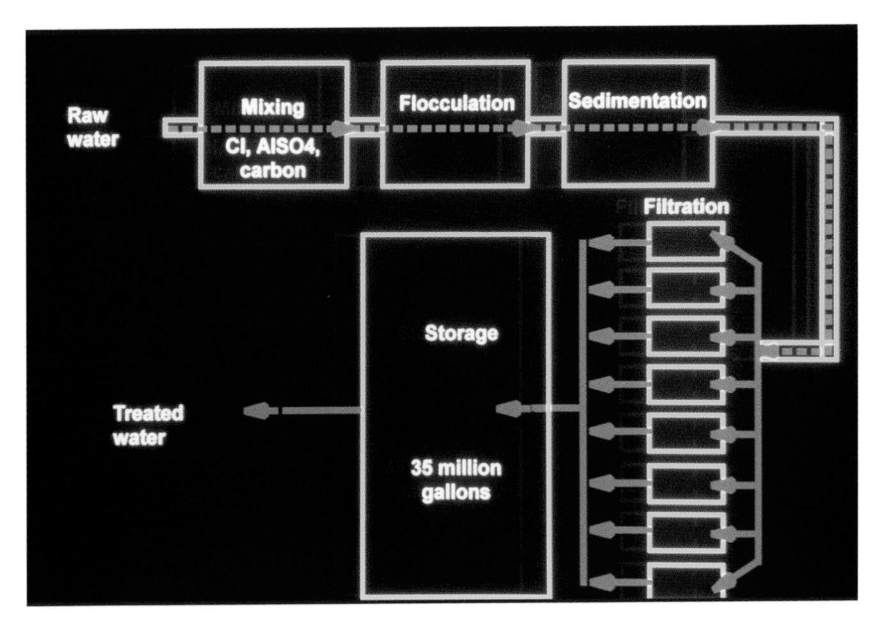

FIGURE 2-12 Depiction of the water treatment process used in the northern and southern Milwaukee Water Works water treatment plants in early 1993.

In September 1992, both plants changed the type of coagulant used, from the venerable alum to polyaluminum chloride. This was done in response to concern that lead and copper might leach from the aging water distribution infrastructure if the pH was too low, which was more likely to occur if the alum coagulant was used (Addiss et al., 1995).

From January 1983 to January 1993, the turbidity of treated water at the southern plant did not exceed 0.4 nephelometric turbidity unit (NTU). From February to April 1993, the turbidity of treated water at the southern plant did not exceed 0.25 NTU until March 18, when it increased to 0.35 NTU. From March 23 to April 1, the maximal daily turbidity of treated water was consistently 0.45 NTU or higher, with peaks of 1.7 NTU on March 28 and 30, despite an adjustment of the dose of polyaluminum chloride (Figure 2-13; Mac Kenzie et al., 1994b). Although marked improvement in the turbidity of treated water had been achieved by April 1 with the use of polyaluminum chloride, on April 2 the southern plant resumed use of alum instead of polyaluminum chloride as a coagulant. On April 5, the turbidity of treated water increased to 1.5 NTU. During February through April 1993, the northern plant treated water turbidity did not exceed 0.45 NTU (Mac Kenzie et al., 1994b). There was no correlation between the turbidity of treated water and the turbidity or temperature of untreated water. From February

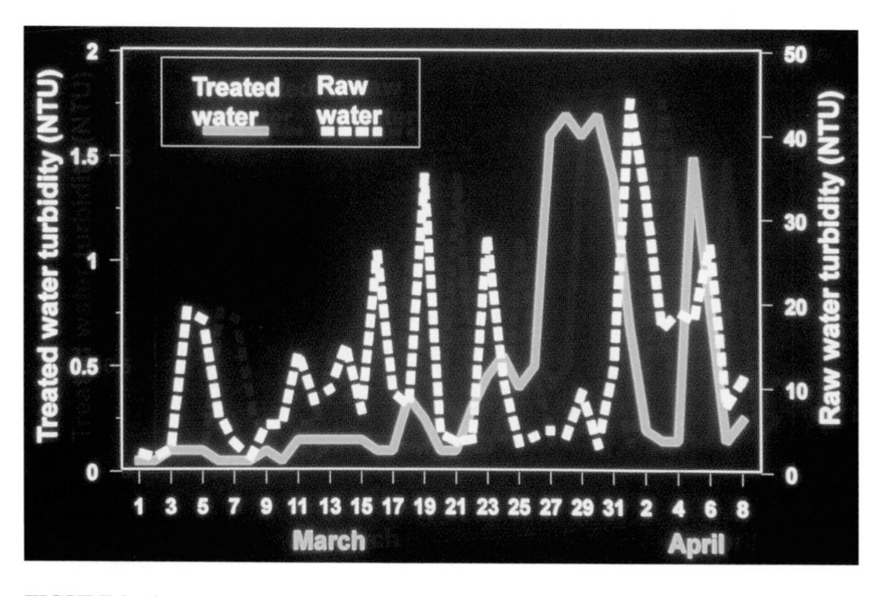

FIGURE 2-13 Maximum daily raw and treated water turbidity at the southern Milwaukee Water Works treatment plant, March-April 1993.
SOURCE: Wisconsin Division of Public Health (unpublished).

through April 1993, samples of treated water from both plants were negative for coliforms and were within the limits established by the Wisconsin Department of Natural Resources for water quality (Mac Kenzie et al., 1994b).

A federal Environmental Protection Agency (EPA) water engineer inspected both Milwaukee water treatment plants and found them to meet existing state and federal water quality standards at the time of the outbreak. However, at the southern plant the water quality data showed a marked increase in turbidity, which reflected poor filtration. The turbidity was measured every eight hours—the minimum amount required by authorities for routine monitoring. Prior to the outbreak, water turbidity was not generally viewed as a potential indicator of protozoan contamination.

The EPA inspector also found that, to decrease the costs of chemicals used in water treatment, the Milwaukee plants recycled water used to backflush and clean their sand filters. This backflushed water (backwash) containing whatever was caught by the sand filter was added to source water coming into the plant rather than being discharged into a sewer. Over time, such recycling of backwash effectively increases the concentration of any contaminant in the water being treated by the plant and increases the risk that the sand filters may not effectively remove the contaminant (Mac Kenzie et al., 1994b).

Weather conditions prior to the outbreak were also very unusual (Addiss et al., 1995; Mac Kenzie et al., 1994b). An extremely high winter snowpack had melted rapidly while the frostline remained high, resulting in high runoff containing greater-than-usual levels of organic material. There was also extraordinarily heavy rainfall during March and April, that exceeded the previous record for the period between March 21 and April 20 (set in 1929) by 30 percent.

At the time of the outbreak, during periods of heavy rain, Milwaukee's storm sewers frequently overflowed. During these periods, sewage was chemically disinfected but otherwise bypassed full sewage treatment (Figure 2-14). Thus, during periods of high flow, the storm sewer and sanitary sewer water that bypassed treatment then emptied into an area within a breakfront on Lake Michigan, just north of the intake for the south water plant as may be seen in Figure 2-15 and further depicted in Figure 2-16.

At the same time, high and frequent northeasterly winds (an unusual wind direction in Milwaukee) probably accentuated the southerly flow of water out of the breakfront and toward the intake for the southern water plant (Figure 2-16). The winds also forced the water within the breakfront closer to the lakeshore, accentuating plumes of storm water and treated sewage that flowed through gaps in the breakfront toward the nearby south plant intake grid (Addiss et al., 1995).

At the southern water plant, personnel lacked experience with dosing the new coagulant in response to spikes in finished water turbidity. By the time the decision was made, on April 2, to resume the use of alum as the coagulant, treated water was already significantly contaminated with *Cryptosporidium* oocysts.

We also became aware of an additional factor that was of potential importance. In early 1993, a university in central Milwaukee was constructing new

FIGURE 2-14 Milwaukee skyline demonstrating confluence of rivers merging just west of the Milwaukee harbor. The Milwaukee Metropolitan Sewage District plant is located on the land just south of the convergence and west of the overpass.
SOURCE: Image courtesy of Kathy Blair.

FIGURE 2-15 Milwaukee River emptying into the Lake Michigan harbor following a period of high flow and attendant creation of a plume. Note the breakfront and the southerly movement of the plume.
SOURCE: Image courtesy of Kathy Blair.

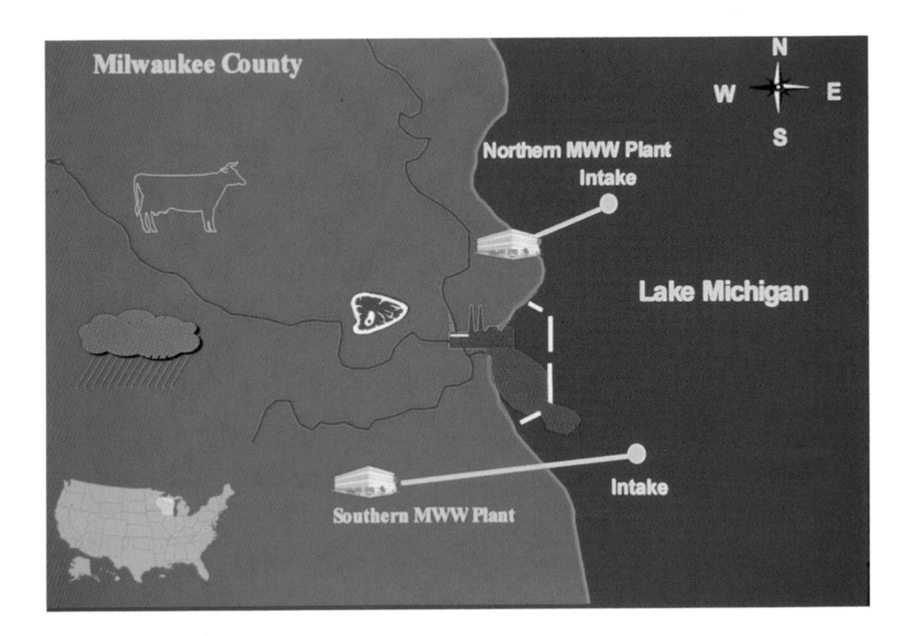

FIGURE 2-16 Location of the three rivers that flow through Milwaukee County, Wisconsin, the breakfront protecting the city of Milwaukee harbor and the northern and southern Milwaukee Water Works water treatment plants and their intake grids, and depiction of creation and southerly flow of a plume toward the southern plant and through a gap in the breakfront near the plant's intake grid in Lake Michigan.

soccer fields. The drainage from these fields was directed into a small storm sewer that had to be connected to a larger main sewer. When construction workers cut into the main sewer to make this connection, they discovered a large impaction of bovine entrails and other waste from a large meatpacking plant located nearby. Ensuing investigation and inspection by city officials revealed a cross connection of a sewer from the abattoir kill floor with the storm sewer. This cross-connection existed for years, and these wastes accumulated over a prolonged time. Following correction of the cross-connection, removal of the impacted wastes and hauling the wastes away occurred in early March. Potentially some of these disrupted wastes could have been discharged through the storm sewer directly into the Menomonee River or directly reach the sewage treatment facility following correction of the cross connection. While it is not clear whether the existence and correction of the cross-connection and clean-up of the sewer influenced this outbreak, it was an issue that was addressed during the investigation.

The Ultimate Source of These *Cryptosporidium* Oocysts: Animals or Humans

As previously noted, as few as 10 *Cryptosporidium hominis* oocysts constituted an ID50 in adult volunteers (Chappell et al., 2006). An infected person persistently excretes billions of oocysts over an extended period. The mounting numbers of people ill with watery diarrhea, each of whom were likely excreting billions of oocysts into sanitary sewers during the course of their illness, placed rapidly increasing demand on the MWW water treatment system and perpetuated an explosive cycle of *Cryptosporidium*-related oocyst ingestion, illness, and oocyst amplification. Additionally, oocysts can remain infective in moist environments for two to six months (Fayer, 2004). The opportunity for infection in this outbreak was, therefore, both inordinately high and sustained.

From the random digit dialing survey, we determined that the highest attack rates of watery diarrhea occurred in people aged 30 to 39 years (Mac Kenzie et al., 1994b). These tended to be working adults, many of whom were commuting from lower risk to higher risk places. Using age-related results of the random digit dialing survey we estimated an attack rate of 18 to 20 percent among children under the age of 10 years, and less than 15 percent among independent-living individuals over 70 years of age (Mac Kenzie et al., 1994b).

Subsequently, the frequency of asymptomatic infection among exposed Milwaukee residents was demonstrated to be very high. Although there was no reliable serologic test for *Cryptosporidium* at the time of the outbreak, stool and serum samples were collected from volunteers during this period. Sera from children who had blood tests for lead levels around the time of the outbreak were preserved; later, when a serologic test for *Cryptosporidium* became available, these sera were tested (McDonald et al., 2001). The serologic study in children revealed that the prevalence of anti-cryptosporidium antibodies from southern Milwaukee increased from 7 percent prior to the outbreak to approximately 80 percent after the outbreak, indicating that most children were infected and many infections were asymptomatic. These data also supported prior studies implicating the southern plant. Interestingly, the prevalence of anti-cryptosporidium antibody in Milwaukee children tested in 1998 was 7 percent, indicating transmission had returned to baseline. The results demonstrated that the Milwaukee outbreak affected considerably more people than we had previously estimated (McDonald et al., 2001).

Efforts were made to obtain large-volume specimens of stool from volunteers with acute onset of diarrheal illness during the outbreak, store the specimens in potassium dichromate, and hopefully maintain viable oocysts for isolation and subsequent analysis. Only five specimens were obtained, including three from patients with AIDS—one of which was obtained in 1996 from a patient with chronic infection who initially was infected during the 1993 outbreak (Peng et al., 1997). CDC investigators purified oocysts from isolates obtained from stools of four of the volunteers. Approximately one million oocysts from each specimen were orally administered to two-day-old calves or four- to six-day-old BALBc or severe combined immunodeficient (SCID) mice. Further, the oocysts were ruptured, parasitic DNA was harvested, specific fragments were amplified, and the DNA was sequenced and analyzed. None of the isolates established infections in calves and mice, suggesting these were not bovine strains. Isolates in the overall study could be divided into two genotypes of, at that time, *Cryptosporidium parvum*, on the basis of genetic polymorphism at one locus; the four Wisconsin isolates were similar to isolates observed only in humans that were noninfective in cows and mice (genotype 1). The other genotype (genotype 2) was infective in calves or mice (Peng et al., 1997). Thus, as noted by the authors of the study, the genotypic and experimental infection data from the four isolates examined suggest a human rather than bovine source. However, the results come from the analysis of only four samples from a massive outbreak, and the degree to which these samples are representative of the entire outbreak remains uncertain.

In later studies, *C. parvum* genotype 1 became known as *C. hominus* (Morgan-Ryan et al., 2005). Infection with a strain that was human adapted rather than bovine adapted is consistent with the massive numbers of human illnesses and asymptomatic human infections noted in this outbreak.

Peng et al. noted that possible sources of Lake Michigan's contamination with *Cryptosporidium* oocysts included cattle along two rivers that flow into the Milwaukee Harbor, slaughterhouses, and human feces (Addiss et al., 1995; Mac Kenzie et al., 1994b; Peng et al., 1997). At that time cattle had been the most commonly implicated source of water contamination in *Cryptosporidium* outbreaks outside the United States, but not conclusively within the United States (Peng et al., 1997). Measures for preventing water contamination have in some cases included the removal of cattle from watershed areas in or around municipalities. If, however, sewer overflows and inadequate sewage treatment are the primary source of water contamination in urban settings where anthroponotic[7] cycles were maintained, focusing only on cattle could fail to eliminate a very important source of infection (Peng et al., 1997). In the Milwaukee outbreak, the latter point is very important because of the combined sewer overflows (prolonged high flow interval in March and April 1993) and attendant inadequate sewage treatment and the anthroponotic nature of the outbreak.

Lessons Learned

Among the many lessons learned from the 1993 Milwaukee *Cryptosporidium* outbreak, the following lessons and needs stand out:

1. *Consistent application of stringent water quality standards.* At the time of the outbreak, drinking water was regulated either by the EPA or by

[7]Transmission from human to human and potentially from human to animal.

individual states, as was the case in Wisconsin, and the MWW water treatment and quality testing results were in compliance with all state and federal standards. Existing state and federal standards for treated water were insufficient to prevent this outbreak (Mac Kenzie et al., 1994b). Moreover, it is important to use measures of turbidity in treated water as an indicator of potential contamination rather than viewing turbidity as an aesthetic measure of clarity. Consistent application of stringent water quality standards was needed. More stringent federal water quality standards, which had been under development for several years, were implementled shortly after this massive waterborne outbreak (Addiss et al., 1995). The vastly improved attention to monitoring and to the quality of water filtration is a powerful impact of this investigation.

2. *Application of technical advances to monitor water safety and minimize the amount of inadequately filtered water to the public.* The post-filtration turbidity (and particle counts, if possible) of treated water should be monitored continuously for each filter to detect changes in filtration status. Alarm systems for each of the filters and particle counting devices *are* available to detect spikes in particles in the size range inclusive of *Cryptosporidium* oocysts and facilitate rapid filter shutdown and diversion of potentially contaminated water when thresholds are reached (Mac Kenzie et al., 1994b).
3. *Testing of source and finished water for Cryptosporidium.* This was needed to detect risk for an outbreak and to determine when the water was safe to drink afterward. At the time of the outbreak, the sampling process for such testing was difficult and lengthy, and it was not standardized. Improved means of sampling and testing source and finished water for *Cryptosporidium* were needed.
4. *Environmental studies.* A coordinated plan was needed to investigate the environment following a waterborne outbreak of *Cryptosporidium* infection, but not many such events had occurred. Thus, we had to overcome considerable challenges, particularly regarding designing, funding, and mobilizing appropriate studies relevant to human health. There needed to be equipment and trained human capacity for rapid deployment of specimen collection followed by prompt testing.
5. *Surveillance. Cryptosporidium* infection was not a reportable public health condition at the time of the outbreak. Watery diarrhea proved to be a good case definition for *Cryptosporidium* infection in an outbreak setting; a more refined clinical case definition was necessary to detect sporadic cases. The random-digit dialing surveys were very valuable in assessing the scope and progress of this large community outbreak; and nursing home surveillance, as described, was very effective (Mac Kenzie et al., 1994b; Proctor et al., 1998). It would have been useful to have a surveillance system in place to analyze consumer complaints to the water

authority before the outbreak as this spike in complaints to the MMW was very striking; these might have focused attention on the unusual turbidity of treated water that preceded the outbreak (Proctor et al., 1998).

6. *Testing of human stool and serum.* Because of the time and expense involved, generally only patients with HIV infection, particularly those with AIDS, were routinely tested for *Cryptosporidium* at the time the outbreak occurred. In addition to delay in determination of the cause of the outbreak, the infrequent use of these tests likely contributed to delay in outbreak recognition. Improved assays were clearly needed. The striking data from the serologic testing of children (McDonald et al., 2001) demonstrated the value of serologic assays to assess background occurrence and the magnitude and impacts of outbreaks *Cryptosporidium* infection.
7. *Routine assays for Cryptosporidium.* Physicians, other clinicians, and public health officials clearly needed to broaden and sustain the index of suspicion for *Cryptosporidium* infection (Mac Kenzie et al., 1994b). This was challenging because of added costs of testing and the limited assays available at that time.
8. *Communication.* To our advantage, we had good interagency communication and worked closely with communities of individuals at greatest risk. For example, the AIDS service organization in Milwaukee had access to over 700 case patients, and we could monitor morbidity and mortality in that population (Frisby et al., 1997). As a result of the outbreak, we developed targeted public health messages and shared them with other health departments. However, due to insufficient understanding of the pathogen and its public health effects, we lacked guidelines for governmental response to findings of oocysts, increased turbidity of finished water, and elevated particle counts in finished water. During our investigation, we worked to establish interagency coordination to remedy this situation.
9. *The media.* Electronic and print media were essential to communicating risk and delivering other important public health messages during the outbreak, and in particular facilitated public inquiry by setting up phone banks. The *Milwaukee Journal* and the *Milwaukee Sentinel* jointly produced an issue in Spanish to inform a large segment of non-English speakers about the outbreak, and they maintained a help line in Spanish as well. The media published treated water turbidity data on a regular basis, which was especially helpful to individuals with HIV infection and AIDS. Daily news conferences and televised updates occurred through the lifting of the boil water advisory on April 14 and related articles appeared daily in the news for weeks.
10. *Other lessons.* The many outbreak-related studies conducted during the time of the outbreak yielded other lessons on the clinical spectrum of

cryptosporidiosis, epidemiologic features of cryptosporidiosis, effectiveness of control measures in specific subpopulations, effectiveness of preventive measures, and the economic impact of such a massive outbreak. Recurrence of diarrhea after a period of apparent recovery was documented frequently—in 39 percent of persons with laboratory-confirmed *Cryptosporidium* infections (Mac Kenzie et al., 1995b), a finding that has implications for patient counseling. A study among HIV-positive persons found that the severity of illness, but not the attack rate, was significantly greater in persons with HIV infection (Frisby et al., 1997). Among young children attending daycare centers, asymptomatic or minimally symptomatic *Cryptosporidium* infection was frequent (Cordell et al., 1997), a fact that may have contributed to several outbreaks associated with recreational water in other parts of the state several months after the Milwaukee outbreak (CDC, 1993; Mac Kenzie et al., 1995a). Surveillance studies revealed the potential usefulness of monitoring sales of over-the-counter antidiarrheal drugs as an early indicator of community-wide outbreaks (Proctor et al., 1998); the effectiveness of control measures and the absence of drinking water as a risk factor for the relatively low level of transmission during the post-outbreak period (Osewe et al., 1996); the importance of testing more than one stool specimen (Cicirello et al., 1997); the usefulness of death certificate review for estimating outbreak-related mortality (Hoxie et al., 1997); and the effectiveness of point-of-use water filters with pore diameters of less than 1 micron (Addiss et al., 1996). Additionally, a detailed cost analysis revealed the enormous economic impact of the outbreak (Corso et al., 2003).

Conclusions and Outcomes

The 1993 Milwaukee cryptosporidiosis outbreak was the largest documented waterborne disease outbreak in the United States. *Cryptosporidium* oocysts in untreated water from Lake Michigan that entered the plant were inadequately removed by the coagulation and filtration process at the Milwaukee southern water treatment plant. Water quality standards were inadequate to prevent this outbreak. There was a lack of laboratory testing for *Cryptosporidium*, which delayed recognition of the microbial etiology of the outbreak. While the environmental source of the oocysts in this outbreak is not specifically known, limited data from genotyping of *Cryptosporidium* DNA from a small number of human stool specimens obtained during the outbreak supports the hypothesis that the environmental source of the oocysts was human. Nonetheless, there was a confluence of important factors that contributed to the occurrence of this massive outbreak (Mac Kenzie et al., 1994b). How oocysts ultimately made their way from sewers into water feeding the intake of the southern water treatment plant (e.g., by sewage overflows related to heavy rains, cross-connections, or inadequate treatment at the sewage treatment plant located at the mouth of the Milwaukee River and facilitated by unusual wind conditions) is unknown.

Based on these conclusions and the opinions of EPA staff and other consultants, the MWW instituted continuous turbidity monitoring in all of its filters. They also put alarms on the filters to enable automatic shutdown if turbidity reached a threshold level, and they set and achieved the goal of maintaining turbidity at a very low level. The MWW modified and improved water treatment procedures and adopted very stringent water quality standards. Substantial enhancement of the filter beds in both treatment plants occurred and the new filter media was installed. The intake grid for raw water entering the southern plant was moved considerably further eastward. These efforts resulted in continuous production by the MWW of high-quality treated water with mean turbidities of 0.01 NTU (Mac Kenzie et al., 1994a). Furthermore, recognizing that *Cryptosporidium* oocysts are highly resistant to chlorine, the City of Milwaukee constructed ozonization plants at each treatment facility. Ozone disrupts the oocyst cell wall prior to disinfection.

When we studied turbidity events among Wisconsin surface water treatment plants over a 10-year period, we discovered other sites with similar challenges. For example, during the months of February through April, turbidity events occur frequently on Lake Michigan; these events affect all treatment plants that use water from this lake (Wisconsin Division of Public Health, unpublished data).

We advocated increasing *Cryptosporidium* testing of stools from persons with watery diarrhea and made *Cryptosporidium* infection a reportable condition (Hayes et al., 1989). Annually, the DPH receives about 450 reports of *Cryptosporidium* infection: some are recreational, some are connected with agriculture, but rarely do they originate in Milwaukee (Wisconsin Division of Public Health, unpublished data). We also advocated including *Cryptosporidium* testing in federal rules that stipulated the collection of information on both raw water sources and finished water (Juranek et al., 1995).

The massive Milwaukee waterborne *Cryptosporidium* outbreak, and the resulting modification of the city's water treatment facilities, has received attention from water authorities worldwide. The attendant events and actions have been very instructive.

PREVENTION IS PAINFULLY EASY IN HINDSIGHT: FATAL *E. COLI* O157:H7 AND *CAMPYLOBACTER* OUTBREAK IN WALKERTON, CANADA, 2000

Steve E. Hrudey, Ph.D.[8]
University of Alberta

Elizabeth J. Hrudey[8]
University of Alberta

Summary

In May 2000, a comfortable rural community of about 4,800 people in Canada's largest province (Ontario) experienced an outbreak of waterborne disease that killed seven people and caused serious illness in many others. The contamination was ultimately traced to a source that had been identified 22 years earlier as a threat to the drinking water system, but no remedial action was taken to manage the public health risk. The operators of the system were oblivious to this danger and the regulators responsible for safe drinking water largely overlooked the problems that existed. Even as the outbreak unfolded the regulatory response was slow and unfocused, suggesting a serious loss of capacity to regulate drinking water safety had occurred in Ontario.

Introduction[9]

Walkerton, located about 175 km northwest of Toronto, Ontario, Canada, experienced serious drinking water contamination in May 2000. The facts of this account are drawn from the Walkerton Inquiry (the Inquiry), a $9 million public inquiry into this disaster called by the Ontario Attorney General (O'Connor, 2002a). This disaster has resulted in a complete overhaul of the drinking water regulatory system in Ontario.

What Happened in Walkerton

Walkerton was served by three wells in May of 2000, identified as Wells 5, 6, and 7. Well 5 was located on the southwest edge of the town, bordering adjacent farmland. It was drilled in 1978 to a depth of 15 m with 2.5 m of overburden and protective casing pipe to 5 m depth (O'Connor, 2002a). The well was completed in fractured limestone with the water-producing zones ranging from 5.5 to 7.4 m

[8]Analytical and Environmental Toxicology, Faculty of Medicine and Dentistry, University of Alberta, Edmonton, AB, T6G 2G3, Canada.

[9]Much of the following text was reprinted with permission from Hrudey (2006a).

depth and it provided a capacity of 1.8 ML/d that was able to deliver ~56 percent of the community water demand. Well 5 water was to be chlorinated with hypochlorite solution to achieve a chlorine residual of 0.5 mg/L for 15 minutes contact time.

Well 6 was located 3 km west of Walkerton in rural countryside and was drilled in 1982 to a depth of 72 m with 6.1 m of overburden and protective casing to 12.5 m depth (O'Connor, 2002a). An assessment after the outbreak determined that Well 6 operated from seven producing zones with approximately half the water coming from a depth of 19.2 m. This supply was judged to be hydraulically connected to surface water in an adjacent wetland and a nearby private pond. Well 6 was disinfected by a gas chlorinator and provided a nominal capacity of 1.5 ML/d that was able to deliver 42 to 52 percent of the community water demand (O'Connor, 2002a).

Well 7, located approximately 300 m northwest of Well 6, was drilled in 1987 to a depth of 76.2 m with 6.1 m of overburden and protective casing to 13.7 m depth (O'Connor, 2002a). An assessment following the outbreak determined that Well 7 operated from three producing zones at depths greater than 45 m with half the water produced from below 72 m. A hydraulic connection discovered between Well 6 and Well 7 reduced the security of an otherwise good-quality groundwater supply. Well 7 was also disinfected by a gas chlorinator and provided a nominal capacity of 4.4 ML/d that was able to deliver 125 to 140 percent of the community water demand (O'Connor, 2002a).

From May 8 to May 12, Walkerton experienced ~134 mm of rainfall, with 70 mm falling on May 12. This was unusually heavy, but not record, precipitation for this location. Such rainfall over a 5-day period was estimated by Environment Canada to happen approximately once in 60 years (on average) for this region in May (Auld et al., 2004). The rainfall of May 12, which was estimated by hydraulic modeling to have occurred mainly between 6 PM and midnight, produced flooding in the Walkerton area.

Stan Koebel, the general manager of the Walkerton Public Utilities Commission (PUC), was responsible for managing the overall operation of the drinking water supply and the electrical power utility. From May 5 to May 14, he was away from Walkerton, in part to attend an Ontario Water Works Association meeting. He had left instructions with his brother Frank, the foreman for the Walkerton PUC, to replace a nonfunctioning chlorinator on Well 7. From May 3 to May 9, Well 7 was providing the town with unchlorinated water in contravention of the applicable provincial water treatment requirements.

From May 9 to 15, the water supply was switched to Wells 5 and 6. Well 5 was the primary source during this period, with Well 6 cycling on and off, except for a period from 10:45 PM on May 12 until 2:15 PM on May 13 when Well 5 was shut down. Testimony at the Inquiry offered no direct explanation about this temporary shutdown of Well 5. No one admitted to turning Well 5 off and the supervisory control and data acquisition (SCADA) system was set to keep Well 5 pumping. Flooding was observed near Well 5 on the evening of May 12 because of the heavy rainfall that night, but why or how Well 5 was shut down for this period remains unknown.

On May 13 at 2:15 PM, Well 5 resumed pumping. That afternoon, according to the daily operating sheets, foreman Frank Koebel performed the routine daily checks on pumping flow rates and chlorine usage, and measured the chlorine residual on the water entering the distribution system. He recorded a daily chlorine residual measurement of 0.75 mg/L for Well 5 treated water on May 13 and again for May 14 and 15. Testimony at the Inquiry indicated that these chlorine residual measurements were never made and that all the operating sheet entries for chlorine residual were fictitious. The monitoring data were typically entered as either 0.5 mg/L or 0.75 mg/L for every day of the month.

On Monday, May 15, Stan Koebel returned and early in the morning turned on Well 7, presumably believing that his instruction to install the new chlorinator had been followed. When he learned a few hours later that it had not, he continued to allow Well 7 to pump into the Walkerton system, without chlorination, until Saturday, May 20. Well 5 was shut off at 1:15 PM on May 15, making the unchlorinated Well 7 supply the only source of water for Walkerton during the week of May 15. Because the Well 5 supply was ultimately determined to be the source of pathogen contamination of the Walkerton drinking water system, the addition of unchlorinated water into the system from Well 7 would have failed to reduce the pathogen load by any means other than dilution.

PUC employees routinely collected water samples for bacteriological testing on Mondays. Samples of raw and treated water were to be collected from Well 7 that day along with two samples from the distribution system. Although samples labeled *Well 7 raw* and *Well 7 treated* were submitted for bacteriological analyses, the Inquiry concluded that these samples were not taken at Well 7 and were more likely to be representative of Well 5. Stan Koebel testified that PUC employees sometimes collected their samples at the PUC shop, located nearby and immediately downstream from Well 5, rather than traveling to the more distant wells (~3 km away) or to the distribution system sample locations.

During this period, a new water main was being installed (the Highway 9 project). The contractor and consultant for this project asked Stan Koebel if they could submit their water samples from this project to the laboratory being used by the Walkerton PUC for bacteriological testing. Stan Koebel agreed and included three samples from two hydrants for the Highway 9 project. On May 1, the PUC began using a new laboratory for bacteriological testing, a lab the PUC had previously used only for chemical analyses.

The first set of samples submitted to the new laboratory was taken on May 1, but was improperly submitted with inadequate sample volumes for the analyses requested and discrepancies between the written documentation and numbers of samples sent. No samples were submitted by the PUC for May 8. The May 15 samples taken by the PUC repeated the problems with inadequate sample volumes and discrepancies in the paperwork.

Early on the morning of Wednesday, May 17, the lab phoned Stan Koebel to advise him that all of the water main construction project samples taken May 15 were positive for *E. coli* and total coliforms, and that the distribution system samples were also contaminated. Because these tests indicated only the presence or absence of indicator bacteria, it was not possible to estimate the numbers of indicator bacteria in each sample. Only the sample labeled *Well 7 treated* was analyzed by the membrane filtration method. The latter procedure would normally allow a bacterial count to be determined, but in this case the sample was so contaminated that it produced an overgrown plate with bacterial colonies too numerous to count (both total coliforms and *E. coli* > 200 / 100 mL). The Inquiry concluded that this sample was most likely mislabeled and was more likely representative of the water from Well 5 entering the distribution system.

The new laboratory was not familiar with the "expectations" (not regulatory requirements at that time) to report adverse microbial results to either the Ministry of Environment (MOE) or the responsible Medical Officer of Health (MOH). Accordingly, this lab reported these adverse sample results only to the PUC General Manager, Stan Koebel. In turn, he advised the consultant for the Highway 9 project contractor that their samples had failed so they would need to rechlorinate, flush and re-sample to complete the project.

On Thursday, May 18, the first signs of illness were becoming evident in the health-care system. Two children, a seven-year-old and a nine-year-old, were admitted to the hospital in Owen Sound, 65 km from Walkerton. The first child had bloody diarrhea and the second developed bloody diarrhea that evening. The attending pediatrician, Dr. Kristen Hallett, noted that both children were from Walkerton and attended the Mother Theresa School in Walkerton. Bloody diarrhea is a notable symptom for serious gastrointestinal infection, particularly infection with *Escherichia coli* O157:H7. Accordingly, Dr. Hallett submitted stool samples from these children to evaluate that diagnosis. By May 18, at least 20 students were absent from the Mother Theresa School.

By Friday, May 19, the outbreak was evident at many levels. Twenty-five children were now absent from the Mother Theresa School and 8 children from the Walkerton public school were sent home suffering from stomach pain, diarrhea, and nausea. Three residents of the Maple Court Villa retirement home and several residents of the BruceLea Haven long-term care facility developed diarrhea, two with bloody diarrhea. A Walkerton physician had examined 12 or 13 patients suffering from diarrhea to that time. Dr. Hallett first notified the Bruce-Grey-Owen Sound Health Unit, the responsible public health agency for Walkerton with its main office in Owen Sound, of the emerging problems on May 19. She expressed concerns to Health Unit staff that Walkerton residents were telling her something was "going on" in Walkerton, and the receptionist from the Mother

Theresa School advised that the parent of one student stated that something was wrong with the town's water supply.

An administrator at the Mother Theresa School called James Schmidt, the public health inspector at the Walkerton office of the Health Unit, to report the 25 children absent. She noted that some were from Walkerton, others from adjacent rural areas, and that the ill students were from different grades and classrooms. She suspected the town's water supply. In contrast, the Health Unit officials suspected a foodborne basis for the outbreak, by far the most common cause of such diseases. Nonetheless, James Schmidt placed a call to Stan Koebel in the early afternoon of May 19. By the time he called, the chlorinator had been installed on Well 7 so that it was now supplying chlorinated water to Walkerton's distribution system. According to James Schmidt, Stan Koebel was asked whether anything was wrong with Walkerton's water and he advised Schmidt that "everything's okay" (J. Schmidt, Inquiry Transcript of Evidence, December 15, 2000, p. 172). By the time of that conversation, Stan Koebel had been faxed the adverse microbial results from the Highway 9 project, the distribution system, and the sample labeled *Well 7 treated* two days earlier.

Later that afternoon, David Patterson, an administrator of the Health Unit based in Owen Sound, called Stan Koebel to advise him of public concerns about the water. Patterson asked whether anything unusual had happened in the water system. Stan Koebel mentioned that there was water main construction under way near the Mother Theresa School, but made no mention of the adverse bacteriological results or of operating Well 7 from May 3 to 9 and from May 15 to 19 without a chlorinator.

The Inquiry concluded that Stan Koebel's lack of candor seriously hampered the Health Unit's early investigation of and response to the outbreak. Because patients had bloody diarrhea, health officials suspected the outbreak was caused by *E. coli* O157:H7. The failure by PUC personnel to mention any problems with the Walkerton water system allowed health officials to continue with their misinformed search for a foodborne cause of the outbreak. At that time, Health Unit personnel were not aware that any outbreaks of this disease had occurred in a chlorinated drinking water system. Earlier waterborne outbreaks of *E. coli* O157:H7—Cabool, Missouri; Alpine, Wyoming; and Washington County, New York—involved unchlorinated drinking water (Hrudey and Hrudey, 2004). Stan Koebel's reassurances about the water's safety kept the Health Unit staff pursuing a foodborne cause. However, the emerging outbreak, with cases distributed across a wide geographic region and across the population from very young and very old, was making any foodborne explanation increasingly improbable.

Suspicions about the safety of the water were spreading in the community. The BruceLea Haven nursing home, where a number of patients had become ill, began to boil water on the initiative of the nursing staff despite the absence of any public health warnings. Some citizens, including Robert MacKay, an employee of the Walkerton PUC, also began to boil their water on Friday, May 19. After his conversations with health officials that afternoon, in which he reassured them about the water, Stan Koebel increased the chlorination level at Well 7. He also began to flush the distribution system through a hydrant near the Mother Theresa School and subsequently at other hydrants throughout the system until May 22.

By Saturday, May 20, on a long holiday weekend, the outbreak was straining the Walkerton hospital with more than 120 calls from concerned residents, more than half of whom complained of bloody diarrhea. After the Owen Sound hospital determined that a stool sample from one of the children admitted on May 18 was presumptive positive for *E. coli* O157:H7, the health unit notified other hospitals in the region because this pathogen may cause hemolytic uremic syndrome (HUS). This was important because antidiarrheal medication or antibiotics, which might be prescribed for diarrhea, can worsen the condition of patients infected with this pathogen, so emergency staff had to avoid dispensing such medication.

David Patterson asked James Schmidt to contact Stan Koebel again to determine the current chlorine residual levels in the water and to receive reassurance that the water system would be monitored over the holiday weekend. Koebel assured Schmidt that there were measurable levels of chlorine residual in the distribution system, leading health officials to believe that the water system was secure.

Early on Saturday afternoon, David Patterson contacted Dr. Murray McQuigge, the local Medical Officer of Health who was out of town during the onset of the outbreak, to advise him of the emerging outbreak. By that time, several people in Walkerton were reporting bloody diarrhea and ten stool samples had been submitted for pathogen confirmation. Dr. McQuigge advised that any further cases diagnosed with *E. coli* O157:H7 should be interviewed for more details, and he returned that evening to Owen Sound.

David Patterson called Stan Koebel to advise him that a local radio station was reporting that Walkerton water should not be consumed. Patterson wanted Koebel to call the radio station to correct this report and reassure the public about the safety of the Walkerton water supply, but Koebel was apparently reluctant to comply with this request. Patterson asked again whether anything unusual had occurred in the water system and Koebel again failed to report the adverse microbiological results from the May 15 samples or that Well 7 had been operating with no chlorination.

Robert MacKay, a PUC operator who had been on sick leave from the PUC, began to suspect something was wrong with Walkerton's water. He had learned from Frank Koebel that the samples from the Highway 9 project had failed testing. MacKay phoned the Spills Action Centre (SAC) of the MOE anonymously to report his concerns and provide a contact number at the PUC for the MOE to call about the Walkerton water system. In the early afternoon of Saturday, May 20, Christopher Johnston, the MOE employee who received MacKay's anonymous call, phoned Stan Koebel to find out if there were problems with the system.

Johnston understood from this conversation with Stan Koebel that any problems with bacteriological results had been limited to the Highway 9 mains replacement project some weeks earlier, but that chlorine residual levels were satisfactory as of May 19. MacKay, now experiencing diarrhea himself, placed another call to the MOE number that evening to find out what was being done. MacKay was advised that Stan Koebel had been contacted, but that MacKay's concern about drinking water safety was really a matter for the Ministry of Health. This feedback from the MOE was wrong: the MOE was designated as the lead agency for drinking water regulation in Ontario. MacKay was provided with a phone number for the wrong regional health office, eventually leading him to call back to the SAC. This time, the MOE SAC staff person agreed to contact the nearest MOE office, in Owen Sound, with a request to investigate the matter.

The outbreak continued to expand. By Sunday, May 21, there were more than 140 calls to the Walkerton hospital and two more patients admitted to the Owen Sound hospital. A local radio station interviewed Dr. McQuigge on Sunday morning and subsequently reported on the noon news that Dr. McQuigge believed that drinking water contamination was an unlikely source of this outbreak. At about that time, the Health Unit was advised that the first presumptive *E. coli* O157:H7 had been confirmed and that another patient sample, presumptive for *E. coli* O157:H7, was being tested for confirmation. David Patterson and Dr. McQuigge conferred with their staff about these results and decided to issue a boil water advisory at 1:30 PM on Sunday, May 21. The notice, hastily drafted by David Patterson, stated:

> The Bruce-Grey-Owen Sound Health Unit is advising residents in the Town of Walkerton to boil their drinking water or use bottled water until further notice. The water should be boiled for five minutes prior to consumption. This recommendation is being made due to a significant increase in cases of diarrhea in this community over the past several days.
>
> Although the Walkerton PUC is not aware of any problems with their water system, this advisory is being issued by the Bruce-Grey-Owen Sound Health Unit as a precaution until more information is known about the illness and the status of the water supply.
>
> Anybody with bloody diarrhea should contact his or her doctor or the local hospital.

This notice was provided only to the local AM and FM radio stations; additional publicity by the television station or by direct door-to-door notification was not pursued. According to the report subsequently prepared on the outbreak with the assistance of Health Canada (BGOSHU, 2000), a community survey showed that only 44 percent of respondents were aware that the Health Unit had issued a boil water advisory on May 21 and only 34 percent heard the announcement on the radio. In retrospect, Health Unit personnel acknowledged that the community could have been more effectively notified. However, given Stan Koebel's consistent reassurance about the safety of the Walkerton water system, the Health Unit's caution in attributing the outbreak to the local drinking water at this emerging stage of the outbreak is understandable.

After issuing the boil water advisory, Dr. McQuigge notified the MOE SAC that there was an *E. coli* outbreak in Walkerton. In exchange, the SAC advised Dr. McQuigge about the anonymous calls about adverse results for the Walkerton water system. The Health Unit updated the MOE SAC that there were now 2 confirmed cases of *E. coli* O157:H7 and 50 cases of bloody diarrhea. The MOE called Stan Koebel to discuss the situation; Koebel once again failed to report the adverse samples from May 15 (reported to him on May 17). During his Inquiry testimony, Stan Koebel responded to a question about whether he had deliberately avoided disclosing these results during his conversation with MOE personnel by answering: "I guess that's basically the truth and I was waiting on the Ministry of the Environment to call from the Owen Sound office with further confirmation" (S. Koebel, Inquiry. Transcript of Evidence, December 20, 2000, p. 108).

The Health Unit established a strategic outbreak team to deal with the emergency. Local public institutions were to be notified about the boil water advisory, but two high-risk institutions, the BruceLea Haven nursing home and the Maple Court Villa retirement home, were inadvertently missed. The Walkerton hospital had been reassured about the safety of the water until that afternoon and had not taken any measures to address water safety. In fact, hospital staff had been advising those caring for patients with diarrhea to provide ample fluids to maintain patient hydration, advice that caused many ill patients to be additionally exposed to contaminated Walkerton tapwater.

Once notified of the problems, the hospital was forced to find an alternative safe water and ice supply, shut off its public fountains, and discard any food prepared or washed with Walkerton tap water. By that evening, the Health Unit had notified provincial health officials of the outbreak and requested the assistance of major hospitals in London (over 150 km distant) and Toronto (over 200 km distant) in treating Walkerton residents and the assistance of Health Canada in conducting an epidemiological investigation.

By Monday, May 22, the Health Unit had received reports of 90 to 100 cases of *E. coli* infection. Phillip Bye, the regional MOE official in Owen Sound, who had been notified the previous evening about the outbreak, did not initiate a MOE investigation, even after being advised about the large number of cases of *E. coli* infection and that the Health Unit suspected the Walkerton water system. Only after being contacted later that day by Dr. McQuigge, who stressed the urgency of the situation, did the regional MOE initiate an investigation by sending environmental officer James Earl to Walkerton to meet first with the Health Unit before meeting Stan Koebel. The Health Unit advised Earl about the "alarming" number of illnesses and said that Health Unit investigations failed to reveal any plausible foodborne cause, making the water system highly suspect. David Patterson asked Earl to obtain any microbiological test results from the PUC for the previous two weeks. Earl was also informed of the anonymous call. He surmised, without any supporting evidence, that intentional contamination might be possible. When Earl interviewed Stan Koebel and asked about any unusual events of the previous two weeks, Koebel did not tell him about the adverse bacteriological results for May 15 or the operation of Well 7 without a chlorinator. However, Koebel provided Earl with a number of documents, including the May 17 report (results for May 15).

James Earl returned to Owen Sound and reviewed these documents that evening. Although Earl noted the result showing high *E. coli* numbers for the water system, he did not report this alarming evidence to his supervisor, Phillip Bye, or the Health Unit at that time. James Earl apparently believed that the boil water advisory eliminated any urgency concerning the revelation about adverse microbial results for Walkerton's drinking water supply.

In the meantime, the Health Unit began to plot an outbreak curve that revealed an apparent peak of disease onset for May 17, suggesting a most likely date of contamination between May 12 and 14. They also plotted the residence locations for those who were infected. This plot revealed that cases were distributed all across the area served by the Walkerton water distribution system. By that evening, the Health Unit was convinced this was a waterborne outbreak, even though they had not yet been provided with the adverse results for May 15.

On Monday, May 22, the first victim of the outbreak, a 66-year-old woman, died. Subsequently, a 2-year-old child who visited Walkerton on Mother's Day (May 14) and consumed only one glass of water, died on Tuesday, May 23. Ultimately, 5 more deaths to total 7, 27 cases (with a median age of 4) of HUS, a life-threatening kidney condition that may subsequently require kidney transplantation, and 2,300 cases of gastrointestinal illness were attributed to consuming Walkerton water. Stool cultures from victims confirmed exposure to *E. coli* O157:H7, *Campylobacter jejuni*, and other enteric pathogens (BGOSHU, 2000). Longer term health consequences have been both documented (longer term gastrointestinal symptoms, irritable bowel syndrome, arthritis symptoms and albuminuria; Garg et al., 2006a, 2008a,b; Marshall et al., 2006) and found absent (renal sequelae among non-HUS cases; Garg et al., 2006b).

Well 5 was providing drinking water to Walkerton during the period of most likely contamination (May 12 to 14) according to the SCADA system. Well 5 was located close to two farms posing a water contamination risk from manure (Figure 2-17). The original hydrogeology report for Well 5, written in 1978, recognized the risk of contamination from nearby farms, having found that fecal coliforms appeared after 24 hours during pump testing (O'Connor, 2002a).

As illness emerged in the community, the Koebel brothers remained convinced that water was not to blame and they continued to drink the water. In the

FIGURE 2-17 Location of Walkerton Well 5 near farms to south and west.
SOURCE: Adapted from original photo taken for the Walkerton Inquiry by Constable Marc Bolduc, Royal Canadian Mounted Police, used with permission.

past, they had often consumed Well 5 water before chlorination because they did not recognize the danger of pathogen contamination.[10]

Direct Causes of the Walkerton Outbreak

The immediate direct cause of failure was that organic loading from manure contamination arising from a nearby farm barnyard overwhelmed the fixed chlorine dose that was used by the Walkerton PUC, leaving no disinfection capacity to inactivate the pathogens entering the distribution system. If the chlorine residual had been monitored as it should have been by the PUC operators, this problem would have been immediately evident, but no valid chlorine residuals were measured during the critical period (around May 12) after contamination was washed into the shallow Well 5.

[10]End of Hrudey (2006a) text.

Despite very clear and unambiguous findings in the first report by Justice O'Connor concerning the roles and responsibilities of the operators in failing to prevent this outbreak, the public record on the operators' roles and responsibilities became confused in December 2004 with the criminal conviction and sentencing of the Koebel brothers. They were charged with breach of trust, uttering forged documents, and common nuisance for their roles in the Walkerton outbreak. The prosecution agreed to a plea bargain, dropping the more serious charges in return for guilty pleas to common nuisance. Stan Koebel was sentenced to one year in jail and Frank Koebel was sentenced to 9 months of house arrest.

The trial and sentences were controversial. On one side, the admissions by the Koebel brothers of failing to perform monitoring and treatment, withholding adverse monitoring results, and falsifying operating records clearly pointed blame their way. However, the severe and systemic deficiencies in the operator training and regulatory systems of the Ontario government led the Walkerton Inquiry to find that the Koebel brothers were not solely responsible for the failures, nor were they the only ones who could have prevented the disaster.

The criminal trial in Ontario Superior Court increased confusion by means of a statement of "facts" agreed to by the prosecution to secure the guilty pleas. This agreed statement of "facts" contained inaccurate elements that were in direct contradiction of the well-documented findings of the Walkerton Inquiry. In this case the prosecution was obliged to gather evidence without reliance on evidence collected by the Inquiry, but their investigation was directed to the same set of facts, and it should have been able to reach the same conclusions as the Inquiry was able to document. The statement of "facts" attested to by the prosecution and the defense cited an epidemiologist as its sole authority to find that, even had the Koebel brothers increased the chlorine level in Walkerton's water system dramatically, that action "would not have prevented this tragedy." The epidemiologist acknowledged not being a specialist in disinfection during testimony at the Walkerton Inquiry. The prosecution concluded from the epidemiologist's opinion: "It therefore cannot be said that the criminal conduct of Stan Koebel and Frank Koebel . . . their failure to properly monitor, sample and test the well water . . . was, in law, a significant contributing cause of the deaths and injuries."

The operators were to ensure the chlorine residual was measured daily. Yet the Inquiry found "virtually all of the entries on the 1999 daily operating sheets are false. Fictitious entries in the daily operating sheets continued until the outbreak in May 2000." Inquiry Commissioner O'Connor observed: "One of the purposes of measuring chlorine residual is to determine whether contamination is overwhelming the disinfectant capacity of the chlorine." Accordingly, he found: "The scope of the outbreak would very likely have been substantially reduced if the Walkerton PUC operators had measured chlorine residuals at Well 5 daily, as they should have, during the critical period when contamination was entering the system." At least eight days without valid chlorine residual monitoring passed between the contamination influx and the boil water advisory issued by the health unit, after illness was already widespread (Figure 2-18).

The Walkerton PUC operators, if properly trained and acting competently, should have been alerted by the heavy rains and the obvious flooding that occurred on May 12. They should have commenced more frequent checking of their chlorine dosage and residuals. Even if they had stuck with the limited monitoring schedule required by the regulator, they should have recognized that the low to non-existent chlorine residual they would have measured on May 13 was cause for alarm. The chlorine dose should have been increased immediately to try to achieve a satisfactory chlorine residual. In this particular case, because Well 7 was able to provide the entire supply for Walkerton, the operators should have shut down Well 5 immediately, knowing its vulnerability. The Well 7 option was compromised by Frank Koebel's failure to follow Stan Koebel's instructions to replace a defective chlorinator on Well 7, an action that was not taken until May 19.

Once the operators became aware that water with no disinfection had entered the distribution system, the water storage should have been dosed with chlorine solution and the mains flushed. The regulator and local health authorities should have been notified of the problem and the actions being taken. If adequate chlo-

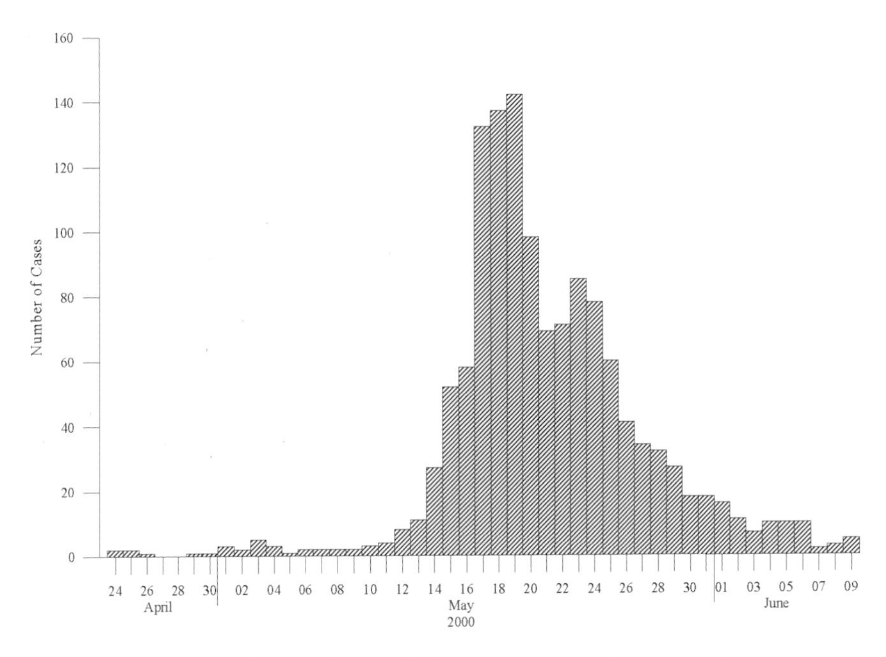

FIGURE 2-18 Outbreak curve for the Walkerton epidemic of gastroenteritis.
SOURCE: Adapted from BGOSHU (2000) and Hrudey and Hrudey (2004).

rine residual could not have been restored, at a minimum a boil water advisory should have been issued immediately. These actions could have and should have all been completed in the first 24 hours after the May 13 morning sample, for chlorine residual would have identified the problem.

[11]Even if these steps did not eliminate the consumption of contaminated water entirely, they would have substantially reduced the exposure of Walkerton consumers and the resulting health impacts. As it was, contaminated water was distributed for a full week longer than necessary. If the policies adopted in 1994 by the Ontario Chlorination Bulletin had been applied to Walkerton as they should have been, this vulnerable shallow well would have been equipped with a continuous chlorine residual analyzer. The continuous monitor should have been established in a fail-safe mode that would automatically shut off Well 5 when the chlorine residual fell below the set point for minimum effective disinfection.

The Walkerton operators should have realized their system was vulnerable to catastrophic failure. It was totally reliant on a single chlorination barrier that was not fail safe. The requirement for at least a second barrier (source protection) was identified more than 20 years earlier, but was never implemented. Water system operators must be able to recognize the threats to their system contrasted with the capability of their system to cope. They have a personal responsibility to ensure deficiencies are identified, made known to management, and effectively remediated. Pending necessary improvements, increased vigilance is required by operators together with contingency plans to cope with periods of stress.

The Koebel brothers lacked the training and expertise to identify the vulnerability of Well 5 and the need for additional safety barriers. They had been certified by a "grand-parenting" process that failed to provide them with the training needed to do their jobs properly. Their experience allowed them to handle the mechanical requirements of their jobs, but they lacked any understanding of water quality or associated public health risks. The Koebel brothers did not intend to harm their fellow citizens through their flawed practices. In fact, they continued to drink the water even as the outbreak was unfolding.[12]

[13]There were many potential direct causes for this outbreak, including new water main construction, fire events, main breaks and repairs, contamination of treated water storage, cross connections, flooding, and human sewage or sewage sludge contamination of the wells. Despite the diversity of possible causes, the Inquiry found consistent and convincing evidence that this outbreak was caused by contamination from cattle manure being washed from an adjacent farm into the shallow Well 5 on or about May 12 because of the heavy rainfall that day.

---

[11]The following text was reprinted from Hrudey and Walker. 2005. *Opflow* 31(6) with permission from the American Water Works Association.

[12]End Hrudey and Walker (2005) text.

[13]The following text was reprinted from Hrudey and Hrudey (2004) with permission from IWA Publishing.

Consequently, the following explanations will focus on the evidence for and understanding of that specific cause. Other plausible causes will be mentioned only briefly. However, under different circumstances, each of these could have caused or contributed to an outbreak.[14]

Well 5 (Figure 2-17) was identified as being vulnerable to surface contamination from the time it was first commissioned. The hydrogeologist who conducted the original assessment of this well wrote in his commissioning report:

> The results of the bacteriological examination indicate pollution from human or animal sources, however, this was not confirmed by the chemical analyses. The supply should definitely be chlorinated and the bacteria content of the raw and treated water supply should be monitored. The nitrate content should also be observed on a regular basis. . . .
>
> The Town of Walkerton should consider establishing a water-protection area by acquiring additional property to the west and south in the vicinity of Well four [now 5]. Shallow aquifers are prone to pollution in farming and human activities should be kept away from the site of the new well as far as possible (Wilson [1978] as reported in November 8, 2000, W. I. Transcript; evidence of J. Budziakowski).

[15]Pump testing on this well in 1978 demonstrated that bacteriological contamination occurred within 12 to 24 hours of initiating pumping, reaching a peak of 12 fecal coliforms per 100 mL after 48 hours. During the well's first two years, the MOE conducted a number of inspections that revealed continuing concerns for surface contamination. These concerns included the nearby agricultural activities; the shallowness of the well with its relatively thin overburden (the layer of soil between the surface and the aquifer); observed fluctuations in turbidity; bacteriological monitoring indicating fecal contamination; and elevated pumping levels in concert with spring thaw and early rainfall suggesting rapid surface recharge of the shallow aquifer. In 1980, the bacteriological samples of the raw water at Well 5 reached as high as 260 total coliforms per 100 mL and 230 fecal coliforms per 100 mL, with 4 out of 42 samples that year showing bacterial contamination. Because none of the chlorine disinfected samples from Well 5 showed bacterial contamination, the poorer quality raw water seems to have been accepted despite the obvious signs of surface contamination. Turbidity measurements were found to be occasionally elevated (up to 3.5 NTU) and to fluctuate well beyond what would be expected from a secure groundwater source.

Unfortunately, the concerns about surface contamination influencing the raw water at Well 5 appeared to have been forgotten in the MOE files during the 1980s

[14]End Hrudey and Hrudey (2004) text.

[15]The following text was reprinted from Hrudey and Hrudey (2004) with permission from IWA Publishing.

when no inspections were performed. However, the investigation by Golder Associates, Ltd. (Golder, 2000) after the outbreak confirmed that Well 5 was definitely under the influence of surface contamination, as the early water quality monitoring indicators had so clearly revealed. In a dramatic demonstration, a shallow pond (~10 cm deep) adjacent to Well 5 went dry within 30 minutes after the pump test commenced, and a deeper nearby pond dropped 27 cm over 36 hours of pumping. Furthermore, subsequent tracer tests conclusively demonstrated the hydraulic connection between the surface pond and the producing zone of Well 5. In fact, there were multiple entry points to the shallow aquifer feeding Well 5, possibly including point source breaches of the overburden by fencepost holes, improperly abandoned wells (none were located) and sand or gravel lenses. The investigations after the outbreak did not confirm the exact route of contamination entry into Well 5, but the relevant experts at the Inquiry agreed that the overall evidence for contamination of Well 5 was entirely consistent and the most plausible explanation for the outbreak.

The epidemiologic evidence and the timing of illness in the community strongly suggested that contamination occurred on or about May 12. Well 5 was the major source of water to Walkerton between May 10 and 15, with intermittent contributions from Well 6. The heavy rainfall experienced by Walkerton on May 12 peaked between 6 pm and midnight. Bacteriological sampling data were limited and were confounded by the inaccurate labeling practiced by PUC personnel. Given the location of the PUC shop in the distribution system downstream of Well 5, combined with the documented poor practices of the PUC operators, it was likely that the May 15 sample labeled *Well 7 treated* was actually taken at the PUC shop and represented Well 5 water entering the distribution system. This sample, the one that Stan Koebel concealed from health authorities, was heavily contaminated with greater than 200 *E. coli* per 100 mL.

A number of samples were collected by the local Health Unit, the MOE and the PUC between May 21 and 23. All Well 5 samples were positive for both total coliforms and *E. coli* while neither Well 6 nor Well 7 samples were positive for either bacterial indicator. A June 6 sample taken from the spring adjacent to Well 5 had a count of 80 *E. coli* per 100 mL. Pump tests were done at two monitoring wells near Well 5, including one on the adjacent farm, in late August 2000. After the 32-hour pump test, *E. coli* counts climbed to 12,000 per 100 mL on the monitoring well 225 m west of Well 5 and to 900 per 100 mL on the monitoring well 105 m west-northwest of Well 5. Dr. R. Gillham, the hydrogeology expert called by the Inquiry, concluded that a large area of the shallow aquifer supplying Well 5 had been heavily contaminated.

In addition to the reasonably consistent circumstantial evidence implicating Well 5 as the primary, if not sole, source of microbial contamination of Walkerton's water supply, there was reasonably compelling evidence linking the bacterial pathogens that caused the human illness with cattle and manure samples from the farm near Well 5. Dr. A. Simor, the Inquiry's expert on medical microbiology, described how pathogens were characterized in the laboratory (A. Simor, W. I. Transcript, February 26, 2000, pp. 142-146). Three methods were used to gain more evidence about the specific strains of pathogens identified: phage typing, serotyping and pulsed-field gel electrophoresis (PFGE).

The first method exploits the ability of certain viruses to infect bacteria. These viruses are named bacteriophages—phages for short. Different bacteria are susceptible to infection by different phages, so exposing a strain of bacteria to a range of different phages can be used to type that strain for its susceptibility to phage attack. That pattern of phage susceptibility can be used to distinguish one strain of bacteria from another strain of the same species.

The second method, serotyping, relies on detecting specific antigens on the exterior of a bacterial cell. These include O antigens that characterize components of the bacterial cell walls and H antigens that characterize the flagella (the whip-like tails that bacteria use for motion). For example, the name *E. coli* O157:H7 refers to the strain of *E. coli* with the 157 antigen in the cell wall and the 7 antigen in the flagellum. Individual strains of *Campylobacter* species, such as *C. jejuni,* can also be characterized by serotyping.

The third method, PFGE, looks at the molecular properties of the DNA found in a bacterial strain. Because the DNA provides the genetic material that causes specific strains of a bacterial species to be distinct, evaluating and comparing the DNA of individual strains provides a relatively direct method for identifying specific strains. In this procedure, DNA is extracted from the bacterial cell and is cut at chosen locations using specific enzymes to yield DNA fragments of varying size. These fragments are separated on a gel plate by electrophoresis to yield a pattern of bands distributed according to the relative size of the fragments. The resulting pattern can be interpreted in terms of the original DNA structure to compare with DNA from different strains. Identical strains will have identical DNA fragment patterns, while the patterns of closely related strains may differ in only a few fragments. Dr. Simor's expert opinion at the Inquiry (A. Simor, W. I. Transcript, p. 160, February 26) was that strains differing by six or fewer DNA fragment bands are considered genetically related in the context of a common source for an outbreak. These advanced methods were used to compare pathogens recovered from cattle manure with those from infected humans.

By August 31, 2000, in the follow-up investigation, the outbreak team working for the Health Unit had identified 1,730 cases as suspected cases (BGOSHU, 2000). Following contact attempts by phone or mail, 80 percent of contacts were judged to have an illness related to exposure to Walkerton municipal water, and 1,346 cases met the definition adopted for the investigation. "A case was defined as a person with diarrhea, or bloody diarrhea; or stool specimens presumptive positive for *E. coli* O157 or *Campylobacter spp.* or HUS between April 15 to June 30. For the purposes of attributing cases to the water system, a primary case was defined as a person who had exposure to Walkerton water. A secondary case was defined as a person who did not have any exposure to Walkerton water but had exposure to a primary case as defined above. A person was classified as unknown if their exposure status was not indicated" (BGOSHU, 2000).

Of these cases, 675 had submitted stool samples for culture, yielding 163 positive for *E. coli* O157:H7, 97 positive for *C. jejuni*, 7 positive for *C. coli* and 12 positive for both *E. coli* O157:H7 and *Campylobacter*. The outbreak curve is plotted in Figure 2-18.

The second peak in the epidemic curve (Figure 2-18) has been discussed as possibly representing the second of two types of infection that occurred, with *C. jejuni* and with *E. coli*. Another possibility that was not discussed is that the second peak occured on May 23, the date that Dr. McQuigge gave his first press conference on the outbreak. The resulting high profile media coverage that day might have anchored May 23 in the memories of some victims when they responded to the survey performed later to determine the date of onset of illness for each case.

Various cultures were also done on environmental samples, allowing some comparison with the pathogens causing illness. The Health Unit collected samples from 21 sites in the Walkerton distribution system on May 21 and collected raw and treated water from Well 5 on May 23. Concurrent samples taken at Wells 6 and 7 showed neither total coliforms nor *E. coli*. Two of the distribution system sites remained positive for total coliforms and *E. coli* over several days. One of the distribution system sites, along with cultures from the May 23 raw and treated water samples from Well 5, was analyzed by PCR, another molecular diagnostic technique. This technique is able to amplify DNA from a sample to allow extremely sensitive detection for specific genes that may be present. Using PCR, these samples, representing a contaminated location in the Walkerton distribution system and Well 5, all showed the same genes for O157, H7 and the specific verotoxin, VT2.

Working with Health Canada and the Ontario Ministry of Agriculture and Food, the Health Unit undertook livestock sampling on farms within a 4 km radius of each of Wells 5, 6, and 7 between May 30 and June 13 (BGOSHU, 2000). They obtained livestock fecal samples from 13 farms and found human pathogens (mainly *Campylobacter*) in samples from 11. On two farms, both *C. jejuni* and *E. coli* O157:H7 were found. These farms were selected for further sampling on June 13. The results are summarized in Table 2-1. Farm 1 was located in the vicinity of Wells 6 and 7 and Farm 2 was located within sight of Well 5 (Figure 2-17).

The most telling features of these typing efforts are revealed in Table 2-2, which compares the strain characteristics from the cattle fecal samples at the two farms with the cultures from human cases infected with *E. coli* O157:H7 or *Campylobacter spp*. Details of the extensive strain typing work that was done have now been published by Clark et al. (2003).

TABLE 2-1 Culture Results from Two Farms Resampled on June 13

	Number Positive	
Pathogens	Farm 1	Farm 2
E. coli O157:H7	2	6
C. jejuni/coli	8	9
Both	2	—

SOURCE: Derived from data reported in BGOSHU (2000) and reprinted from Hrudey and Hrudey (2004) with permission from IWA Publishing.

TABLE 2-2 Pathogen Strain Typing Comparison Between Human Cases and Cattle Fecal Samples at Farms 1 and 2

		Cattle Fecal Samples	
	Human Cases	Farm 1	Farm 2
Total individuals tested	675	20	38
E. coli O157:H7 positive	163	4	6
Phage type 14	147 (90% of +)	0	6 (100% of +)
Phage type 14a	3 (2% of +)	4 (100% of +)	0
PFGE pattern A	150 (92% of +)	0	6 (100% of +)
PFGE pattern A1	2 (1.2% of +)	4 (100% of +)	0
PFGE pattern A4	2 (1.2% of +)	0	1 (17% of +)
Verotoxin VT2	majority	–	6 (100% of +)
Campylobacter spp. positive	105	8	9
Phage type 33	56 (53% of +)	0	9 (100% of +)
Other phage types 2, 13, 19var, 44, 77		8 (100% of +)	0

SOURCE: Derived from data reported in BGOSHU (2000) and reprinted from Hrudey and Hrudey (2004) with permission from IWA Publishing.

These results do not provide absolute confirmation that manure from Farm 2 was responsible for contamination of the Walkerton water supply for a number of reasons. The cattle samples were taken in mid-June, about a month after the suspected date of contamination, and it is not possible to be certain that cattle on Farm 2 were infected on May 12. Likewise, the DNA typing by PFGE must be recognized as much less certain than DNA typing used in human forensic analysis. Because bacteria reproduce by binary fission, each progeny cell is a clone of its parent (i.e., each progeny cell has identical DNA to the parent cell, so individual cells are not genetically unique). However, the DNA makeup of bacteria changes rapidly because their rapid rate of reproduction allows genetic mutation through a number of mechanisms that alters their DNA quickly compared with humans. In total, the level of evidence for this outbreak is far more compelling than the quality and level of evidence that has historically been available for outbreak investigations. The main features suggesting that Farm 2, located near Well 5, was the primary source of pathogens that caused the outbreak are the match of the phage type 14 for *E. coli* O157:H7, with PFGE pattern A and with phage type 33 for the *Campylobacter spp*.

Finding that raw water in Well 5 was contaminated by pathogens detected in cattle manure from a nearby farm does not explain how this contamination was allowed to cause the disastrous disease outbreak in the community. The water produced by all the wells serving Walkerton was supposed to be chlorinated continuously to achieve a chlorine residual of 0.5 mg/L for 15 minutes of contact time (ODWO). This level of disinfection would have provided a concentration-contact time (CT) value of 7.5 mg/L-min. That CT value is more than 150 times greater than literature CT values of 0.03-0.05 mg-min/L for 99 percent inactivation of *E. coli* O157:H7 and more than 80 times greater than a CT value of 0.067 to 0.090 mg-min/L for 99.99% inactivation of *E. coli* (Hoff and Akin, 1986; Kaneko, 1998). Clearly, the specified level of chlorine residual and contact time was not operative for Well 5 in May 2000. If it had been, inactivation of the *E. coli* pathogen greater than 99.99 percent would have been achieved, a level of protection that is certainly not consistent with the magnitude of the outbreak that occurred in Walkerton.[16]

Indirect Causes of the Walkerton Outbreak

Although the PUC operators were obviously culpable for their misdeeds of omission and commission, they were clearly not the sole cause of this disaster. The overwhelming impression that follows from reviewing what happened in Walkerton is that complacency was pervasive at many levels. The majority of the discussion of the cause of this disaster in the Part 1 Walkerton Inquiry report (O'Connor, 2002a) was devoted to evaluating the contributions from many other parties to this disaster. Those interested in pursuing the institutional failures are referred to the Inquiry report. A brief summary of those institutional failures is provided below.

The Ontario Ministry of Environment

- The original certificate of approval was issued in 1978 for Well 5 without including any formal operating conditions to deal with the hazards that were apparent from the pump testing at the commissioning of this well.

[16]End Hrudey and Hrudey (2004) text.

- In 1994 the MOE adopted a policy to require continuous chlorine residual monitoring for vulnerable shallow wells (under the influence of surface contamination) but the MOE failed to implement policy this for existing vulnerable wells like Well 5.
- There was no follow-up on deficiencies identified during rare inspections including one that was conducted in 1998 which identified some of the deficiencies in the monitoring practices of the PUC operators.
- The false records and clear deficiencies in performance of the Walkerton operators were not recognized by the MOE.
- The MOE showed surprisingly little institutional knowledge about drinking water safety.

The Walkerton Public Utilities Commission

- The PUC placed total confidence in their General Manager and it took no apparent interest or responsibility for oversight of PUC operations.
- The PUC ignored an adverse MOE inspection report in 1998 and demonstrated no interest in assuring that the PUC was responding to the deficiencies identified.
- The PUC maintained a substantial financial surplus while not making investments in improving the water system.
- The PUC took an adversarial stance with the Health Unit when the outbreak occurred.

The Government of Ontario

- The Ontario government slashed the budget of the MOE drastically (>40 percent) over the previous five years with little evidence of concern for consequences to public health.
- Responsibility for drinking water safety was largely offloaded to individual communities without providing effective assistance for these communities to handle that responsibility.
- Water testing was removed from the provincial laboratory without adding any regulatory requirement for mandatory reporting of adverse results by the private labs to the MOE or the local health units.

Preventing Drinking Water Outbreaks: Turning Hindsight into Foresight

The challenge in preventing drinking water disasters like Walkerton is to learn from the experience of such disasters. There have been a surprising number of drinking water outbreaks in affluent countries and many have features in common with Walkerton (Hrudey and Hrudey, 2004). The task is essentially one of turning hindsight into foresight. This requires a risk management approach that demands a committed focus on prevention, as evidenced by:

- Informed vigilance, is actively promoted and rewarded.
- Understanding of the entire water system, from source to consumer's tap, its challenges and limitations, is promoted and actively maintained.
- Effective, real-time treatment process control is the basic operating approach.
- Fail-safe multi-barriers are actively identified and maintained at a level appropriate to the challenges facing the system.
- Close calls are documented and used to train staff about how the system responded under stress and to identify what measures are needed in future.
- Operators, supervisors, lab personnel and management all understand that they are entrusted with protecting the public's health and are committed to honoring that responsibility above all else.
- Operational personnel are afforded the status, training, and remuneration commensurate with their responsibilities as guardians of the public's health.
- Response capability and communication are enhanced.
- An overall continuous improvement, total quality management mentality will pervade the organization (Hrudey and Hrudey, 2004).

Justice O'Connor concluded in his second Inquiry report (O'Connor, 2002b) that "Ultimately, the safety of drinking water is protected by effective management systems and operating practices, run by skilled and well-trained staff." Ontario has committed to implementing substantial improvements in the scope and quality of operator training. Operators can prevent a disaster like Walkerton if they assure the following key elements are established and followed (Hrudey and Walker, 2005):

- Operators must understand their system, including the major contamination hazards it faces in relation to the safety barriers and their capabilities for assuring safe water.
- Operators must develop guidance limits for monitoring parameters that are able to detect abnormal conditions, based on knowing their system.
- Operators must watch for and recognize signals for abnormal conditions (i.e., increase in turbidity or chlorine demand and drop of chlorine residual).
- Operators must work with management to anticipate plausible abnormal conditions and plan effective responses well before a serious incident occurs, including appropriate notification of regulatory authorities. Pre-

paredness should support but does not replace the need for thoughtful analysis and problem-solving as events unfold.

- Operators must recognize when they are facing a problem that is beyond their understanding or training and call for assistance.
- Operators' understanding of their system should include recognition of any inherent vulnerability that needs improvement to reduce contamination risks.
- Operators need to be prepared to take ownership of problems and lead efforts to ensure that their managers fully understand the existence of such problems that must be rectified.

Concluding Comments

The Walkerton outbreak provides a strong argument for the multiple barrier approach for assuring safe drinking water. This needs to be based on a preventive risk management approach like that described in the Australian Drinking Water Guidelines (NHMRC, 2004) or the WHO water safety plan approach (WHO, 2004) that is firmly grounded in a quality management approach that is founded on a thorough understanding of the risks facing any particular system (Hrudey, 2004).

Because outbreaks of disease caused by drinking water remain comparatively rare in North America, particularly in contrast with the developing world, complacency about the dangers of waterborne pathogens can easily occur. Yet, the source of waterborne disease in the form of microbial pathogens is an ever present risk because these pathogens are found in human fecal waste and in fecal wastes from livestock, pets or wildlife, making any drinking water source at risk of contamination before or after treatment (Hrudey, 2006b).

SEAS AND GOTUZZO REFERENCES

Clemens, J., M. J. Albert, M. Rao, F. Qadri, S. Huda, B. Kay, F. P. van Loon, D. Sack, B. A. Pradhan, and R. B. Sack. 1995. Impact of infection by *Helicobacter pylori* on the risk and severity of endemic cholera. *Journal of Infectious Diseases*171(6):1653-1656.

Colwell, R. R., M. L. Tamplin, P. R. Brayton, A. L. Gauzens, B. D. Tall, D. Herrington, M. M. Levine, S. Hall, A. Huq, and D. A. Sack. 1990. Environmental aspects of *Vibrio cholerae* in transmission of cholera. In *Advances in Research on Cholera and Related Diarrheas*, vol. 7, edited by R. B. Sack, and Y Zinnaka. Tokyo: KTK Scientific.

Colwell, R. R., A. Huq, M. S. Islam, K. M. A. Aziz, M. Yunus, N. Huda Khan, A. Mahmud, R. Bradley Sack, G. B. Nair, J. Chakraborty, D. A. Sack, and E. Russek-Cohen. 2003. Reduction of cholera in Bangladeshi villages by simple filtration. *Proceedings of the National Academy of Sciences* 100(3):1051-1055.

Franco, A. A., A. D. Fix, A. Prada, E. Paredes, J. C. Palomino, A. C. Wright, J. A. Johnson, R. McCarter, H. Guerra, and J. G. Morris, Jr. 1997. Cholera in Lima, Peru, correlates with prior isolation of *Vibrio cholerae* from the environment. *American Journal of Epidemiology* 146(12):1067-1075.

Gotuzzo, E., J. Cieza, L. Estremadoyro, and C. Seas. 1994. Cholera: lessons from the epidemic in Peru. *Infectious Disease Clinics of North America* 8(1):183-205.

León-Barúa, R., S. Recavarren-Arce, E. Chinga-Alayo, C. Rodríguez-Ulloa, D. N. Taylor, E. Gotuzzo, M. Kosek, D. Eza, and R. H. Gilman. 2006. *Helicobacter pylori*-associated chronic atrophic gastritis involving the gastric body and severe disease by *Vibrio cholerae*. *Transactions of the Royal Society of Tropical Medicine and Hygiene* 100(6):567-572.

Lipp, E. K., A. Huq, and R. R. Colwell. 2002. Effects of global climate on infectious disease: the cholera model. *Clinical Microbiology Reviews* 15(4):757-770.

Pascual, M., X. Rodó, S. P. Ellner, R. Colwell, and M. J. Bouma. 2000. Cholera dynamics and El Niño-southern oscillation. *Science* 289(5485):1766-1769.

Salazar-Lindo, E., C. Seas, and D. Gutierrez. 2008. The ENSO and cholera in South America: what we can learn about it from the 1991 cholera outbreak. *International Journal of Environment and Health* 2(1):30-36.

Seas, C., J. Miranda, A. I. Gil, R. Leon-Barua, J. Patz, A. Huq, R. R. Colwell, and R. B. Sack. 2000. New insights on the emergence of cholera in Latin America during 1991: the Peruvian experience. *American Journal of Tropical Medicine and Hygiene* 62(4):513-517.

WHO (World Health Organization). 2008. Cholera. *Weekly Epidemiological Record* 83(31):269-284.

Zuckerman, J. N., L. Rombo, and A. Fisch. 2007. The true burden and risk of cholera: implications for prevention and control. *Lancet Infectious Diseases* 7(8):521-530.

DAVIS REFERENCES

Addiss, D. G., W. R. Mac Kenzie, N. J. Hoxie, M. S. Gradus, K. A. Blair, M. E. Proctor, J. J. Kazmierczak, W. L. Schell, P. Osewe, H. Frisby, H. Cicirello, R. L. Cordell, J. B. Rose, and J. P. Davis. 1995. Epidemiologic features and implications of the Milwaukee cryptosporidiosis outbreak. In *Protozoan parasites and water*, edited by W. B. Betts, D. Casemore, C. Fricker, H. Smith, and J. Watkins. Cambridge, UK: The Royal Society of Chemistry. Pp. 19-25.

Addiss, D. G., R. S. Pond, M. Remshak, D. Juranek, S. Stokes, and J. P. Davis. 1996. Reduction of risk of watery diarrhea with point-of-use water filters during a massive outbreak of waterborne *Cryptosporidium* infection in Milwaukee, 1993. *American Journal of Tropical Medicine and Hygiene* 54(6):549-553.

CDC (Centers for Disease Control and Prevention). 1994. *Cryptosporidium* infections associated with swimming pools—Dane County, Wisconsin, 1993. *Morbidity and Mortality Weekly Report* 43(31):561-563.

Chappell, C. L., P. C. Okhuysen, R. C. Langer-Curry, G. Widmer, D. E. Akiyoshi, S. Tanriverdi, and S. Tzipori. 2006. *Cryptosporidium hominis:* experimental challenge of healthy adults. *American Journal of Tropical Medicine and Hygiene* 75(5):851-857.

Cicirello, H. G., K. S. Kehl, D. G. Addiss, M. J. Chusid, R. I. Glass, J. P. Davis, and P. L. Havens. 1997. Cryptosporidiosis in children during a massive waterborne outbreak, Milwaukee, Wisconsin: clinical, laboratory, and epidemiologic findings. *Epidemiology and Infection* 119(1):53-60.

Cordell, R. L., D. G. Addiss, P. Thor, J. Theurer, R. Lichterman, S. Ziliak, F. Steurer, D. D. Juranek, and J. P. Davis. 1997. Impact of a massive waterborne cryptosporidiosis outbreak on child care facilities in metropolitan Milwaukee, Wisconsin. *Pediatric Infectious Diseases Journal* 16(7):639-644.

Corso, P. S., M. H. Kramer, K. A. Blair, D. G. Addiss, J. P. Davis, A. C. Haddix. 2003. The cost of illness in the waterborne *Cryptosporidium* outbreak, Milwaukee, Wisconsin. *Emerging Infectious Diseases* 9(4):426-431.

Current, W. L., N. C. Reese, J. V. Ernst, W. S. Bailey, M. B. Heyman, and W. M. Weinstein. 1983. Human cryptosporidiosis in immunocompetent and immunodeficient persons: studies of an outbreak and experimental transmission. *New England Journal of Medicine* 308(21):1252-1257.

D'Antonio, R. G., R. E. Winn, J. P. Taylor, T. L. Gustafson, W. L. Current, M. M. Rhodes, G. W. Gary, Jr., and R. A. Zajac. 1985. A waterborne outbreak of cryptosporidiosis in normal hosts. *Annals of Internal Medicine* 103(6 Pt 1):886-888.

Du Pont, H. L., C. L. Chappell, C. R. Sterling, P. C. Okhuysen, J. B. Rose, and W. Jakubowski. 1995. The infectivity of *Cryptosporidium parvum* in health volunteers. *New England Journal of Medicine* 332(13):855-859.

Fayer, R. 2004. *Cryptosporidium*: a water-borne zoonotic parasite. *Veterinary Parasitology* 126(1-2):37-56.

Frisby, H. R., D. G. Addiss, W. J. Reiser, B. Hancock, J. M. Vergeront, N. J. Hoxie, and J. P. Davis. 1997. Clinical and epidemiologic features of a massive waterborne outbreak of cryptosporidiosis among persons with human immunodeficiency virus (HIV) infection. *Journal of Acquired Immune Deficiency Syndromes and Human Retrovirology* 16(5):367-373.

Gallagher, M. M., J. L. Herndon, I. J. Nims, C. R. Sterling, D. J. Grabowski, and H. F. Hull. 1989. Cryptosporidiosis and surface water. *American Journal of Public Health* 79(1):39-42.

Hayes, E. B., T. D. Matte, T. R. O'Brien, T. W. McKinley, G. S. Logsdon, J. B. Rose, B. L. Ungar, D. M. Word, P. F. Pinsky, M. L. Cummings, M. A. Wilson, E. G. Long, E. S. Hurwitz, and D. D. Juranek. 1989. Large community outbreak of cryptosporidiosis due to contamination of a filtered public water supply. *New England Journal of Medicine* 320(21):1372-1376.

Hoxie, N. J., J. P. Davis, J. M. Vergeront, R. D. Nashold, and K. A. Blair. 1997. Cryptosporidiosis-associated mortality following a massive waterborne outbreak in Milwaukee, Wisconsin. *American Journal of Public Health* 87(12):2032-2035.

Joseph, C., G. Hamilton, M. O'Connor, S. Nicholas, R. Marshall, R. Stanwell-Smith, R. Sims, E. Ndawula, D. Casemore, P. Gallagher, and P. Harnett. 1991. Cryptosporidiosis in the Isle of Thanet: an outbreak associated with local drinking water. *Epidemiology and Infection* 107(3):509-519.

Juranek, D. D., D. G. Addiss, M. E. Bartlett, M. J. Arrowood, D. G. Colley, J. E. Kaplan, R. Perciasepe, J. R. Elder, S. E. Regli, and P. S. Berger. 1995. Cryptosporidiosis and public health: workshop report. *Journal of the American Water Works Association* 87(9):69-80.

Leland, D., J. McAnulty, W. Keene, and G. Stevens. 1993. A cryptosporidiosis outbreak in a filtered-water supply. *Journal of the American Water Works Association* 85(6):34-42.

Mac Kenzie, W. R., D. G. Addiss, and J. P. Davis. 1994a. *Cryptosporidium* and the public water supply. *New England Journal of Medicine* 331(22):1529-1530.

Mac Kenzie, W. R., N. J. Hoxie, M. E. Proctor, M. S. Gradus, K. A. Blair, D. E. Peterson, J. J. Kazmierczak, D. G. Addiss, K. R. Fox, J. B. Rose, and J. P. Davis. 1994b. A massive outbreak in Milwaukee of *Cryptosporidium* infection transmitted through the public water supply. *New England Journal of Medicine* 331(3):161-167.

Mac Kenzie, W. R., J. J. Kazmierczak, and J. P. Davis. 1995a. Cryptosporidiosis associated with a resort swimming pool. *Epidemiology and Infection* 115(3):545-553.

Mac Kenzie, W. R., W. L. Schell, K. A. Blair, D. G. Addiss, D. E. Peterson, N. J. Hoxie, J. J. Kazmierczak, and J. P. Davis. 1995b. Massive outbreak of waterborne *Cryptosporidium* infection in Milwaukee, Wisconsin: recurrence of illness and risk of secondary transmission. *Clinical Infectious Diseases* 21(1):57-62.

McDonald, A. C., W. R. Mac Kenzie, D. G. Addiss, M. S. Gradus, G. Linke, E. Zembrowski, M. R. Hurd, M. J. Arrowood, P. J. Lammie, and J. W. Priest. 2001. *Cryptosporidium parvum*–specific antibody responses among children residing in Milwaukee during the 1993 waterborne outbreak. *Journal of Infectious Diseases* 183(9):1373-1379.

Meisel, J. L., D. R. Perera, C. Meligro, and C. E. Rubin. 1976. Overwhelming watery diarrhea associated with *Cryptosporidium* in an immunosuppressed patient. *Gastroenterology* 70(6):1156-1160.

Morgan-Ryan, U. M., A. Fall, L. A. Ward, N. Hijawi, I. Sulaiman, R. Payer, R. C. Thompson, M. Olson, A. Lal, and L. Xiao. 2002. *Cryptosporidium hominis* n. sp. (Apicomplexa: Cryptosporidiidae) from *Homo sapiens*. *Journal of Eukaryotic Microbiology* 49(6):433-440.

Nime, F. A., J. D. Burek, D. L. Page, M. A. Holscher, and J. H. Yardley. 1976. Acute enterocolitis in a human being infected with the protozoan *Cryptosporidium*. *Gastroenterology* 70(4):592-598.

Osewe, P., D. G. Addiss, K. A. Blair, A. Hightower, M. L. Kamb, and J. P. Davis. 1996. Cryptosporidiosis in Wisconsin: a case-control study of post-outbreak transmission. *Epidemiology and Infection* 117(2):297-304.

Peng, M. M., L. Xiao, A. R. Freeman, M. J. Arrowood, A. A. Escalante, A. C. Weltman, C. S. L. Ong, W. R. Mac Kenzie, A. A. Lal, and C. B. Beard. 1997. Genetic polymorphism among *Cryptosporidium parvum* isolates: evidence of two distinct transmission cycles. *Emerging Infectious Diseases* 3(4):567-573.

Proctor, M. E., K. A. Blair, and J. P. Davis. 1998. Surveillance data for waterborne illness: an assessment following a massive waterborne outbreak of *Cryptosporidium* infection. *Epidemiology and Infection* 120(1):43-54.

Richardson, A. J., R. A. Frankenberg, A. C. Buck, J. B. Selkon, J. S. Colbourne, J. W. Parsons, and R. T. Mayon-White. 1991. An outbreak of waterborne cryptosporidiosis in Swindon and Oxfordshire. *Epidemiology and Infection* 107(3):485-495.

HRUDEY AND HRUDEY REFERENCES

Auld, H., D. MacIver, and J. Klaassen. 2004. Heavy rainfall and waterborne disease outbreaks: the Walkerton example. *Journal of Toxicology and Environmental Health* 67(20-22):1879-1887.

BGOSHU (Bruce-Grey Owen Sound Health Unit). 2000. *The investigative report of the Walkerton outbreak of waterborne gastroenteritis May-June 2000*, http://water.sesep.drexel.edu/outbreaks/WalkertonReportOct2000/REPORT_Oct00.PDF (accessed April 20, 2009).

Clark, C. G., L. Price, R. Ahmed, D. L. Woodward, P. L. Melito, F. G. Rodgers, F. Jamieson, B. Ciebin, A. Li, and A. Ellis. 2003. Characterization of waterborne outbreak-associated *Campylobacter jejuni*, Walkerton, Ontario. *Emerging Infectious Diseases* 9(10):1232-1241.

Garg, A. X., J. Marshall, M. Salvadori, H. R. Thiessen-Philbrook, J. Macnab, R. S. Suri, R. B. Haynes, J. Pope, and W. Clark. 2006a. A gradient of acute gastroenteritis was characterized, to assess risk of long-term sequelae after drinking bacterial contaminated water. *Journal of Clinical Epidemiology* 59(4):421-428.

Garg, A. X., W. F. Clark, M. Salvadori, H. R. Thiessen-Philbrook, and D. Matsell. 2006b. Absence of renal sequelae after childhood *Escherichia coli* O157: H7 gastroenteritis. *Kidney International* 70(4):807-812.

Garg, A. X., J. E. Pope, H. R. Thiessen-Philbrook, W. Clark, and J. Ouimet. 2008a. Arthritis risk after acute bacterial gastroenteritis. *Rheumatology* 47(2):200-204.

Garg, A. X., M. Salvadori, J. M. Okell, H. R. Thiessen-Philbrook, R. S. Suri, G. Filler, L. Moist, D. Matsell, and W. F. Clark. 2008b. Albuminuria and estimated GFR 5 years after *Escherichia coli* O157: H7 hemolytic uremic syndrome: an update. *American Journal of Kidney Diseases* 51(3):435-444.

Golder. 2000. *Report on hydrogeological assessment, bacteriological impacts, Walkerton town wells, municipality of Brockton, county of Bruce, Ontario*. Golder Associates Ltd, Walkerton Inquiry Part 1 Exhibit 259.

Hoff, J. C., and E. W. Akin. 1986. Microbial resistance to disinfectants: mechanisms and significance. *Environmental Health Perspectives* 69:7-13.

Hrudey, S. E. 2004. Drinking water risk management principles for a total quality management framework. *Journal of Toxicology and Environmental Health A* 67(20-22):1555-1566.

———. 2006a. Fatal disease outbreak from contaminated drinking water in Walkerton, Canada. *Case Studies in Environmental Engineering and Science*. Champaign, IL: Association of Environmental Engineering and Science Professors.

———. 2006b. Waterborne outbreak of cryptosporidiosis in North Battleford, Canada. *Case Studies in Environmental Engineering and Science*. Champaign, IL: Association of Environmental Engneering and Science Professors.

Hrudey, S. E., and E. J. Hrudey. 2004. *Safe drinking water—lessons from recent outbreaks in affluent nations*. London: IWA.

Hrudey, S. E., and R. Walker. 2005. Walkerton—5 years later. Tragedy could have been prevented. *Opflow* 31(6):1, 4-7.

Kaneko, M. 1998. Chlorination of pathogenic *E. coli* O157. *Water Science and Techology* 38(12):141-144.

Marshall, J. K., M. Thabane, A. X. Garg, W. F. Clark, M. Salvadori, and S. M. Collins. 2006. Incidence and epidemiology of irritable bowel syndrome after a large outbreak of bacterial dysentery. *Gastroenterology* 131(2):445-450.

NHMRC (National Health and Medical Research Council). 2004. *Australian drinking water guidelines*. National Health and Medical Research Council. Canberra, http://www.nhmrc.gov.au/publications/synopses/eh19syn.htm (accessed April 20, 2009).

O'Connor, D. R. 2002a. *Report of the Walkerton inquiry. Part 1. The events of May 2000 and related issues*. Toronto, The Walkerton Inquiry, http://www.attorneygeneral.jus.gov.on.ca/english/about/pubs/walkerton (accessed April 20, 2009).

———. 2002b. *Report of the Walkerton inquiry. Part 2. A strategy for safe water*. Toronto, The Walkerton Inquiry, http://www.attorneygeneral.jus.gov.on.ca/english/about/pubs/walkerton (accessed April 20, 2009).

WHO (World Health Organization). 2004. Water safety plans, Chapter 4. *WHO guidelines for drinking water quality*. Geneva: WHO. Pp. 54-88, http://www.who.int/water_sanitation_health/dwq/gdwq3_4.pdf (accessed April 20, 2009).

3

Vulnerable Infrastructure and Waterborne Disease Risk

OVERVIEW

This chapter highlights an assortment of vulnerabilities in water and sanitation infrastructure and the various means used to assess their potential consequences on scales ranging from local to global. The first paper, by workshop speaker Michael Beach and coauthors from the Centers for Disease Control and Prevention's (CDC's) National Center for Zoonotic, Vector-Borne, and Enteric Diseases, demonstrates that the United States, despite its relatively light burden of waterborne disease, is home to a deteriorating public drinking water distribution system, increasing numbers of unregulated private water systems, and a limited, passive waterborne disease surveillance system. Beach and colleagues discuss major national trends in waterborne disease dynamics as detected by the CDC's Waterborne Disease Outbreak Surveillance System (WBDOSS) and identify emerging needs in waterborne disease prevention and control, which include a deeper understanding of the ecology of waterborne disease as it pertains to drinking water distribution systems, safe water reuse programs, and an estimate of the burden of waterborne disease *in toto* to advocate for, as well as inform, active surveillance efforts.

Climate change presents a serious challenge to safe water availability worldwide, for numerous reasons summarized by presenter Joan Rose of Michigan State University in the chapter's second paper. In this context, she evaluates the findings of key studies relating health, climate, and water quality, and identifies critical questions for future research. Such studies have pursued three main lines of inquiry:

1. Relationships between extreme weather events and outbreaks of waterborne disease;
2. Associated changes in fecal bacterial concentrations in water and climate factors; and
3. Quantitative assessments of the relationship between various environmental factors (e.g., infrastructure and climate) and transmission risk for specific waterborne pathogens.

The essay concludes with a summary of critical needs that must be met in order to predict the effects of climate change on waterborne disease.

The subsequent contribution, by speaker Kelly Reynolds of the University of Arizona and Kristina Mena of the University of Texas-Houston, expands on a topic introduced by Rose: quantitative microbial risk assessment of waterborne disease. Reynolds and Mena observe that human pathogens make difficult subjects for risk assessment due to their "relatively low prevalence and infectious dose, specific virulence characteristics, and variably susceptible populations"; the vast diversity of water systems in use around the globe amplifies that challenge in the case of waterborne pathogens.

Following a description of microbial risk assessment methodologies for waterborne disease, the authors review representative studies (most of which were conducted in the United States) that describe drinking water contamination and the role of the water distribution system in spreading waterborne disease, as well as that played by premise plumbing and the biofilms present therein. They discuss the potential for improving risk assessment science by taking full advantage of the complementary relationship between epidemiological and forecasting studies, and also with increasingly accurate mathematical models and improved monitoring capacity.

That final, essential component of assessing risk for waterborne disease—pathogen monitoring—was the subject of a presentation in the same workshop session by Mark Sobsey of the University of North Carolina at Chapel Hill, entitled "Current Issues and Approaches to Microbial Testing of Water: Applicability and Use of Current Tests in the Developing World." While clearly beneficial in industrialized countries, water testing is "essential" to providing safe water in developing countries, Sobsey observed. Water quality data informs the selection of promising sources for drinking water and appropriate treatments to ensure its safety, as well as the classification of existing sources for the purposes of studying their health effects. Unfortunately, he observed, most water tests are not accessible, are too complicated, or are too costly for use in developing countries.

Sobsey described the ideal microbial water test for low-resource settings as portable, self-contained, lab-free, electricity-free, low cost, globally available, able to support data communication, and capable of educating and mobilizing stakeholders, especially youth, to improve public health. These goals eventually may be met through a variety of approaches and options but are currently limited mainly to culturing *E. coli* with enhanced detection. In the future, Sobsey predicted, culture-free and direct methods for detecting waterborne pathogens would predominate. These tests could be performed at ambient or body temperature (on Petri films or absorbent pads, or in small volumes of liquid, that could be incubated in a pocket or armpit). They would display simple, picture-based, shareable results.

Progress toward developing such a test is being made by the Aquatest Project, an international, multidisciplinary consortium led by the University of Bristol, United Kingdom (Aquatest, 2009). "The idea was to . . . develop a low-cost test that would be accessible and affordable for the developing world . . . sort of like a home pregnancy or glucose test," Sobsey explained. Following a successful feasibility study, the project is now in its second phase: a four-year, $13 million-plus project funded by the Bill and Melinda Gates Foundation to develop a test for *E. coli,* field-test it in India and South Africa, and prepare to deploy it on a global basis.

Sobsey concluded his presentation with the following recommendations to build on Aquatest and support continued development of microbial water tests for developing countries:

- Engage a wider network of collaborators and donors;
- Experiment with various test formats;
- Explore target microbes other than *E. coli*;
- Consider potential uses of testing results by a range of sectors (e.g., water science and engineering, health, and development);
- Link test development to waterborne disease epidemiology and quantitative risk assessment; and
- Use test as a tool for education and policy making.

THE CHANGING EPIDEMIOLOGY OF WATERBORNE DISEASE OUTBREAKS IN THE UNITED STATES: IMPLICATIONS FOR SYSTEM INFRASTRUCTURE AND FUTURE PLANNING

Michael J. Beach, Ph.D.[1]
Centers for Disease Control and Prevention

Sharon Roy, M.D., M.P.H.[2]
Centers for Disease Control and Prevention

Joan Brunkard, Ph.D.[2]
Centers for Disease Control and Prevention

Jonathan Yoder, M.P.H., M.S.W.[2]
Centers for Disease Control and Prevention

Michele C. Hlavsa, R.N., M.P.H.[2]
Centers for Disease Control and Prevention

The timing for this presentation is fortuitous since it is September 23, 2008, the eve of the 100th anniversary of the addition of chlorine to the Jersey City, New Jersey drinking water supply—the first time chlorine was added to water to kill microbes and improve water quality at an American drinking water treatment plant. This centennial reminds us, as we explore current challenges in providing safe drinking water in this country, of the pivotal role that inclusion of filtration and disinfection in water treatment plants had in reducing the burden of waterborne diseases in the United States (Cutler and Miller, 2005).

Since 1971, the Centers for Disease Control and Prevention (CDC) in collaboration with the U.S. Environmental Protection Agency (EPA) and the Council for State and Territorial Epidemiologists (CSTE) has tracked epidemiological trends in waterborne disease in the United States through the national Waterborne Disease and Outbreak Surveillance System[3] (WBDOSS). The WBDOSS receives investigative information on individual cases and outbreaks of waterborne disease from public health departments in states, territories, and the Freely Associated States (composed of the Republic of the Marshall Islands, the Federated States

[1]Corresponding author. Associate Director for Healthy Water, National Center for Zoonotic, Vector-Borne and Enteric Diseases, Centers for Disease Control and Prevention, 4770 Buford Highway, F-22, Atlanta, Georgia 30341; E-mail: mbeach@cdc.gov; Tel: 770-488-7763; Fax: 770-488-7761.

[2]Parasitic Diseases Branch, Division of Parasitic Diseases, National Center for Zoonotic, Vector-Borne and Enteric Diseases.

[3]Information on the WBDOSS can be accessed at http://www.cdc.gov/healthywater/statistics/wbdoss/index.html.

of Micronesia, and the Republic of Palau; formerly parts of the United States-administered Trust Territories of the Pacific Islands). Although initially designed to collect data about drinking water outbreaks in the United States, the WBDOSS now captures outbreaks associated with drinking water, recreational water, and nonrecreational water that is not intended for drinking or where the intended use is unknown. Annual or biennial surveillance summaries of the data have been published by CDC since the system's inception in 1971.[4] This system is now the primary source of data on waterborne disease outbreaks (including those caused by pathogens, chemicals, and toxins) associated with ingestion, contact, or inhalation of drinking water, recreational water, or water not intended for drinking (i.e., cooling towers, industrial use) occurring within the United States. The WBDOSS has documented a wide range of outbreaks of waterborne illnesses including acute gastrointestinal illness (AGI), infections of the skin, ear, eye, respiratory tract, urinary tract, wounds, and neurological system. These include outbreaks of AGI caused by a variety of pathogens such as *Campylobacter* (Vogt et al., 1982), *Cryptosporidium* (CDC, 2007b, 2008b; Mac Kenzie et al., 1994; Wheeler et al., 2007), *E. coli* O157:H7 (McCarthy et al., 2001; Swerdlow et al., 1992), norovirus (Parshionikar et al., 2003; Podewils et al., 2007), *Giardia* (Katz et al., 2006; Kent et al., 1988), *Salmonella* (Angulo et al., 1997), and *Shigella* (CDC, 2001; Iwamoto et al., 2005). Other nonenteric illness outbreaks have also been documented in the United States and include illnesses such as *Pseudomonas*-related dermatitis/folliculitis and outer ear infections (CDC, 1982; Gustafson et al., 1983; Yoder et al., 2008a), adenovirus-related pharyngoconjunctival fever (D'Angelo et al., 1979; Turner et al., 1987), legionellosis (i.e., Legionnaire's disease and Pontiac fever; Benin et al., 2002; Burnsed et al., 2007; Fields et al., 2001), echovirus-related aseptic meningitis (CDC, 2004), primary amebic meningoencephalitis (CDC, 2008a; Visvesvara et al., 1990), hepatitis A (Bergeisen et al., 1985; Mahoney et al., 1992), leptospirosis (Morgan et al., 2002), and conditions caused or exacerbated by waterborne chemicals or toxins (for example, there are apparent links between bronchial health effects and chloramines, which are volatile irritants formed when nitrogenous waste such as urine or sweat is oxidized by hypochlorous acid used to disinfect swimming pools) (Bowen et al., 2007; CDC, 2007a, 2009; Kaydos-Daniels, 2008; Weisel et al., 2008).

Trends in Drinking Water-Associated Disease Outbreaks

Over the course of its existence, WBDOSS surveillance has revealed four major trends in drinking water-related outbreaks that reflect the positive impact

---

[4]All WBDOSS surveillance summaries of data from 1971 to the latest summary can be found electronically on CDC's Healthy Water website at http://www.cdc.gov/healthywater/statistics/wbdoss/surveillance.html.

of regulation in improving drinking water safety in the United States, as well as where gaps in regulation exist:

- Public drinking water system-related disease outbreaks have decreased, reflecting the positive impact of national regulations (e.g., the Safe Drinking Water Act of 1974[5] and its amendments in 1986[6] and 1996[7]; primary drinking water standards set in 1985; the Surface Water Treatment rule of 1989[8]), as well as improved water system practices (Figure 3-1). However, this apparent correlation of decreasing outbreaks with improved regulation underscores how disease prevention efforts must be maintained and improved. This includes a continued emphasis on enforcing and improving existing regulation as new data become available or new pathogens emerge, implementation of new regulation as needed, source water protection, drinking water infrastructure investment, and other efforts responsible for the gains made to this point.
- The proportion of surface water-related disease outbreaks has declined in relation to groundwater-related disease outbreaks. Many regulations, including the Surface Water Treatment Rule of 1989, have focused on improving treatment of public drinking water supplies using surface water sources (e.g., rivers, lakes). It is therefore not surprising that while surface water-related disease outbreaks have decreased (Figure 3-2), groundwater-

[5]The Safe Drinking Water Act of 1974 put into motion a new national program to reclaim and ensure the purity of the water we consume. Under the Act, each level of government, every local water system, and the individual consumer have well-defined roles and responsibilities. But both the opportunity and the challenge of implementing the Act begins with EPA (for more information, see http://www.epa.gov/history/topics/sdwa/07.htm).

[6]The 1986 amendment created a demonstration program to protect aquifers from pollutants, mandated state-developed critical wellhead protection programs, required the development of drinking water standards for many contaminants now unregulated, and strengthened EPA's enforcement powers in dealing with recalcitrant water systems and underground injection well operators. It also imposed a ban on lead-content plumbing materials. Studies have found that excessive levels of lead in drinking water can harm the central nervous system in humans, especially children. The measure also provides substantial new authority to EPA to enforce the law including increased civil and criminal penalties for violations (for more information, see http://www.epa.gov/history/topics/sdwa/04.htm).

[7]The 1996 amendments established a strong new emphasis on preventing contamination problems through source water protection and enhanced water system management. This emphasis transformed the previous law, with its largely after-the-fact, regulatory focus, into a truly environmental statute that can better provide for the sustainable use of water by the nation's public water systems and their customers. The states are central, creating and focusing prevention programs and helping water systems improve operations and avoid contamination problems (for more information, see http://www.epa.gov/ogwdw/sdwa/theme.html).

[8]The Surface Water Treatment Rule of 1989 was designed to prevent waterborne diseases caused by viruses, Legionella, and Giardia intestinalis. These disease-causing microbes are present at varying concentrations in most surface waters. The rule requires that water systems filter and disinfect water from surface water sources to reduce the occurrence of unsafe levels of these microbes (for more information, see http://epa.gov/ogwdw/therule.html#Surface).

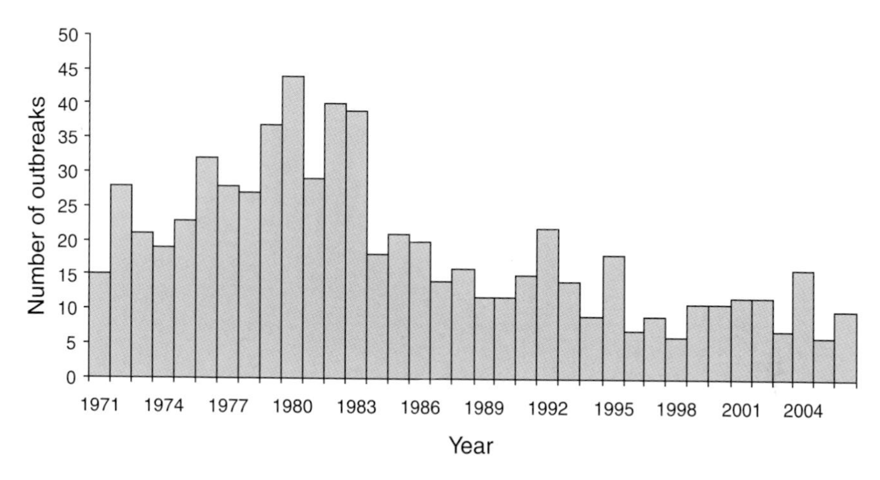

FIGURE 3-1 Number of reported waterborne-disease outbreaks in public drinking water systems—United States, 1971-2006 (N = 680).
SOURCE: CDC, unpublished WBDOSS data.

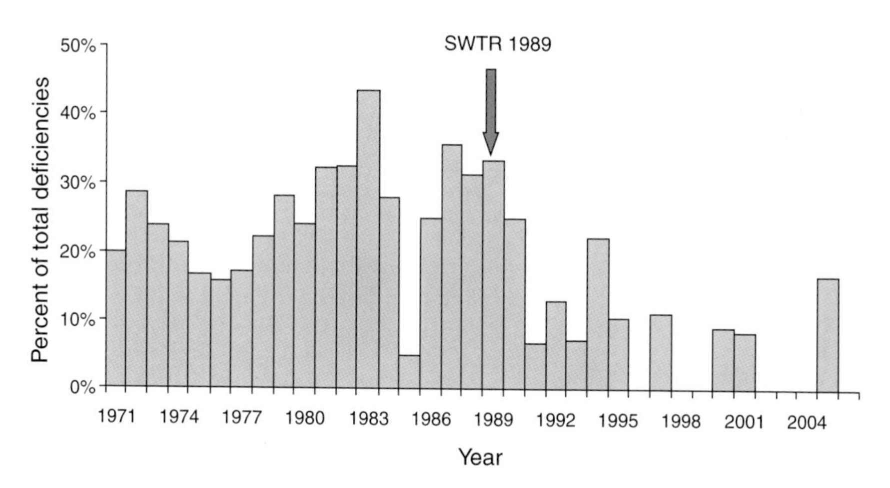

FIGURE 3-2 Proportion of deficiencies in public drinking water systems associated with untreated or improperly treated surface water—United States, 1971-2006. Deficiency = antecedent event or situation that results in exposure of persons to a disease-causing agent or agents. May be single or multiple deficiencies associated with each outbreak. SWTR = Surface Water Treatment Rule.
SOURCE: CDC, unpublished WBDOSS data.

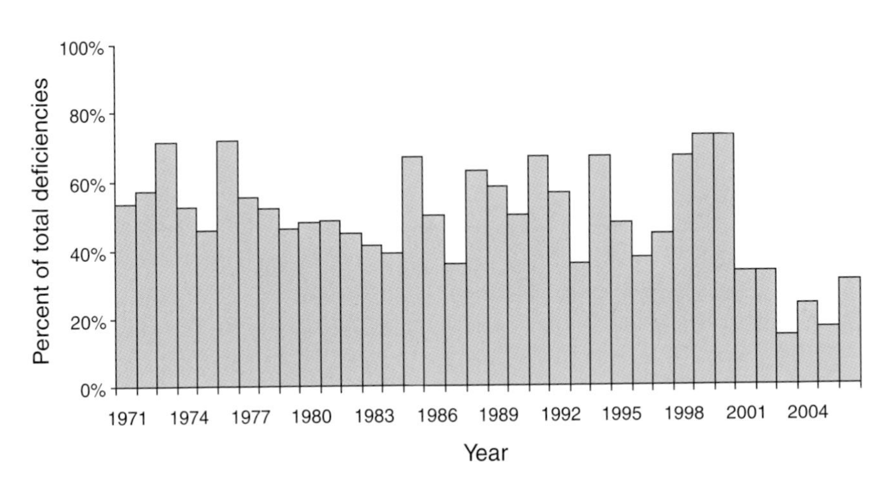

FIGURE 3-3 Proportion of deficiencies in public drinking water systems associated with untreated or improperly treated ground water—United States, 1971-2006. Deficiency = antecedent event or situation that results in exposure of persons to a disease-causing agent or agents. May be single or multiple deficiencies associated with each outbreak.
SOURCE: CDC, unpublished WBDOSS data.

related disease outbreaks have continued to be reported (Figure 3-3). CDC is hopeful that the Groundwater Rule,[9] published by EPA in the *Federal Register* in November 2006 (EPA, 2006), will eventually produce a decline in groundwater-related disease outbreaks similar to the results observed after enactment of the Surface Water Treatment Rule.

- Individual or unregulated water systems represent an important gap in waterborne disease prevention. Approximately 15.6 million households—about 12 percent of U.S. households—receive their water from private wells or small well water systems serving fewer than 25 people that are not regulated by EPA (U.S. Census Bureau, 2008). Although some of the smaller systems may be partially regulated by the state, private residential wells go unregulated. Generally, private well owners are not legally compelled to test or treat their drinking water, or to maintain the system to any standards. As a result, testing and maintenance schedules are likely to be less than optimal, resulting in increased vulnerability for well contamina-

[9]The purpose of the rule is to reduce disease incidence associated with disease-causing microorganisms in drinking water. The rule established a risk-based approach to target groundwater systems that are vulnerable to fecal contamination. Groundwater systems that are identified as being at risk of fecal contamination must take corrective action to reduce potential illness from exposure to microbial pathogens. The rule applies to all systems that use groundwater as a source of drinking water (for more information, see http://www.epa.gov/safewater/disinfection/gwr/regulation.html and EPA, 2006).

tion. The potential ramifications for the health of children drinking private well water recently prompted an American Academy of Pediatrics policy statement providing recommendations for inspection, testing, and remediation for wells providing drinking water for children (AAP, 2009). As Figure 3-4 demonstrates, an increasing proportion of reported waterborne disease outbreaks are associated with use of individual private wells.

- *Legionella* is a continuing threat. Although the outbreak that led to the identification of *Legionella* as a pathogen occurred in 1976 (Fraser et al., 1977), only Pontiac fever, primarily associated with hot tub exposure, was reported to WBDOSS until 2001. In 2001, the system began capturing data on outbreaks of Legionnaires' disease. In the latest surveillance summary (Yoder et al., 2008b), which covers drinking water–associated disease outbreaks reported from 2005 to 2006, half of all reported drinking water–related disease outbreaks were attributed to *Legionella* (Figure 3-5). Until *Legionella* outbreaks were included in the WBDOSS, AGI was the predominant type of illness associated with waterborne-disease outbreaks. *Legionella*, a thermophilic bacterium, colonizes and amplifies in premise plumbing systems (hot water heaters, taps, shower heads) as well as other sources of water (e.g., recreational hot tubs, cooling towers, etc.) and is transmitted by inhaling aerosols containing the bacterium (Fields et al., 2002). It can cause fatal pneumonia in vulnerable populations, such

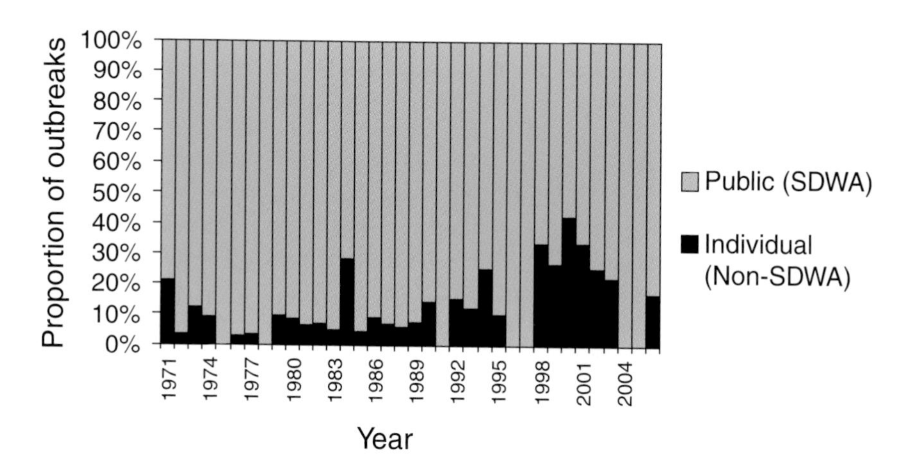

FIGURE 3-4 Percentage of waterborne-disease outbreaks in public and individual drinking water systems—United States, 1971-2006 (N = 762). Excludes 18 outbreaks occurring in multiple system types at the same time, bottled water, bulk water purchase, and unknown system types. SDWA: drinking water systems covered by the Safe Drinking Water Act; Non-SDWA: drinking water systems not covered by the Safe Drinking Water Act.
SOURCE: CDC, unpublished WBDOSS data.

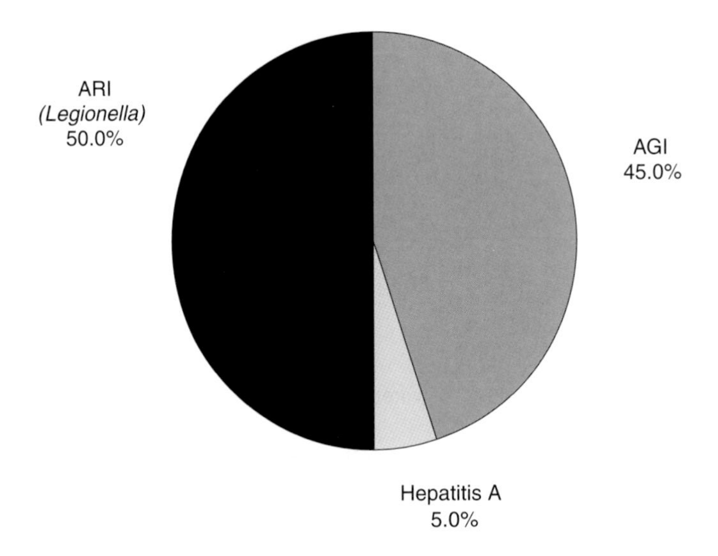

FIGURE 3-5 Percentage of waterborne-disease outbreaks associated with drinking water use, by illness and etiology—United States, 2005-2006 (N = 20); ARI: acute respiratory illness; AGI: acute gastrointestinal illness.
SOURCE: Yoder et al. (2008b).

as residents of health-care facilities and nursing homes. *Mycobacterium avium* also appears to be filling the same ecologic habitats colonized by *Legionella*, and it too has been associated with waterborne disease outbreaks (Falkinham, 2003). This underscores the need to focus on contamination of drinking water after it leaves regulated infrastructure, enters a building, or emerges at its point of use. Threats to drinking water safety from premise plumbing or public health challenges resulting from other uses of water (i.e., cooling towers, hot tubs) represent an important target for waterborne disease prevention.

Limitations of Waterborne Disease Surveillance

While the WBDOSS has been useful in elucidating the aforementioned trends and, therefore, highlighting areas of emerging public health need, this surveillance system has a number of critical limitations. First and foremost, the WBDOSS is a passive system based on outbreak reports from state and local public health agencies that do not necessarily actively track waterborne disease outbreaks. In many instances, waterborne disease outbreaks go unrecognized and therefore are neither investigated nor reported to CDC; thus, the WBDOSS provides, at best, an underestimate of waterborne disease outbreak occurrence and trends. Waterborne disease outbreaks tend to be reported more consistently by health departments with adequate resources for investigation; therefore, the geographic distribution of reported outbreaks is unlikely to be representative of the true geographic distribution of outbreaks.

Another shortcoming of WBDOSS is that it does not collect data on endemic waterborne disease. At this time, there are no reliable estimates of the total burden of disease for these illnesses in the United States. Several reviews of existing epidemiologic studies of drinking water use have produced preliminary estimates ranging from 4 million to 33 million cases of AGI per year that result from drinking water supplied by public water systems (Colford et al., 2006; Messner et al., 2006). However, these estimates need to be refined as better data and improved methods of estimating endemic waterborne disease become available. Furthermore, these estimates do not include the full scope of waterborne illness in the United States (e.g., illness other than AGI), illness in the 15.6 million households served by private wells (U.S. Census Bureau, 2008), and illness in the more than 55 million swimmers using recreational water six or more times a year in the United States (U.S. Census Bureau, 2009).

Emerging Challenges

Drinking Water Distribution System Infrastructure

Public drinking water systems in the United States supply 34 billion gallons per day of drinking water to approximately 87 percent of U.S. households (NRC, 2006; U.S. Census Bureau, 2008). The nation's drinking water infrastructure contains more than 50,000 community water systems that rely on water treatment plants and distribution systems that include over one million miles of pipes plus associated pumps, valves, storage tanks, reservoirs, meters, and fittings (NRC, 2006). Most of the pipes in these systems will reach the end of their expected lifespan within the next 30 years; some are already overdue for replacement, as attested by water main breaks and the appearance of sinkholes in cities across the United States. Water pipes typically are located on top of sewers and sewer pipes. When water mains break, there is potential for contamination of the entire system. Based on a simple Google search for the term "boil water advisory," hundreds to perhaps thousands of such incidents may occur in the United States each year. Such breaches in the water infrastructure result in costly repairs, increase the risk of water supply contamination, and impose a huge burden on local utilities and public health or environmental agencies. EPA has estimated that $276.8 billion will be needed over the next 20 years to repair and replace aging infrastructure, including $183.6 billion for transmission and distribution systems (EPA, 2005).

Although the total number of drinking water-associated disease outbreaks has declined, the proportion of outbreaks caused by deficiencies in public water distribution systems (before the building point-of-entry) has been fairly consistent over time, accounting for approximately 10 percent of drinking water outbreaks associated with public water systems since 1971 (CDC, unpublished data). However, this proportion would be expected to increase if infrastructure is not replaced on schedule. A recent epidemiologic study assessing the potential health risks associated with low pressure events in distribution systems found a relative risk of 1.58 (95 percent confidence interval: 1.1-2.3) for AGI following water main breaks or maintenance work in seven community water systems across Norway (Nygård et al., 2007). According to the 2006 National Research Council (NRC) report, *Drinking Water Distribution Systems: Assessing and Reducing Risks*, the water distribution system is the one remaining component of U.S. public water systems yet to be adequately addressed to reduce waterborne disease outbreaks; thus, the committee recommended conducting epidemiological studies that specifically target the distribution system component of waterborne diseases (NRC, 2006).

The Total Coliform Rule (TCR), which was intended to protect distribution systems by testing for coliform bacteria, has been recently revised by EPA (2009a). Research and information collection priorities with respect to drinking water distribution systems in the revised Total Coliform Rule/Distribution System Advisory Committee (TCRDSAC) include the following (EPA, 2009b):

- Understand the role of cross-connections and backflow in system contamination.
- Learn how storage facility design, operation, and maintenance can influence distribution systems and lead to contamination.
- Identify the best methods for installing and repairing water mains in order to reduce contamination.
- Assess the role of intrusions due to pressure conditions and physical gaps in causing contamination.
- Understand the significance of biofilm formation as an agent of waterborne disease, and identify effective controls for biofilm and microbial growth in water distribution systems.
- Study the role of nitrification in promoting bacterial blooms that contaminate water systems.
- Determine safe methods for removing scale and sediment from drinking water distribution system components.

Clearly the potential health impact of not replacing the nation's drinking water infrastructure is high; investing funds over the next 20 years to upgrade and replace aging infrastructure is a public health imperative.

Climate Change, Severe Weather Events, and Safe Water Availability

The Arctic is the "canary in a coal mine" of climate change; the 4 million people in the Arctic are already experiencing the dramatic effects of rising temperatures (Arctic Council and IASC, 2004). These include thawing permafrost and reduced pack ice that has led to massive coastal erosion, damaged water and wastewater infrastructure (some water systems have been washed away), breached waste lagoons, saltwater intrusion into water supplies, and increased organic loads in source water that challenge water treatment procedures. Habitat changes (e.g., increased ocean temperature, northern treeline movement) are projected to influence the distribution of zoonotic diseases that may impact water sources and human health (e.g., northerly beaver migration and giardiasis). Increasing ocean temperatures also seem tied to a recent *Vibrio parahaemolyticus* AGI outbreak associated with oyster consumption: this extended by 1,000 km the northernmost documented source of oysters contaminated with *Vibrio parahaemolyticus* (McLaughlin et al., 2005).

Extreme weather events such as hurricanes, floods, and drought have debilitated water treatment and distribution systems in many parts of the United States and elsewhere around the world, limiting residents' access to safe water. Extensive flooding in the midwestern United States in 1993 resulted in widespread and long-term contamination of wells in nine states (CDC, 1998). In 2008, severe drought limited water availability from Georgia's Lake Lanier, the main source of drinking water for the city of Atlanta. Severe weather (e.g., heavy precipitation, floods) appears to also be associated with increased reporting of waterborne outbreaks (Curriero et al., 2001), and the potential impact of severe weather on waterborne disease is part of a larger discussion about potential climate change impacts on public health (Patz et al., 2000, 2008; Rose et al., 2001).

One means by which the increasing impact of climate change and severe weather events on water availability can be addressed is by reusing wastewater. Indirect potable reuse—so-called "toilet-to-tap"—is not a popular option with the public. However, it is being utilized in places like Orange County, California, where up to 70 million gallons of highly treated wastewater per day are being injected into a groundwater aquifer to serve as a barrier to saltwater intrusion and to augment groundwater supplies needed for provision of municipal drinking water. In Gwinnett County, Georgia, a pipeline will soon deliver 60 million gallons of highly treated wastewater per day into drought-depleted Lake Lanier, Atlanta's main water supply.

The majority of water used in households does not flow through toilets but through showers, bathtubs, clothes washing machines, and sinks. If plumbing systems were designed to separate this "graywater" from wastewater (i.e., toilets), graywater derived from these sources would be likely to have a low risk of pathogen transmission and could be "recycled" for nonpotable household uses such as irrigation, exterior washing, and toilet flushing. Public awareness and education about water reuse is key to improving public perception and understanding about the range of water reuse options that will likely play a bigger role in supplying both potable and nonpotable water in the coming decades.

With increasing regional drought conditions, emerging conflict surrounding water rights and access, and competing demand for scarce water supplies, water reuse decisions are going to be a reality for more places in the future. To prepare for these decisions, we need new epidemiologic studies and methodological approaches to accurately measure the health effects of water reuse so that a scientific knowledge base is available for use by environmental, public health, and government officials. These decisions must balance the objectives of increasing water availability and protecting public health while being transparent to the general public.

Recreational Water Use

Perhaps lessons learned in preventing drinking water–associated disease can guide public health and the aquatics sector in combating dramatic changes in recreational water–associated disease in the United States. While the incidence of the drinking water–associated disease outbreaks has decreased over the last several decades (Yoder et al., 2008b), WBDOSS reporting has demonstrated a dramatic increase in the number of AGI outbreaks attributable to recreational water use in the United States (Figure 3-6). This increase is primarily due to an increased number of outbreaks associated with use of public swimming pools and other disinfected venues such as water parks and interactive fountains. The leading etiologic agent, *Cryptosporidium*, is extremely chlorine resistant, which allows the parasite to bypass the primary barrier to protecting swimmers from pathogens,

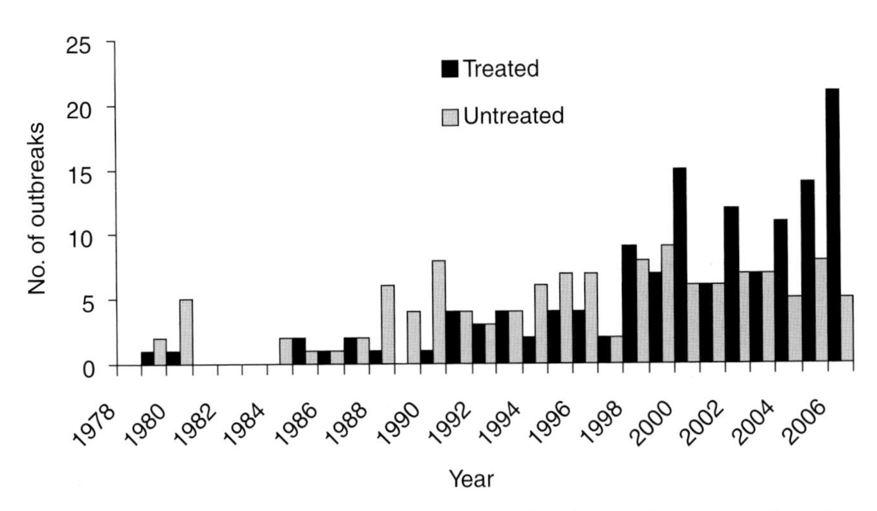

FIGURE 3-6 Number of recreational water-associated outbreaks of acute gastrointestinal illness (n = 259), by water type and year—United States, 1978-2006.
SOURCE: Yoder et al. (2008a).

chlorination (Korich et al., 1990; Shields et al., 2008). As a result, the parasite accounted for 68.3 percent of disinfected venue-associated outbreaks from 1997 to 2006 (Yoder et al., 2008a). In addition, enteric infections such as *Salmonella* and *Shigella* have also been linked to aquatic recreational activities despite traditional thinking about other exposures (e.g., food, child care centers; Denno et al., 2009). In natural, nondisinfected water (lakes, rivers, marine beaches), no barriers to transmission of pathogens are used. As a result, multiple studies from around the globe indicate that natural, nondisinfected water swimmers are at increased risk for AGI compared to nonswimmers (Prüss, 1998; Wade et al., 2006, 2008). Other low incidence but severe outcome recreational water-associated infections such as those caused by *Naegleria fowleri* (primary amoebic meningoencephalitis; CDC, 2008a; Visvesvara et al., 1990) and *Vibrio* species (Yoder et al., 2008a) have also been reported to CDC. In the United States, waterborne disease associated with recreational water use likely results in a high level of morbidity. These challenges require a concerted effort to conduct sound research, develop and evaluate appropriate interventions and strategies that address emerging issues such as chlorine-resistant microbes, and create a science-based regulatory framework that promotes healthy and safe recreational water use.[10]

Summary and the Path Forward

Data suggest that the epidemiology of waterborne disease is changing in the United States. WBDOSS has documented decreasing surface water outbreaks, likely the result of drinking water regulation; increasing numbers of recreational water outbreaks, particularly associated with the chlorine-resistant parasite *Cryptosporidium*; the increasing importance of outbreaks associated with groundwater and private systems; the continued importance of *Legionella* as a pathogen and premise plumbing issues; and the ramifications of severe weather and potential climate change. New complexities have arisen that include an aging drinking water infrastructure, severe weather effects, water recycling, and recreational water use.

To our detriment, those of us who study waterborne disease tend to compartmentalize water. We attend drinking water meetings, recreational water meetings, water conservation meetings, and so on, but we too often fail to look at water-

[10]All regulation of public pools in the United States is set at the state or local level. The lack of a uniform federal standard for public pool design, construction, operation, and maintenance has been recognized as a challenge for prevention of recreational water illnesses and injuries. Consequently, CDC is sponsoring a national effort, currently led by New York State, to develop the national Model Aquatic Health Code (MAHC). The objective of this knowledge-based and scientifically supported model health code is to reduce injuries and the transmission of pathogens at disinfected aquatic facilities by making the best available standards and practices available for voluntary adoption by state and local health agencies. More information is available at http://www.cdc.gov/healthyswimming/MAHC/model_code.htm.

borne disease *in toto*. This challenge could be met by expanding discussions to include partners across the waterborne disease and water use spectrum. These discussions should involve all water uses (e.g., drinking, recreational, etc.), non-enteric as well as enteric disease, chemical challenges, water quantity as well as quality issues, and the implications of climate variability (e.g., drought, flood). This "One Water" concept would strengthen our ability to address the full magnitude and burden of waterborne disease and better reflect the full scope of effects of water on public health.

In order to address the threat of waterborne disease over the long term, it will be essential to assess all types of water use, study the impact of both microbial and chemical contamination, and investigate the sources of human- and animal-specific contamination. This will require multiyear intervention studies, laboratory methods development, and health-effects studies, involving state, local, and federal agencies, as well as academic researchers in the United States. FoodNet,[11] an active surveillance system for foodborne illness in the United States, provides a model for such an effort, which could be used for designing a public health network for addressing key national questions related to waterborne disease. The development of "WaterNet" with foundational and sustained funding could create or renew key strategic partnerships with multiple groups across a spectrum of disciplines, including epidemiology, clinical medicine, microbiology, laboratory, engineering, environmental and behavioral sciences, and health communications. Such a diversity of expertise will be necessary to identify and tackle key issues to reduce the burden of waterborne disease. The first step toward this goal should be to produce an accurate estimate of the overall waterborne disease burden, to be used as both a benchmark and an advocacy tool for resource allocation and priority setting. The specific objectives of "WaterNet" should be the following:

- Determine the size of the problem.
  - Include all water types and uses (drinking, recreational, other) and both enteric and nonenteric illness.
  - Assess the prevalence of waterborne pathogens and chemicals in small water systems and premise plumbing.
- Increase investment.
  - Invest in building a science-based foundation of data for decision making.
- Enhance disease detection.
  - Improve detection, investigation, and reporting.
- Measure health effects of waterborne contaminants.
  - Invest in research on the health effects associated with drinking water infrastructure.
  - In particular, study chronic effects of chemical exposure.

[11] See http://www.cdc.gov/FoodNet/.

- Improve contaminant detection and removal.
  - Improve laboratory and epidemiologic capacity to sample, detect, and characterize waterborne contaminants in humans, animal reservoirs, and the environment.
  - Improve environmental tracking of contaminants.
  - Develop and evaluate methods for removing or inactivating contaminants.
- Promote access to water-related public health information.
  - Improve consumer knowledge and understanding of waterborne disease and prevention.
- Develop prevention plans.
  - Test appropriate prevention interventions.
  - Develop and test adaptive strategies for water scarcity issues including conservation and reuse.
  - Develop sound, science-based public health policy to prevent waterborne disease.

No natural resource is more fundamental to public health than water, and in the United States—where we are fortunate to have one of the safest water supplies in the world—we too often take it for granted. This paper has outlined some of the current water-related public health issues we face, but in the coming decades we will confront new and more intractable water and public health challenges. These include drought, decreased water availability and deteriorating water quality, aging water infrastructure, climate change impacts, chemical contamination, and the potential emergence of newly identified waterborne disease pathogens, which may be more difficult to combat because of chlorine or drug resistance. Preparing our public health system to address these challenges, as would occur if the proposed "WaterNet" platform were implemented, would provide essential information needed to prevent waterborne disease and strengthen public health protection of U.S. water resources in the future.

HEALTH, CLIMATE CHANGE, AND WATER QUALITY

Joan B. Rose, Ph.D.[12]
Michigan State University

Introduction

The Intergovernmental Panel on Climate Change (IPCC) has predicted that global climate change will increase the threat to human health, ecosystems, and

[12]Homer Nowlin Chair in Water Research, Michigan State University, 13 Natural Resources, E. Lansing, MI 48824. Phone: 517-432-4412; Fax: 517-432-1699; E-mail: rosejo@msu.edu.

socioeconomic conditions (IPCC, 2007). Direct human health impacts are likely to be associated with increased extremes in the weather and climate patterns (e.g., temperature and precipitation) and disasters such as weather phenomena including floods, hurricanes, and tornados. But the role of climate change in the spread of infectious diseases is one area of human health which, due to the complex interactions between the environment (e.g., land and water) and people, and the high variability, is not well understood. Waterborne diseases, in particular, are influenced by distinct climate and weather factors that affect water pollution at water basin and watershed scales. In an attempt to improve our understanding and mitigate the potential for disease spread at a reasonably large geographic scale and at the community level, we will need to investigate further the relationships between health, climate, and water.

Water resources and water-related disasters are key areas of concern in regard to climate and health. The recent IPCC report focused on the changes that might be expected, the causes, the projected effects, the need for adaptation and mitigation, and finally provided a long-term view of predictions. The relationship between climate and water resources has undergone extensive assessment (Roberson et al., 2008) and the key climate predictions that will likely affect water include:

- increased warming over land and at most high northern latitudes;
- contraction of snow cover;
- increased frequency of hot temperature extremes, heat waves, and heavy precipitation;
- precipitation increased at high latitudes and decreased at most subtropical land regions;
- increased annual river runoff and water availability at high latitudes and decreases in some dry regions in the midlatitudes and the tropics; and
- decreased water resources in many semi-arid areas (western United States).

In addition, as a result of the increase in atmospheric concentrations of greenhouse gases, the frequency, intensity, and duration of extreme weather events is predicted to change. For example, more hot days, heat waves, and heavy precipitation events are expected. The risks of floods and droughts in many regions would increase. Global annual precipitation is also projected to increase, although both increases and decreases in annual precipitation are projected at the regional scale. One of the necessary next steps beyond the assessment of extremes and water quantity changes is linking these changes to changes in water quality.

Our ability to understand the effects of climate fluxes on the changes in water quality, which may in turn affect the likelihood of waterborne diseases, moves the field closer toward building predictive approaches that can be used for science and evidence-based management strategies. The research on water quality-health-climate interactions has focused on three main lines of evidence:

(1) human disease and climate, (2) changes in fecal indicators (pollution) in water associated with climate factors, and (3) changes in actual pathogen loading. And finally, research has begun to address adaptation and mitigation, primarily focused on water treatment.

Human Disease and Climate: Waterborne Diseases

Some of the strongest evidence now emerging is the association of precipitation as an extreme event with drinking waterborne disease outbreaks in the United States and Canada. About 50 percent of the outbreaks in the United States were statistically associated with extreme rainfall (at the 95 percentile compared to the 25-year average) and in Canada, rainfall events greater than the 93rd percentile were associated with outbreaks at a relative odds factor of 2.283 (95 percent [CI] = 1.216-4.285; Curriero et al., 2001; Thomas et al., 2006). Both surface water (Milwaukee, WI) and ground waters (Walkerton, Ontario; Put-In-Bay, Ohio) have been involved, and temporal differences have been suggested to be linked to the transport phenomenon (Auld et al., 2004; Curriero et al., 2001; Fong et al., 2007; Mac Kenzie et al., 1994).

Globally, extreme rainfall remains the largest cause of both direct and indirect effects on human health and well-being. Flooding is the most frequent natural weather disaster (causing 30 to 46 percent of natural disasters in 2004-2005), affecting over 70 million people worldwide each year (Hoyois et al., 2007). The most common illnesses associated with floods described in the literature are diarrhea, cholera, hepatitis (jaundice), leptospirosis, and typhoid. Unusual illnesses such as tetanus have also been reported. The etiological agents identified include *Cryptosporidium spp.,* hepatitis A virus, hepatitis E virus, *Leptospira spp., Salmonella typhi,* and *Vibrio cholera.* Cholera in particular has been directly associated with flooding in Africa.

In 1998, flooding in West Bengal was followed by a severe outbreak of cholera with 16,000 cases (Sur et al., 2000). Studies of flooding in Bangledesh in 1988, 1998, and 2004 confirmed that cholera was the most prevalent pathogen associated with flooding—increasing by almost 20-fold (Schwartz et al., 2006). Also increasing were other fecal-oral pathogens including rotavirus *Shigella, Salmonella,* and *Giardia* (Schwartz et al., 2006). In Central America, Hurricane Mitch affected a large geographical area incluing Nicaragua, Guatemala, Honduras, El Salvador, and Belize. An estimated six feet of rain drowned crops, leading to food shortages. Gastrointestinal and respiratory diseases were rampant throughout the affected area. In Guatemala, the country most affected by cholera, the average weekly number of cases before Hurricane Mitch (January-October 1998) was 59, whereas after Hurricane Mitch (November 1998) the average number of cases per week was 485 (Figure 3-7; PAHO, 1998).

There are many reports of devastating "outbreaks" associated with flooding. On May 5, 2005, outbreaks of waterborne diseases were reported as the death toll

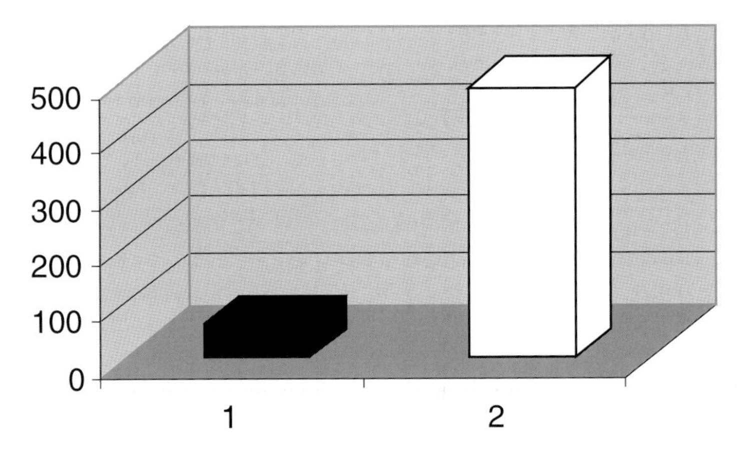

FIGURE 3-7 Cholera cases (1) pre and (2) post Hurricane Mitch in Guatemala in 1998.
SOURCE: Based on data in PAHO (1998).

rose into the hundreds in southeastern Ethiopia after flooding and heavy rains. In March 2006, a cycle of drought and flood in Malawi's southern and central regions was reported to aggravate a cholera outbreak with many deaths reported. "Malawi is dealing with three crises at the same time: food shortages, floods, and cholera," said Dr. Eliab Some, who heads UNICEF's health and nutrition team in Malawi (IRIN News, 2006). Over 4,000 cases of cholera, a disease associated with poor sanitation, lack of hygiene, and lack of access to potable water, have been recorded over the past three months, mostly in Malawi's southern region.

During the Tsunami of 2004, in the case of Aceh Province, many communities reported diarrhea as the main cause of morbidity (85 percent of the cases were in children under five years of age), but neither increases in mortality nor outbreaks of cholera or other potentially epidemic diseases were reported (Brennen and Rimba, 2005). In some towns, more than two-thirds of the population died at the time of impact, almost 100 percent of homes were destroyed, and 100 percent lacked access to clean water and sanitation. To a large extent, the Australian army and other groups are to be credited with rapidly deploying environmental health teams to swiftly implement public health measures, including provision of safe drinking water, proper sanitary facilities, and mosquito control measures. Widespread fecal pollution of the surface waters was shown, yet the saltiness of the potable water supply after the disaster made much of the water unpalatable. Wells were vulnerable, perhaps to other etiological agents of fecal origin including viruses, and *Shigella,* with greater probability of infection than *Vibrio*, thus leading to the widespread diarrhea.

After Hurricane Katrina, on the U.S. Gulf Coast, the lack of what was perceived or anticipated as an epidemic disease (thousands of cases) may have been due to inadequate assessment and reporting (it was acknowledged that the public health response was grossly inadequate), dilution of sewage wastes, and rapid die-offs of pathogens of concern from the sewage-laden waters (due to the salinity and high temperatures). However, to imply that no infectious disease outbreaks occurred would be far from the truth. *Vibrio vulnificus* caused at least 22 wound-associated infections and killed five people (CDC, 2005a). In addition, a norovirus outbreak occurred among Katrina evacuees in Houston, Texas (CDC, 2005b). Disease reports of diarrheal illness steadily increased and jumped from 3 per 1,000 resident-days to 20 per 1,000 resident-days, within 5 days of exposure in the effected area (for evacuees in shelters). Reports of diarrheal illness peaked at 43.5 cases per 1,000 resident-days after an outbreak in a shelter was reported after day 7. Thus, there is much speculation about how much was or was not due to exposure to contaminated floodwaters (Cookson et al., 2008).

Water Quality Changes in Fecal Pollution Indicators and Pathogens Associated with Climate

In order to build an understanding of the public health risks associated with climate, the development of models of water quality changes are needed. However, initial studies have been without appropriate resolution in time, space, or hazard identification, and thus an adequate characterization of the quality, transport, fate, and variability of the microbial hazards is limited. There is no doubt that storm waters are laden with *Escherichia coli* and other fecal indicator bacteria; however, often the studies do not help to address risk of infectious disease transmission. Studies that provide good quantitative data at adequate spatial and temporal scales can begin to elucidate the interconnections between sources of the contaminants and risk of human exposure. Initial studies on El Niño in Florida demonstrated that over a 20-year time frame the impacts of such a climate signal on degrading water quality (based on fecal coliform indicator data) could be ascertained for the winter, spring, and fall months only (as the seasonal summer rains diminished any discernable change in either precipitation, river flows, and water quality; Lipp et al., 2001b). In addition, during the 1998 El Niño, increases in all fecal indicators, as well as virus pollution from presumably septic tanks, were related to the El Niño signal, which increased flow in two main rivers, impacting the Gulf of Mexico in Charlotte Harbor, Florida (Figure 3-8; Lipp et al., 2001a). Human viruses were only detected at the marine sites during this event. The other factor identified in this degradation of water quality associated with fecal pollution was water temperature. Thus, virus survival and dispersion were enhanced by low water temperatures and rainfall or river flow, respectively.

Kistemann et al. (2002) have shown that not only do fecal indicators increase but the waterborne parasites *Cryptosporidium* and *Giardia* also increased significantly with rainfall and, thus, microbial loading associated with key rain events could be determined for the watershed. Interestingly enough, their assessment

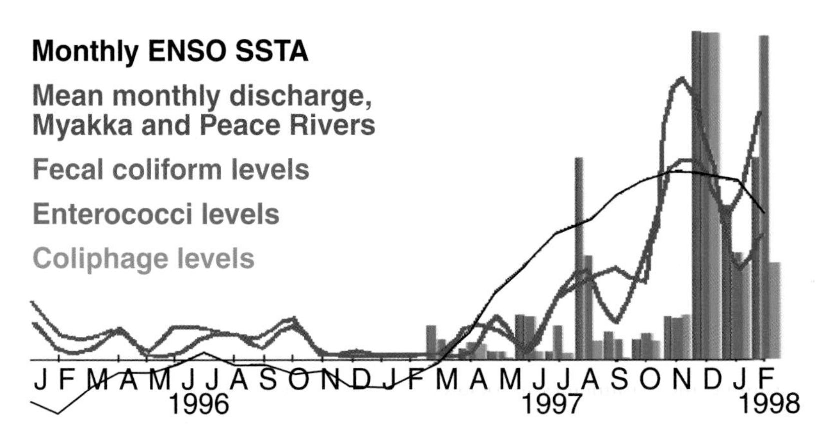

FIGURE 3-8 Changes in water quality associated with septic tanks and the 1998 El Niño. SOURCE: Based on data in Lipp et al. (2001a).

also demonstrated that the sources of the pathogens need to be present in the watershed for the predictions to hold true. While this seems obvious many scientific investigations fail to address the variability in pathogen types and concentrations in relationship to the sources, the climate event, and the disease outcomes. For example, studies after Hurricane Floyd showed significant water quality impacts via fecal indicators and in areas with high concentrations of pig farms that illnesses associated with specific etiological agents such as Adenovirus, *Cryptosporidium, Giardia, Toxoplasma,* and *Helicobacter* were not increased after the flooding even though unidentified illnesses did increase (from 5.1 to 11 outpatient visits per month). Key zoonotic pathogens that might have been present in the hogs, such as *Salmonella* and *Campylobacter*, were not investigated (Setzer et al., 2004). Although one might expect *Cryptosporidium* to be zoonotic, this disease is primarily associated with very young animals and more often with calves and lambs.

While progress is being made on modeling of nitrates and other chemical constituents in floods and under key environmental conditions including land-use models, more studies are need on pollution-associated microbial infectious agents. The best work to date focuses on the naturally-occurring *Vibrios* associated with plankton in coastal waters (Lipp et al., 2002). These types of ecologically-based models can also be useful for pathogens; however, much more data and information will be needed for the future.

Quantitative Microbial Risk Assessment Approaches

It is clear that the poor and developing countries will suffer the most severe consequences of climate and health changes. Here, I would suggest that sewage and human fecal pollution is the source; that climate is the driver of exposure; and, thus, one might hypothesize that the wastewater and drinking water infrastructure, the types of pathogens at any given time in the community wastewater, and the characteristics of the flooding would drive the risk of disease. In order to examine the parameters associated with the risk of disease transmission, a Quantitative Microbial Risk Assessment (QMRA) approach can be used for an extreme flooding event in a developing country where untreated sewage is mixed with water supply. This uses a standard procedure of hazard identification, exposure assessment, dose response relationship, and risk characterization (Haas et al., 1999).

The pathogens *Vibrio cholerae*, *Salmonella typhi*, *Cryptosporidium parvum*, hepatitis A virus, and rotovirus were considered for the assessment (T. Shibata and J. B. Rose, QMRA for floods and disease, personal communication). Probability of infection models were used (Haas et al., 1999) and assumptions based on published data on (1) occurrence in wastewater, (2) survival in the environment, (3) dilution during flooding, and (4) rates of ingestion were used to undertake a risk estimate. Crystal Ball®[13] was used to develop a Monte Carlo simulation and to explore uncertainty analysis in health risk assessment. Figure 3-9 shows the preliminary outputs of the risk assessment. The high risk estimates for the viruses and parasites and the flat line over the first week demonstrate the impact of the high potency (infectivity) of these microbes and their enhanced survival compared with *Vibrio*, which has a greater potency than *Salmonella.* It should be noted that risks above 10^{-1} could be generally notable as outbreaks or increased rates of illness, while risks below that would likely only be observed with large epidemiological studies. This analysis suggests that higher risks could extend out 10 to 30 days for viruses and parasites compared to bacteria.

Based on the literature, sewage concentrations for *Vibrio* are highly variable and were reported less so for hepatitis and *Cryptosporidium.* Thus, dilution played a larger role in defining the uncertainty of the risk for those pathogens. Over time, depending on the decay rates used, die-off in the environment played a significant role in defining the risk.

Use of a QMRA begins to demonstrate the need for critical information to inform the relative level of risk. One can also begin to examine the relative

[13]Crystal Ball® (Oracle Corporation, Redwood Shores, CA) is a statistical software package in which probabilistic simulations can be developed and run.

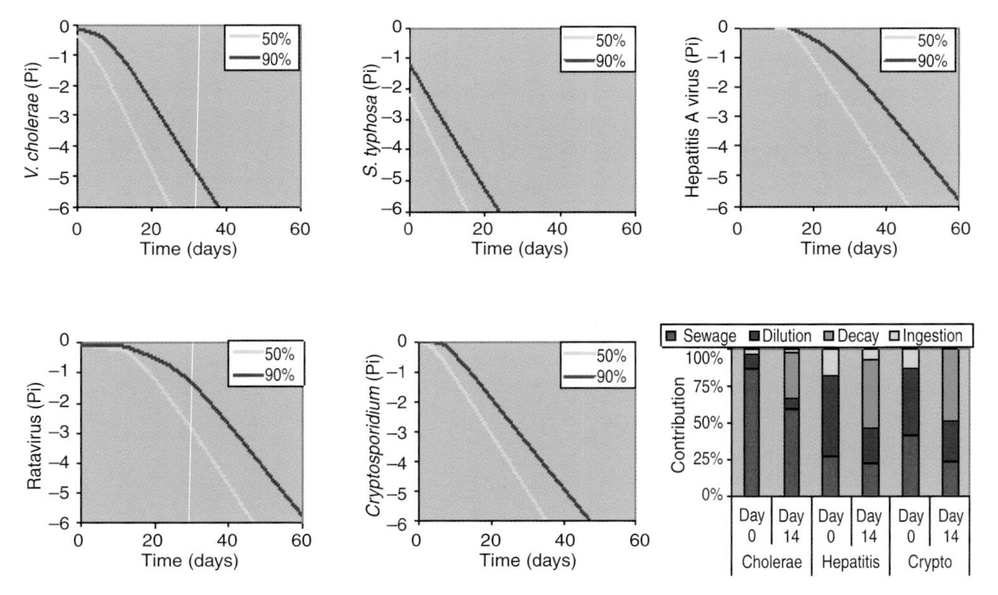

FIGURE 3-9 Probability of infection from 10^{-1} to 10^{-6} for five pathogens over 60 days with 50th (yellow line) and 90th (blue line) percentiles of risk displayed. Bar graph demonstrates the uncertainty analysis associated with sewage concentrations of the pathogens (red bar), dilution (blue bar), decay or survival (green bar), and ingestion rates for cholera, hepatitis, and *Cryptosporidium* at Day 0 and Day 14.

reduction in risk if interventions are used during flood events where sewage contamination is suspected.

Critical Research Needs

Schwartz et al. (2006) are to be commended for specifically defining the terms *epidemic* and *flooding* during their study; most do not. The term *outbreak* may or may not be used, regardless of the definition (greater than one case of disease from a common exposure, including venues [e.g., cruise ships, nursing homes, restaurants]; source [e.g., water, food, or an event]; picnic, flood). For example, the increase in *Vibrio* cases following Katrina certainly met this definition; however, the CDC reported that no outbreaks had occurred. Floret et al. (2006) have titled their paper "Negligible Risk for Epidemics After Geophysical Disasters" and they describe 26 large disasters, including 22 earthquakes, 2 volcanic eruptions, and 2 tsunamis. Of these 26 disasters, 19 percent reported either diseases such as giardiasis, pneumonia, Hep A and E respiratory, and/or diarrhea, and 8 percent reported no outbreak. The remaining 73 percent had no report. Of the more than 600 records, very few reported on infectious disease and of 233 articles retrieved from Medline, only 18 (7.7 percent) reported on infectious disease data collection. Epidemics and outbreaks were not defined and "negligible" was not described based on values, economics, or any other measure. Many have suggested that this implies we have overexaggerated the risks and others have suggested that this means we have improved public health response during disasters, while others have argued that there has been inadequate surveillance and reporting.

As seen in Figure 3-10, the risks associated with flooding or drought are far greater than other types of disasters. Therefore, it is clear that climate change must be considered; and given the nature of water, its importance and its widespread geographic impacts, the role of water quality and health must be addressed. Some of the critical needs that must be met in order to predict the effect of climate change on waterborne disease are

- better knowledge of disease incidence and pathogen excretion;
- better assessment of concentrations of pathogens in sewage and other sources;
- better assessment of the vulnerability of pathogen sources (e.g., combined sewer overflows versus septic tanks);
- better monitoring of sewage indicators to gather source, transport, and exposure information (event monitoring), and monitoring of sediments and other reservoirs; and
- more quantitative data for risk assessment.

We will not be able to address the complexities of climate impacts on our water and our health without obtaining better health surveillance information and

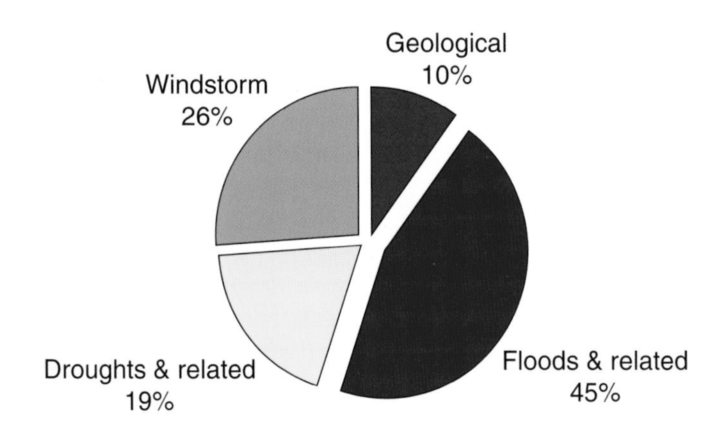

FIGURE 3-10 Percentage of disasters by type, 2000-2004 averages.
SOURCE: Based on data in Hoyois et al. (2007).

water quality data that are temporally and spatially relevant. Thus, as we move forward in investing our own water infrastructure in the United States and develop serious programs to work at the global level, we should also invest in gathering the information and knowledge that will allow us to make informed decisions to assist in adaptation and mitigation strategies.

QUANTITATIVE MICROBIAL RISK ASSESSMENT OF WATERBORNE DISEASE

Kelly A. Reynolds, M.S.P.H., Ph.D.[14]
University of Arizona

Kristina D. Mena, M.S.P.H., Ph.D.[15]
University of Texas-Houston

The Risk Assessment Paradigm

Risk analysis involves (1) the use of risk assessment—characterizing a hazard qualitatively and quantitatively; (2) risk management—incorporating the scientific assessment data with cultural values and ethics to determine appropriate

[14]Environmental and Occupational Health, Mel and Enid Zuckerman College of Public Health, Tucson, AZ 85724.

[15]Public Health, El Paso, TX 79902.

control actions; and (3) risk communication—involving multiple stakeholders in an open discussion of the issues and opportunity to exchange information learned from the risk assessment and control options. Risk analysis is a controversial process where the social importance of specific impacts is not congruent across populations or among individuals. In contrast, risk assessment is a quantifiable concept. Using statistical concepts and mathematical modeling tools, estimates of risk can be related to a range of defined variables of exposure and dose response. Risk and uncertainty can therefore be quantitatively estimated, ultimately leading the risk characterization in populations and individuals.

The basic risk assessment paradigm, published by the National Research Council (1983), provides a framework for the consistent application of risk analysis across multiple disciplines. These specific guidelines were intended to provide consistency to priority setting and the overall regulatory process of the federal government. Through the process of hazard identification, dose-response assessment, exposure assessment, and risk characterization, inferences become science-based and more credible with defined limits and measurable control benefits (Table 3-1).

Although the steps in risk assessment are well defined, less certain is the accuracy of the assumptions we make in the process to estimate data where gaps exist. The need to standardize the risk assessment approach in making quantitative assumptions, while minimizing uncertainties, is apparent across multiple disciplines.

TABLE 3-1 National Research Council Risk Assessment Paradigm

Hazard identification	Identification of hazards, incident scenarios, potential consequences, and agent properties, including factors of virulence, adaptation, resistance, and mutation. Includes consideration of acute and chronic health effects, sensitive populations, and individual immunity.
Dose-response assessment	Characterization of the relationship between dose and incidence of adverse effect in populations exposed to hazards.
Exposure assessment	Estimation of the frequency, amount, and duration of human exposures to agents via determined routes and the determination of the size and nature of the population exposed. Involves consideration of temporal and spatial exposure along with changes in microbial populations.
Risk characterization	Integration of information from hazard identification, dose-response assessment, and exposure assessment to estimate the magnitude (possibilities, probabilities, impacts) of health effects. Aided by tools in mathematical modeling and distribution analysis.

SOURCE: NRC (1983).

Waterborne Disease Estimates

Outbreaks of disease from drinking water are not common in the United States, but they do still occur and can lead to serious acute, chronic, or sometimes fatal health consequences, particularly in sensitive and immunocompromised populations. If properly applied, current protocols in municipal water treatment are effective at eliminating pathogens from water. However, inadequate, interrupted, or intermittent treatment has repeatedly been associated with waterborne disease outbreaks. Drinking water outbreaks exemplify known breaches in municipal water treatment and distribution processes and the failure of regulatory requirements to ensure that water is free of human pathogens.

Numerous epidemiological studies have been conducted over the years to evaluate the role of drinking water in human illness; however, these studies are often criticized for their failure to consider all of the dynamic and complex interactions of the water treatment and distribution process relative to human exposures and eventual health endpoints. Risk assessment allows for the consideration of a wide range of exposure and impact scenarios but lacking is a defined and standard approach across the discipline. Proper study design is of the utmost importance for assuring the development of appropriate and effective policy and regulation.

Drinking Water Contamination

Waterborne disease outbreak data have been collected from the CDC, EPA, and CSTE since 1971. From 1971 to 2006, 841 documented outbreaks were associated with drinking water resulting in 578,829 cases of illness and 87 deaths (Blackburn et al., 2004; Calderon, 2004; Liang et al., 2006; Yoder et al., 2008); however, the true impact of disease is estimated to be much higher. Between 2005 and 2006, 28 outbreaks associated with drinking water were documented in the United States. These outbreaks, caused by human pathogens (including protozoa, viruses, or bacteria) and chemicals or toxins, resulted in 612 documented illnesses and 4 deaths (Yoder et al., 2008).

According to a CDC survey, cross-connections and back-siphonage caused the majority (51 percent) of outbreaks linked to the distribution system from 1971 to 2000, followed by water main contamination (a collective 33 percent) and contamination of storage facilities (16 percent). Data compiled by EPA indicate that only a small percentage of contamination from cross-connections and back-siphonage are reported and that the CDC data underreport known instances of illnesses caused by backflow contamination events. For example, from 1981 to 1998, only 97 of 309 (31 percent) documented incidents were reported to public health authorities (AWWSC, 2002). Of the 97 reported incidences, 75 (77 percent) reported illnesses (4,416 estimated cases); however, only 26 (27 percent) appear in CDC's summaries of waterborne disease outbreaks (reviewed in Reynolds et al., 2008).

National surveys and water utility databases are continuously growing. Recently, for example, several national surveys have documented evidence of viruses in groundwater (reviewed in Reynolds et al., 2008). The newly promulgated Ground Water Rule requires that sanitary surveys are conducted by December 31, 2012, for most community water systems (CWSs) and by 2014 for CWSs with outstanding performance and for all non-CWSs to help identify deficiencies that may lead to impaired water quality. Source water monitoring for indicator microbes, corrective actions for systems with significant deficiencies or source water fecal contamination, and compliance monitoring are further required. Using a risk-based approach, the EPA targeted groundwater sources at greatest risk of contamination rather than requiring all public water systems with a groundwater source to disinfect. The Ground Water Rule is estimated to reduce waterborne viral illnesses by approximately 42,000 cases each year (23 percent reduction from the current baseline estimate).

Role of the Water Distribution System

The distribution system is a complex network of pumps, pipes, and storage tanks that deliver finished water to end users (reviewed in Reynolds et al., 2008). There are approximately one million miles of distribution system networks in the United States and an estimated 154,000 finished water storage facilities with more than 13,000 miles of new pipes installed each year (AWWA, 2003; Grigg, 2005; Kirmeyer et al., 1994). Approximately 26 percent of the distribution pipes in the United States are in poor condition and the annual number of documented main breaks has significantly increased from about 250 in 1970 to 2,200 in 1989 (AWWSC, 2002). It is estimated that even well-run water distribution systems experience about 25 to 30 breaks per 100 miles of piping per year (Deb et al., 1995). Using a value of 27 main breaks per 100 miles per year, Kirmeyer et al. (1994) estimated 237,000 main breaks per year in the United States; however, variation between utilities is considerable. The public health significance of these breaks in the distribution system is not currently known. These data, and the significance of their inherent spatial and temporal variability, must be considered in future public health studies and risk assessment estimates.

Calderon (2004) conducted a survey from 1991 to 2000 and found that the majority of outbreaks occurred due to a lack of treatment (primarily groundwater) or a treatment failure. During the 2003-2004 survey, approximately 52 percent of the contamination events occurred at the point of use (i.e., premise plumbing) while approximately 42 percent were due to source water contamination, treatment inadequacies, or contamination in the municipal distribution system. Little is known about the extensiveness of distribution system inadequacies and whether they are sporadic or continuously occurring (Lee and Schwab, 2005), but outbreaks have been documented following external contamination in the distribution system despite the presence or requirement of residual disinfectant (Craun and Calderon, 2001; Levy et al., 1998).

Other data that may be important to consider in distribution systems and contaminant exposures include hydraulic integrity, back-siphonage events, and intrusion rates. Water systems commonly lose more than 10 percent of the total water produced through leaks in the pipelines (AWWA and AWWARF, 1992). A survey of 26 water utilities in the United States found that the percent of leakage (unaccounted for water) ranged from less than 10 percent to as high as 32 percent (Kirmeyer et al., 2001). Compromised hydraulic integrity (positive pressure) of water distribution has been associated with worldwide epidemics (reviewed in Lee and Schwab, 2005).

At least 20 percent of distribution mains are reported to be below the water table, but it is assumed that all systems have some pipe below the water table for some time throughout the year, thus providing an opportunity for intrusion of exterior water under low or negative pressure conditions (LeChevallier et al., 2003b). In addition, pipes buried in soil are subject to contamination with fecal indicators and pathogens from the surrounding environment (Karim et al., 2003; Kirmeyer et al., 2001). Even minor pressure fluctuations create back-siphonage, where intrusion rates are estimated at more than one gallon per minute (gpm; LeChevallier et al., 2003a). During power outages, up to 90 percent of nodes have been shown to draw a negative pressure (LeChevallier et al., 2003b).

A survey of water utilities in North America found that 28.8 percent of cross-connections resulted in bacterial contamination (Lee et al., 2003). Negative hydraulic pressure can draw pathogens from the surrounding environment into the water supply, where residual disinfection efficacy is uncertain and variable, depending on the magnitude of such events (Gadgil, 1998; Haas and Trussel, 1998; Trussell, 1999). Little is known about the extensiveness of distribution system inadequacies and whether they are sporadic or continuously occurring (Lee and Schwab, 2005), but outbreaks have been documented following external contamination in the distribution system despite the presence or requirement of residual disinfectant (Craun and Calderon, 2001; Levy et al., 1998).

Decline in residual disinfectant in the distribution system is related to many factors, including the distance traveled, water flow velocity, residence time, age and composition of pipes, and water pressure (Egorov et al., 2002). Although residual chlorine is present in the distribution system of treated water, the levels do not provide significant inactivation of pathogens in intrusion events (Payment, 1999; Snead et al., 1980). More recent modeling studies have evaluated intrusion events at specific locations, with consideration to mixing, contact time, and other distribution system variables, prior to consumption. Under these realistic exposure scenarios, monochloramine disinfectants performed poorly against *Giardia* and *Escherichia coli*. Typical concentrations of chlorine residual (0.5 mg/L) inactivated *E. coli* in simulated sewage intrusion events but were again ineffective for *Giardia* (Baribeau et al., 2005; Propato and Uber, 2004). Intentional contamination events in the distribution system are also a concern where public water supplies are potentially vulnerable to bioterrorism threats.

Premise Plumbing/Biofilms

Tap water is known to contain bacteria not found at the original source, treatment facility, or at other points earlier in the distribution system, indicating the possibility of biofilms in the distribution pipes or in the consumer's tap (Chu et al., 2005; Pepper et al., 2004; Schmeisser et al., 2003). Bacterial colonization of pipes, connections, and faucets positioned along the channels of drinking water distribution, including the utility's distribution system, the homeowner's distribution system (premise plumbing), and fixtures (i.e., faucets and hose connections) in the home, is well documented. Pepper et al. (2004) found that the bacteriological quality of water significantly deteriorates in home plumbing relative to the distribution system, as evidenced by heterotrophic plate count (HPC) bacteria increasing up to five-fold from the distribution outlet to the household tap. Stagnant water in premise plumbing provides an environment where bacteria can grow to values several orders of magnitude higher than the municipal distribution system (Edwards et al., 2005).

Legionella is a concern related to microbial growth in the distribution system. Ten waterborne outbreaks from *Legionella* occurred from 2005 to 2006 (Yoder et al., 2008). *Legionella* is known to grow in the distribution systems of large buildings or institutions (Blackburn et al., 2004) and in premise plumbing (Thomas et al., 2006; Tobin-D'Angelo et al., 2004; Vacrewijck et al., 2005). These organisms remind us of the need to evaluate what specific effect(s) contribute to increased exposure to pathogens in the distribution system where not all systems present the same exposure scenarios and not all exposed populations will experience adverse impacts. *H. pylori* and *Mycobacterium* are also associated with biofilms, water distribution systems, and premise plumbing (Flannery et al., 2006; Park et al., 2001; Pryor et al., 2004).

Epidemiological Research

The true burden of waterborne disease is unknown in the United States, although variable approaches have been used to estimate gastrointestinal illness from waterborne pathogens, including epidemiological studies and exposure analysis. Information is lacking, however, regarding risk estimates that consider gastroenteritis and other illnesses. For example, household intervention trials have been used in an attempt to estimate the baseline of gastrointestinal illness within communities (Colford et al., 2002, 2005; Hellard et al., 2001; Payment et al., 1991, 1997). These epidemiological studies involved randomly designating one group of households as the "intervention group," where members drank water additionally purified with an in-home treatment system, and then another group drank tap water or water that passed through a sham device. In the latter situation, the study is *blinded* meaning that neither group knows during the study whether their in-home device is actually providing treatment.

For all of these trials, the human health end point was gastrointestinal illness, with some variation in the specific definition of the symptomology. All participants were immunocompetent individuals who kept health diaries throughout the study to record symptoms related to gastrointestinal illnesses. The source (surface) waters were reported to have varying levels of microbial contamination. Other factors of inlet water quality and distribution system effects undoubtedly have some level of impact on end-use water quality and contaminant exposures.

The initial study by Payment et al. (1991) concluded that an estimated 35 percent of the gastrointestinal illnesses occurring within the tap water group may be attributable to their drinking water. This was of interest given that the tap water met both Canadian and U.S. regulations. The source water, however, was subject to sewage contamination. One potential limitation in the study design was that persons drinking tap water were not provided a sham treatment device. In other words, the study was not blinded and thus participants may have been more inclined to report poor(er) health symptoms. A follow-up study attempted to evaluate the role of distribution system water quality in gastrointestinal incidence and involved four groups of participants: a tap water group and a bottled purified water group (to address those exposed and unexposed, respectively), and a plant bottled water group and a tap water group using a purge valve (to address distribution system water quality). The attributable risk ranged from 3 percent for the bottled plant water group (rate of illness for this group is the same as for those drinking the bottled purified water), to 12 percent for the tap water group, to 17 percent for those in the tap water group with a purge valve. The investigators concluded that the excess number of gastrointestinal illnesses observed was associated with contamination within the distribution system since the rate of illness of the bottled plant water group was similar to the rate of those drinking bottled purified water. This study was limited by population size since about half of the participants in the bottled plant water group dropped out during the course of the study. Another limitation was that this study, too, was not blinded.

Addressing previous limitations, a blinded household intervention study was conducted in Australia (Hellard et al., 2001). Some participants used a water treatment device that involved an ultraviolet application and filtration while others were given a fake (no treatment) device. Unlike the Payment et al. (1991, 1997) studies, this study showed similar rates of illness of the participants in both the control and the intervention groups. Two more household intervention trials have since been conducted in the United States (Colford et al., 2002, 2005). After first determining that the study they designed could incorporate effective participant blinding, an attributable risk of 0.85 was observed, indicating that 24 percent of the gastrointestinal illnesses could be attributable to tap water.

In a follow-up study done by Colford et al. (2005), the investigators observed no difference in the rate of illness between the group with water treatment at the tap and those without. The authors offered the explanation that perhaps their study resulted in this conclusion due to successful water treatment practices and a well-maintained water distribution system. In addition, it was recognized that water consumption by the participants outside of the home may have had some effect on the study results.

Similarly, community intervention studies have also been conducted to address waterborne gastrointestinal disease risks.[16] Two of these studies (Calderon, 2001; Goh et al., 2005) concluded that a reduction in gastrointestinal illnesses was observed due to the intervention (additional water treatment). A preliminary report from the Kunde et al. (unpublished) study also indicates a decrease in diarrheal illness risk in participants over age 35 following the intervention. Conversely, preliminary data analysis from the Frost study (unpublished) does not indicate a significant difference, however, analysis is reported to be ongoing for both of these aforementioned studies (reviewed in Calderon and Craun, 2006).

The Role of Risk Assessment

Conflicting results in replicate studies, as described earlier, indicate a flawed research design, inconsistent variables, or some other confounding effect. Contamination is not evenly distributed but rather is affected by differences in source water quality, hydraulic flow, mixing, biofilm development, age of distribution system, and many other variables that could explain inconsistent conclusions in epidemiological results. Even climatic events can play a role by taxing treatment plant operations or increasing the chances for environmental intrusion events.

It is not practical to monitor water supplies in real time and at the point of use for all groups of contaminants; thus, episodic and routine contamination events are difficult to predict or identify. Epidemiological studies are often criticized for their lack of being properly controlled or randomized where the strength and validity of studies can be compromised by confounding factors, uncertainty, and inherent biases. Nonetheless, epidemiology is a vital component in determining risk factors for disease in a population. Well-conducted epidemiological studies contribute to improving the forecast of events as is done with quantitative risk assessment by providing more scientific rigor to the risk assessment process. Likewise, the process of quantitative risk assessment can be used to identify specific effects of different types of data over defined, probabilistic ranges that have the greatest impact on health. This information can be used, in return, to better inform epidemiological studies.

[16]See Calderon (2001), Frost et al. (unpublished), Goh et al. (2005), Hellard et al. (2002), Kunde et al. (unpublished), and McConnell et al. (2001).

Published guidance documents emphasize the need for improved data collection during drinking water outbreaks (Emde et al., 2001) and tools linking water utilities and health researchers (Parkin et al., 2006). Water utilities collect a plethora of water quality and distribution data, such as system types, sources, treatments, storage, quality indicator monitoring, residual disinfectants/alternative barriers, distribution flow/management, and consumer complaints. These complex databases have not been wholly integrated into risk assessment or health effects study design or analysis. Unknown are the implications these data have, both separately and in an integrated fashion, for the determination of human health risks from drinking water contaminants. Guidelines are needed to evaluate the robustness of available data relative to informing risk assessment and the study of population health risks.

Recent estimates of waterborne illnesses per year in the United States range from 12 million cases/year (Colford et al., 2006) to 16 million cases/year (Messner et al., 2006) to 19.5 million cases/year (Reynolds et al., 2008). Each estimate utilized a different approach and assumptions to calculate the public health risk. Using a proportional, risk-based approach, Colford et al. (2006) made assumptions under hypothetical scenarios of either poor source water quality/poor water treatment or contamination in a distribution system. The latter resulted in the estimated 11.69 million cases of acute gastrointestinal illnesses per year and a lower estimate of 4.26 million cases per year associated with poor source water quality/poor water treatment. Assumptions include the applicability of the attributable risk percent estimates from the household intervention trials to the entire U.S. population. The authors emphasize that the primary purpose of their estimation of acute gastrointestinal illness incidence, attributable to drinking tap water in the United States, is to demonstrate a methodology that can be improved upon with more data. This is an example of the necessary integration of risk assessment principles and epidemiological surveys to improve the rigor of each science separately.

Quantitative Risk Assessment

Quantitative risk assessment allows for the use of mathematical models to forecast public health impacts. By identifying a hazard and incorporating data such as the dose-response relationship and exposure information, we can estimate the health impact of those exposures. Messner et al. (2006) described a risk assessment approach for estimating the incidence of gastrointestinal disease in the United States due to drinking water (reviewed in Reynolds et al., 2008). Assuming that, for each population served by a community water system, a distribution of incidence rates of acute gastroenteritis can be estimated. This rate can be applied to calculate a national estimate of disease attributable to drinking water in the United States. The authors speculate that the mean incidence of acute gastrointestinal illness attributable to drinking water among community water systems ranges widely due to variations in source water quality, water treatment efficiencies, water quality within a distribution system, and water quality management practices.

Messner et al. (2006) proposed the development of a "risk matrix" to categorize community water systems based on relative microbial risk levels. The authors suggest connecting the information obtained from epidemiological studies regarding the incidence rate of acute gastrointestinal illness to risk factors identified in the epidemiological studies that have been conducted and to other CWSs. The identification of these risk factors (characteristics) should allow for risk-based categorizing of other CWSs that have similar characteristics (therefore assuming that generalizations can be made regarding all U.S. CWSs and the populations they serve). Completing this process involves overcoming several challenges due to the lack of data related to both pathogen occurrence and variation in survivability and infectivity, as well as knowing the actual efficiency of water treatment applications (as opposed to theoretical information). Consideration of specific health-based information from previous studies with factors associated with source water/water treatment quality and distribution system deficiencies along with illness rates is warranted.

Improved Monitoring Data

Recent studies in water quality monitoring exemplify the multidisciplinary and innovative approaches needed to improve monitoring data that can be used to provide a more robust risk estimate. For example, characterization of complex transport phenomena provides necessary information related to assumptions of mixing within a pressurized water distribution network. Using a lab-scaled experimental setup of a distribution system and cross junctions equipped with various pumps and sensors, researchers are evaluating published models on water quality and hydraulic flow behavior of water distribution piping systems under various conditions of hydraulic flow. Studies conclude that previous assumptions of mixing in a distribution system are potentially in error. Minimizing these errors has practical implications for improving exposure estimates, determining water distribution infrastructure resilience, and evaluating impacts of a bioterrorism event (Austin et al., 2008; Kim et al., 2008; Romero-Gomez et al., 2008).

Due to potential contamination in the distribution system and in premise plumbing, monitoring as close to the point of consumption as possible is warranted. Miles et al. (2009) described the use of a monitoring scheme at the neighborhood level for microbial pathogens and water quality indicators in municipal water supplies. Samples were collected from municipal water sources, subject to all federal water quality standards and regulations, post-treatment and distribution. In this study, the current widespread network of water vending machines is utilized by collecting machine filters concentrating microbes from large volumes (thousands of liters) of source water over long time periods (several months). Of the 45 filters from 41 unique sites, 10.4 percent and 6.3 percent tested positive for *E. coli* and infectious enteroviruses, respectively.

Real-time monitoring remains an important goal in ensuring the consistent quality of the water supply. Rapid monitoring schemes are available for detection of viable bacteria and protozoa but not for human viruses. Viruses are known to cause approximately 8 percent of waterborne outbreaks in the United States. In addition, an etiological agent is not identified in about 40 percent of all waterborne outbreaks where the outbreak characteristics frequently indicate a viral agent (reviewed in Reynolds et al., 2008).

Viruses present a particular challenge in monitoring water quality due to their small size, low infectious dose, and lack of universal cell culture. Conventional virus detection methods generally involve collection in the field followed by offsite analysis. Detection methods for viable viruses have high sensitivity and specificity but are costly, time-consuming, labor-intensive, and not applicable for the detection of some of the viruses of concern (i.e., noncytopathogenic viruses). Polymerase chain reaction (PCR) methods provide rapid detection but are complicated by inhibitory compounds in environmental samples and detection of nonviable agents. False positive and false negative results are common with each method (Reynolds et al., 1998). Integrating the two methods (integrated cell culture and PCR) has overcome major flaws of each individual method (Blackmer et al., 2000; Reynolds et al., 1996, 2001). However, this approach is neither automated nor rapid.

Advances in near-real-time monitoring for viruses in water include the use of electrodeposition of viruses onto the surface of optical fibers. Subsequently, infrared spectroscopy is used to characterize and identify captured viruses. Specific and unique components of a virus present distinct vibrational fingerprints in the infrared (Wilhelm et al., 2008), which can be used to identify and quantify the type of virus.

Regulatory Perspective

Quantitative risk assessment is frequently used to formulate policy where acceptable risk limits are estimated from assumed exposure values gained from monitoring data. Current regulatory standards and monitoring requirements, however, do not guarantee the absence of human pathogens in tap water. For example, the Total Coliform Rule, mandating the use of bacterial indicators of water quality, does not predict vulnerability to an outbreak (Craun et al., 2002). In fact, few community and noncommunity water systems that reported an outbreak from the survey period of 1991 to 1998 had violated the coliform standard in the 12-month period prior to the outbreak.

Risk limits are used to provide guidance on acceptable endpoints. The EPA suggests a risk of one case of disease in 10,000 persons exposed to potable water. This guidance is therefore used to conduct mathematical modeling of human health risks from various agents to determine exposure standards or levels of contamination in various media that must not be exceeded. In addition, the necessary efficacy of preventative measures can be quantitatively evaluated to meet the acceptable risk level. For quantitative microbial risk assessment, key pieces of information are needed to be either known or predicted, including the infectious dose response of the pathogen of interest; concentration at which the agent can be found in water; the impact of various water treatment strategies on reducing pathogen infectivity; and pathogenicity factors. Values of exposure and dose-response parameters may correspond to a point estimate of interest.

Typical point estimates may reflect the best calculation of risk or to a maximum reasonable exposure or other justified value. Interval estimates are useful for looking at a range of values. In the latter, parameters are not single values but probability distributions. Risk assessment is subject to large uncertainty and variability. Uncertainty occurs when there is an error in the estimate. For example, measurement errors, reporting errors, or inferences from a small, unrepresentative population lead to uncertainty in the estimate. Variability occurs due to intrinsic heterogenicity, such as differences in population consumption patterns, cultures and ethnicity, dose-response sensitivity, and immune function. The advantage of modeling probabilistically is to propagate these uncertainties through the model. Using distribution analysis, a range of possible outcomes can be assessed and key data contributors identified. A range of outcomes can be reviewed under different conditions of uncertainty and mitigations to evaluate parameters that have the greatest health impacts. This process helps to define when better data can be most valuable and identify parameters that can be influenced by policy.

Monte Carlo analysis is a widely applied tool for risk distribution analysis. Using random numbers in a computational process, a desired output as a function of changing variables (i.e., dose, exposure, survival, etc.) can be estimated for known and assumed distribution inputs. The process can be easily repeated for thousands of trials. As part of the risk characterization process, however, a number—or series of numbers—is needed to inform decisions about acceptable risk, possible risks, risk reduction potentials, and risk management decisions. Questions remain regarding acceptable risk limits, susceptible populations, and the use of conservative, worst-case, protective, or interval estimates.

Recommendations and Conclusions

While the discussion of how to appropriately estimate risks associated with contributing values continues, other factors must also be considered. We have learned that water quality variability affects long-term risk and that exposure is not constant over time. In addition, there is no threshold for infectivity with microbes. Even low-level exposures can have a significant impact on risk over time. Where average doses have been used in risk, the possibility of widespread exposure to low doses and limited exposures to large doses should be considered.

Additional epidemiological data are necessary to prove the plausibility of models, and models should be evaluated with increased inputs related to the mechanisms of the host and pathogen interactions.

In addition to the many variables in water quality, treatment, and distribution, of particular concern are sensitive populations in the United States that are susceptible to higher rates of infections and to more serious health outcomes from waterborne pathogens. These subpopulations include not only individuals experiencing adverse health status, but also those experiencing "normal" life stages (e.g., pregnancy or those very young or old). Acceptable risk goals need to be evaluated for a changing population as persons move through these normal life stages that impact their susceptibility to waterborne illness. Risks may be acute or chronic and sequelae are common outcomes that must be considered (Parkin, 2000). With exposure-related inputs contributing the greatest uncertainty in models, more monitoring is needed to inform risk and minimize uncertainty in risk characterization. Finally, better communication between water quality professionals, public health researchers, and health-care providers is needed to design studies that comprehensively assess the impact of waterborne disease and address the multibarrier approach necessary to preserve water quality.

OVERVIEW REFERENCES

Aquatest. 2009. *Aquatest research programme*, http://www.bristol.ac.uk/aquatest/ (accessed April 15, 2009).

BEACH ET AL. REFERENCES

AAP (American Academy of Pediatrics). 2009. Drinking water from private wells and risks to children. *Pediatrics* 123(6):1599-1605.

Angulo, F. J., S. Tippen, D. J. Sharp, B. J. Payne, C. Collier, J. E. Hill, T. J. Barrett, R. M. Clark, E. E. Geldreich, H. D. Donnell, and D. L. Swerdlow. 1997. A community waterborne outbreak of salmonellosis and the effectiveness of a boil water order. *American Journal of Public Health* 87(4):580-584.

Arctic Council and IASC (International Arctic Science Committee). 2004. *Impacts of a warming arctic*. Cambridge, UK: Cambridge University Press.

Benin, A. L., R. F. Benson, K. E. Arnold, A. E. Fiore, P. G. Cook, L. K. Williams, B. Fields, and R. E. Besser. 2002. An outbreak of travel-associated Legionnaires disease and Pontiac fever: the need for enhanced surveillance of travel-associated legionellosis in the United States. *Journal of Infectious Diseases* 185(2):237-243.

Bergeisen, G. H., M. W. Hinds, and J. W. Skaggs. 1985. A waterborne outbreak of hepatitis A in Meade County, Kentucky. *American Journal of Public Health* 75(2):161-164.

Bowen, A., J. Kile, C. Austin, C. Otto, B. Blount, N. Kazerouni, H.-N. Wong, H. Mainzer, J. Mott, M. J. Beach, and A. M. Fry. 2007. Outbreaks of short-incubation illness following exposure to indoor swimming pools. *Environmental Health Perspectives* 115(2):267-271.

Burnsed, L. J., L. A. Hicks, L. M. Smithee, B. S. Fields, K. K. Bradley, N. Pascoe, S. M. Richards, S. Mallonee, L. Littrell, R. F. Benson, M. R. Moore, and Legionellosis Outbreak Investigation Team. 2007. A large, travel-associated outbreak of legionellosis among hotel guests: utility of the urine antigen assay in confirming Pontiac fever. *Clinical Infectious Diseases* 44(2):222-228.

CDC (Centers for Disease Control and Prevention). 1982. Otitis due to *Pseudomonas aeruginosa* serotype 0:10 associated with a mobile redwood hot tub system—North Carolina. *Morbidity and Mortality Weekly Report* 31(40):541.

———. 1998. *A survey of the quality of water drawn from domestic wells in nine midwest states.* CDC funded multi-state report, http://www.cdc.gov/nceh/hsb/disaster/pdfs/A%20Survey%20of%20the%20Quality%20ofWater%20Drawn%20from%20Domestic%20Wells%20in%20Nine%20Midwest%20States.pdf (accessed March 16, 2009).

———. 2001. Shigellosis outbreak associated with an unchlorinated fill-and-drain wading pool—Iowa, 2001. *Morbidity and Mortality Weekly Report* 50(37):797-800.

———. 2004. Aseptic meningitis outbreak associated with echovirus 9 among recreational vehicle campers—Connecticut, 2003. *Morbidity and Mortality Weekly Report* 53(31):710-713.

———. 2007a. Ocular and respiratory illness associated with an indoor swimming pool—Nebraska, 2006. *Morbidity and Mortality Weekly Report* 56(36):929-932.

———. 2007b. Cryptosporidiosis outbreaks associated with recreational water use—five states, 2006. *Morbidity and Mortality Weekly Report* 56(29):729-32.

———. 2008a. Primary amebic meningoencephalitis—Arizona, Florida, and Texas, 2007. *Morbidity and Mortality Weekly Report* 57(21):573-577.

———. 2008b. Communitywide cryptosporidiosis outbreak—Utah, 2007. *Morbidity and Mortality Weekly Report* 57(36):989-993.

———. 2009. Respiratory and ocular symptoms among employees of a hotel indoor waterpark resort—Ohio, 2007. *Morbidity and Mortality Weekly Report* 58(4):81-85.

Colford, J. M., S. L. Roy, M. J. Beach, A. Hightower, S. E. Shaw, and T. J. Wade. 2006. A review of household drinking water intervention trials and an approach to the estimation of endemic waterborne gastroenteritis in the United States. *Journal of Water and Health* 4(Suppl 2):71-88.

Curriero, F. C., J. A. Patz, J. B. Rose, and S. Lele. 2001. The association between extreme precipitation and waterborne disease outbreaks in the United States, 1948-1994. *American Journal of Public Health* 91(8):1194-1199.

Cutler, D., and G. Miller. 2005. The role of public health improvements in health advances: the twentieth-century United States. *Demography* 42(1):1-22.

D'Angelo, L. J., J. C. Hierholzer, R. A. Keenlyside, L. J. Anderson, and W. J. Martone. 1979. Pharyngoconjunctival fever caused by adenovirus type 4: report of a swimming pool-related outbreak with recovery of virus from pool water. *Journal of Infectious Diseases* 140(1):42-47.

Denno, D. M., W. E. Keene, C. M. Hutter, J. K. Koepsell, M. Patnode, D. Flodin-Hursh, L. K. Stewart, J. S. Duchin, L. Rasmussen, R. Jones, and P. I. Tarr. 2009. Tri-county comprehensive assessment of risk factors for sporadic reportable bacterial enteric infection in children. *Journal of Infectious Diseases* 199(4):467-476.

EPA (Environmental Protection Agency). 2005. *Drinking water infrastructure needs survey and assessment*, http://www.epa.gov/safewater/needssurvey/index.html (accessed February 20, 2009).

———. 2006. National primary drinking water regulations: ground water rule; final rule. *Federal Register* 71(216):65573-65660.

———. 2009a. Agreement in principal. Total coliform rule/distribution system advisory committee. *Federal Register* 74(8):1683-1684.

———. 2009b. *Agreement in principal. Total coliform rule/distribution system advisory committee*, http://www.epa.gov/OGWDW/disinfection/tcr/regulation_revisions_tcrdsac.html#aip (accessed March 16, 2009).

Falkinham, J. O., 3rd. 2003. Mycobacterial aerosols and respiratory disease. *Emerging Infections Diseases* 9(7):763-767.

Fields, B. S., T. Haupt, J. P. Davis, M. J. Arduino, P. H. Miller, and J. C. Butler. 2001. Pontiac fever due to *Legionella micdadei* from a whirlpool spa: possible role of bacterial endotoxin. *Journal of Infectious Disease* 184(10):1289-1292.

Fields, B. S., R. F. Benson, and R. E. Besser. 2002. *Legionella* and Legionnaires' disease: 25 years of investigation. *Clinical Microbiology Reviews* 15(3):506-526.

Fraser, D. W., T. R. Tsai, W. Orenstein, W. E. Parkin, H. J. Beecham, R. G. Sharrar, J. Harris, G. F. Mallison, S. M. Martin, J. E. McDade, C. C. Shepard, and P. S. Brachman. 1977. Legionnaires' disease: description of an epidemic of pneumonia. *New England Journal of Medicine* 297(22):1189-1197.

Gustafson, T. L., J. D. Band, R. H. Hutcheson Jr., and W. Schaffner. 1983. *Pseudomonas* folliculitis: an outbreak and review. *Reviews of Infectious Diseases* 5(1):1-8.

Iwamoto, M., G. Hlady, M. Jeter, C. Burnett, C. Drenzek, S. Lance, J. Benson, D. Page, and P. Blake. 2005. Shigellosis among swimmers in a freshwater lake. *The Southern Medical Journal* 98(8):774-778.

Katz, D. E., D. Heisey-Grove, M. J. Beach, R. C. Dicker, and B. T. Matyas. 2006. Prolonged outbreak of giardiasis with two modes of transmission. *Epidemiology and Infection* 134(5):935-941.

Kaydos-Daniels, S. C., M. J. Beach, T. Shwe, D. Bixler, and J. Magri. 2008. Health effects associated with indoor swimming pools: a suspected toxic chloramine exposure. *Public Health* 122(2):195-200.

Kent, G. P., J. R. Greenspan, J. L. Herndon, L. M. Mofenson, J. A. Harris, T. R. Eng, and H. A. Waskin. 1988. Epidemic giardiasis caused by a contaminated public water supply. *American Journal of Public Health* 78(2):139-143.

Korich, D. G., J. R. Mead, M. S. Madore, N. A. Sinclair, and C. R. Sterling. 1990. Effects of ozone, chlorine dioxide, chlorine, and monochloramine on *Cryptosporidium parvum* oocyst viability. *Applied and Environmental Microbiology* 56(5):1423-1428.

Mac Kenzie, W. R., N. J. Hoxie, M. E. Proctor, M. S. Gradus, K. A. Blair, D. E. Peterson, J. J. Kazmierczak, D. G. Addiss, K. R. Fox, J. B. Rose, and J. P. Davis. 1994 A massive outbreak in Milwaukee of *Cryptosporidium* infection transmitted through the public water supply. *New England Journal of Medicine* 331(3):161-167.

Mahoney, F. J., T. A. Farley, K. Y. Kelso, S. A. Wilson, J. M. Horan, and L. M. McFarland. 1992. An outbreak of hepatitis A associated with swimming in a public pool. *Journal of Infectious Diseases* 165(4):613-618.

McCarthy, T. A., N. L. Barrett, J. L. Hadler, B. Salsbury, R. T. Howard, D. W. Dingman, C. D. Brinkman, W. F. Bibb, and M. L. Cartter. 2001. Hemolytic-uremic syndrome and *Escherichia coli* O121 at a lake in Connecticut, 1999. *Pediatrics* 108(4):E59.

McLaughlin, J. B., A. DePaola, C. A. Bopp, K. A. Martinek, N. P. Napolilli, C. G. Allison, S. L. Murray, E. C. Thompson, M. M. Bird, and J. P. Middaugh. 2005. Outbreak of *Vibrio parahaemolyticus* gastroenteritis associated with Alaskan oysters. *New England Journal of Medicine* 353(14):1463-1470.

Messner, M., S. Shaw, S. Regli, K. Rotert, V. Blank, and J. Soller. 2006. An approach for developing a national estimate of waterborne disease due to drinking water and a national estimate model application. *Journal of Water and Health* 4(Suppl 2):201-240.

Morgan J., S. L. Bornstein, A. M. Karpati, M. Bruce, C. A. Bolin, C. C. Austin, C. W. Woods, J. Lingappa, C. Langkop, B. Davis, D. R. Graham, M. Proctor, D. A. Ashford, M. Bajani, S. L. Bragg, K. Shutt, B. A. Perkins, J. W. Tappero, and Leptospirosis Working Group. 2002. Outbreak of leptospirosis among triathlon participants and community residents in Springfield, Illinois, 1998. *Clinical Infectious Diseases* 34(12):1593-1599.

NRC (National Research Council). 2006. *Drinking water distribution systems: assessing and reducing risks*. Washington, DC: The National Academies Press.

Nygård, K., E. Wahl, T. Krogh, O. A. Tveit, E. Bøhleng, A. Tverdal, and P. Aavitsland. 2007. Breaks and maintenance work in the water distribution systems and gastrointestinal illness: a cohort study. *International Journal of Epidemiology* 36(4):873-880.

Parshionikar, S. U., S. Willian-True, G. S. Fout, D. E. Robbins, S. A. Seys, J. D. Cassady, and R. Harris. 2003. Waterborne outbreak of gastroenteritis associated with a norovirus. *Applied Environmental Microbiology* 69(9):5263-5268.

Patz, J. A., D. Engelberg, and J. Last. 2000. The effects of changing weather on public health. *Annual Review of Public Health* 21:271-307.

Patz, J. A., S. J. Vavrus, C. K. Uejio, and S. L. McLellan. 2008. Climate change and waterborne disease risk in the Great Lakes region of the U.S. *American Journal of Preventive Medicine* 35(5):451-458.

Podewils, L. J., L. Zanardi-Blevins, M. Amundson, D. Itani, A. Burns, M. J. Beach, C. Otto, L. Browne, S. Adams, S. Monroe, V. Hill, C. Lohff, and M.-A. Widdowson. 2007. Outbreak of norovirus illness associated with a swimming pool. *Epidemiology and Infection* 135(5):827-833.

Prüss, A. 1998. Review of epidemiological studies on health effects from exposure to recreational water. *International Journal of Epidemiology* 27(1):1-9.

Rose, J. B., P. R. Epstein, E. K. Lipp, B. H. Sherman, S. M. Bernard, and J. A. Patz. 2001. Climate variability and change in the United States: potential impacts on water- and foodborne diseases caused by microbiologic agents. *Environmental Health Perspectives* 109(Suppl 2):211-221.

Shields, J. M., V. R. Hill, M. J. Arrowood, and M. J. Beach. 2008. Inactivation of *Cryptosporidium parvum* under chlorinated recreational water conditions. *Journal of Water and Health* 6(4):513-520.

Swerdlow, D. L., B. A. Woodruff, R. C. Brady, P. M. Griffin, S. Tippen, D. Donnell, E. Geldreich, B. J. Payne, A. Meyer, Jr., J. G. Wells, K. D. Greene, M. Bright, N. H. Bean, and P. A. Blake. 1992. A waterborne outbreak in Missouri of *Escherichia coli* O157:H7 associated with bloody diarrhea and death. *Annals of Internal Medicine* 117(10):812-819.

Turner, M., G. R. Istre, H. Beauchamp, M. Baum, and S. Arnold. 1987. Community outbreak of adenovirus type 7a infections associated with a swimming pool. *Southern Medical Journal* 80(6):712-715.

U.S. Census Bureau. 2008. *Current housing reports, series H150/07, American housing survey for the United States: 2007*. Washington, DC: U.S. Government Printing Office.

———. 2009. *Statistical abstract of the United States. Recreation and leisure activities: participation in selected sports activities 2006*, www.census.gov/compendia/statab/tables/09s1209.pdf (accessed March 16, 2009).

Visvesvara, G. S., and J. K. Stehr-Green. 1990. Epidemiology of free-living ameba infections. *The Journal of Protozoology* 37:25S-33S.

Vogt, R. L., H. E. Sours, T. Barrett, R. A. Feldman, R. J. Dickinson, and L. Witherell. 1982. *Campylobacter enteritis* associated with contaminated water. *Annals of Internal Medicine* 96(3):292-296.

Wade, T. J., R. L. Calderon, E. Sams, M. Beach, K. P. Brenner, A. H. Williams, and A. P. Dufour. 2006. Rapidly measured indicators of recreational water quality are predictive of swimming-associated gastrointestinal illness. *Environmental Health Perspectives* 14(1):24-28.

Wade, T. J., R. L. Calderon, K. P. Brenner, E. Sams, M. Beach, R. Haugland, L. Wymer, and A. P. Dufour. 2008. A rapid method of measuring recreational water quality demonstrates an enhanced sensitivity of children to swimming associated gastrointestinal illness. *Epidemiology* 19:375-383.

Weisel, C. P., S. D. Richardson, B. Nemery, G. Aggazzotti, E. Baraldi, E. R. Blatchley III, B. C. Blount, K.-H. Carlsen, P. A. Eggleston, F. H. Frimmel, M. Goodman, G. Gordon, S. A. Grinshpun, D. Heederik, M. Kogevinas, J. S. LaKind, M. J. Nieuwenhuijsen, F. C. Piper, and S. A. Sattar. 2008. Childhood asthma and environmental exposures at swimming pools: state of the science and research recommendations. *Environmental Health Perspectives* 117(4):500-507.

Wheeler, C., D. Vugia, G. Thomas, M. J. Beach, S. Carnes, T. Maier, J. Gorman, L. Xiao, M. Arrowood, D. Gilliss, and S. B. Werner. 2007. Outbreak of cryptosporidiosis at a California waterpark: employee and patron roles and the long road towards prevention. *Epidemiology and Infection* 135(2):302-310.

Yoder, J., M. Hlavsa, G. F. Craun, V. Hill, V. Roberts, P. Yu, L. A. Hicks, N. T. Alexander, R. L. Calderon, S. L. Roy, and M. J. Beach. 2008a. Surveillance for waterborne disease and outbreaks associated with recreational water use and other aquatic facility-associated health events—United States, 2005-2006. *Morbidity and Mortality Weekly Report Surveillance Summaries* 57(SS09):1-38.

Yoder, J., V. Roberts, G. F. Craun, V. Hill, L. Hicks, N. T. Alexander, V. Radke, R. L. Calderon, M. C. Hlavsa, M. J. Beach, and S. L. Roy. 2008b. Surveillance for waterborne disease and outbreaks associated with drinking water and water not intended for drinking—United States, 2005-2006. *Morbidity and Mortality Weekly Report Surveillance Summaries* 57(SS09):39-62.

ROSE REFERENCES

Auld, H., D. MacIver, and J. Klaassen J. 2004. Heavy rainfall and waterborne disease outbreaks: the Walkerton example. *Journal of Toxicology and Environmental Health* 67(20-22):1879-1887.

Brennan, R. J., and K. Rimba. 2005. Rapid health assessment in Aceh Jaya District, Indonesia, following the December 26 tsunami. *Emergency Medicine Australasia* 17(4):341-350.

CDC (Centers for Disease Control and Prevention). 2005a. *Vibrio* illnesses after Hurricane Katrina—multiple states, August-September 2005. *Morbidity and Mortality Weekly Report* 54(37):928-931.

———. 2005b. Norovirus outbreak among evacuees from Hurricane Katrina—Houston, Texas, September 2005. *Morbidity and Mortality Weekly Report* 54(40):1016-1018.

Cookson, S. T., K. Soetebier, E. L. Murray, G. C. Fajardo, A. Cowell, C. Drenzek, and R. Hanzlick. 2008. Internet-based morbidity and mortality surveillance among Hurricane Katrina evacuees in Georgia. *Preventing Chronic Disease* 5(4):A133.

Curriero, F. C., J. A. Patz, J. R. Rose, and S. Lele. 2001. The association between extreme precipitation and waterborne disease outbreaks in the United States, 1948-1994. *American Journal of Public Health* 91(8):1194-1199.

Floret, N., J. Viel, F. Mauny, B. Hoen, and R. Piarroux. 2006. Negligible risk for epidemics after geophysical disasters. *Emerging Infectious Diseases* 12(4):543-548.

Fong, T., L. S. Mansfield, D. L. Wilson, D. J. Schwab, S. L. Molloy, and J. B. Rose. 2007. Massive microbiological groundwater contamination associated with a waterborne outbreak in Lake Erie South Bass Island, Ohio. *Environmental Health Perspectives* 115(6):1-9.

Haas, C. N., J. B. Rose, and C. P. Gerba. 1999. *Quantitative microbial risk assessment*. New York: John Wiley and Sons.

Hoyois, P., J.-M. Scheuren, R. Below, and D. Guha-Sapin. 2007. *Annual disaster statistical review numbers and trends, 2006*. Brussels, Belgium: Center for Research on the Epidemiology of Disasters.

IPCC (Intergovernmental Panel on Climate Change). 2007. *Climate change: 2007*. Fourth Assessment Report of the Intergovernmental Panel on Climate Change. New York: Cambridge University Press.

IRIN News. 2006. *MALAWI: Cholera outbreak claims 51*, http://www.irinnews.org/PrintReport.aspx?ReportId=58467 (accessed April 13, 2009).

Kistemann, T., T. ClaBen, C. Koch, F. Dangendorf, R. Fischeder, J. Gebel, V. Bacata, and M. Exner. 2002. Microbial load of drinking water reservir tributaries during extreme rainfall and runoff. *Applied and Environmental Microbiology* 68(5):2188-2197.

Lipp, E. K., R. Kurz, R. Vincent, C. Rodriguez-Palcios, S. R. Farrah, J. B. Rose. 2001a. The effects of seasonal variability and weather on microbial fecal pollution and enteric pathogens in a subtropical estuary. *Estuaries and Coasts* 24(2):266-276.

Lipp, E. K., N. Schmidt, M. E. Luther, and J. B. Rose. 2001b. Determining the effects of El Niño-Southern Oscillation events on coastal water quality. *Estuaries and Coasts* 24(4):491-497.

Lipp, E. K., A. Huq, and R. R. Colwell. 2002. Effects of global climate on infectious disease: the cholera model. *Clinical Microbiology Reviews* 15(4):757-770.

Mac Kenzie, W. R., N. J. Hoxie, M. E. Proctor, M. S. Gradus, K. A. Blair, D. E. Peterson, J. J. Kazmierczak, D. G. Addiss, K. R. Fox, J. B. Rose, and J. P. David. 1994. A massive outbreak in Milwaukee of *Cryptosporidium* infection transmitted through the public water supply. *New England Journal of Medicine* 31(3):161-167.

PAHO (Pan American Health Organization). 1998. Impact of Hurricane Mitch on Central America. *Epidemiological Bulletin* 19(4):1-13.

Roberson, J. A., T. Holmes, and C. A. Lane. 2008. What's on Washington's water agenda for 2008? *Journal of the American Water Works Association* 100(3):67-72.

Schwartz, B. S., J. B. Harris, A. I. Khan, R. C. Larocque, D. A. Sack, M. A. Malek, A. S. Faruque, F. Qadri, S. B. Calderwood, S. P. Luby, and E. T. Ryan. 2006. Diarrheal epidemics in Dhaka, Bangladesh, during three consecutive floods: 1988, 1998, and 2004. *American Journal of Tropical Medicine and Hygiene* 74(6):1067-1073.

Setzer, C., and M. E. Domino. 2004. Medicaid outpatient utilization for water borne pathogenic illness following Hurricane Floyd. *Public Health Reports* 119(5):472-478.

Sur, D., P. Dutta, G. B. Nair, and S. K. Bhattacharya. 2000. Severe cholera outbreak following floods in a northern district of West Bengal. *Indian Journal of Medical Research* 112:178-182.

Thomas, K. M., D. F. Charron, D. Waltner-Toews, C. Schuster, A. R. Maarouf, and J. D. Holt. 2006. A role of high impact weather events in waterborne disease outbreaks in Canada, 1975-2001. *International Journal of Environmental Health Research* 16(3):167-180.

REYNOLDS AND MENA REFERENCES

Austin, R. G., B. van Bloemen Waanders, S. McKenna, and C. Y. Choi. 2008. Mixing at cross junctions in water distribution systems. II. An experimental study. *ASCE Journal of Water Resources Planning and Management* 134(3):295-302.

AWWA (American Water Works Association). 2003. *Water stats 2002 distribution survey* CD-ROM. Denver, CO: AWWA.

AWWA and AWWARF (American Water Works Association Research Foundation). 1992. *Water industry database: utility profiles*. Denver, CO: American Water Works Association and American Water Works Association Research Foundation.

AWWSC (American Water Works Service Company). 2002. *Deteriorating buried infrastructure management challenges and strategies*, http://www.epa.gov/ogwdw000/disinfection/tcr/pdfs/whitepaper_tcr_infrastructure.pdf (accessed February 18, 2009).

Baribeau H., N. L. Pozos, L. Boulos, G. F.Crozes, G. A. Gagnon, S. Rutledge, D. Skinner, Z. Hu, R. Hofmann, R. C. Andrews, L. Wojcicka, Z. Alam, C. Chauret, S. A. Andrews, R. Dumancis, and E. Warn. 2005. *Impact of distribution system water quality on disinfection efficacy*. Denver, CO: AWWARF.

Blackburn R. S., G. F. Craun, J. S. Yoder, V. Hill, R. L. Calderon, N. Chen, S. H. Lee, D. A. Levy, and M. J. Beach. 2004. Surveillance for waterborne-disease outbreaks associated with drinking water—United States, 2001-2002. *Morbidity and Mortality Weekly Report* 53(SS-8):23-45.

Blackmer, F., K. A. Reynolds, C. P. Gerba, and I. L. Pepper. 2000. Use of integrated cell culture-PCR to evaluate the effectiveness of poliovirus inactivation by chlorine. *Applied Environmental Microbiology* 66(5):2267-2268.

Calderon, R. L. 2001. Microbes in drinking water: recent epidemiological research to assess waterborne risks. In *Microbial pathogens and disinfection by-products in drinking water. Health effects and management of* risks, edited by G. F. Craun, F. S. Hauchman, and D. E. Robinson. Washington, DC: ILSI Press. Pp. 137-147.

Calderon, R. L. 2004. *Measuring benefits of drinking water technology: ten years of drinking water epidemiology*. NEWWA Water Quality Symposium, May 20, Boxborough, MA.

Calderon, R. L., and G. F. Craun. 2006. Estimates of endemic waterborne risks from community-intervention studies. *Journal of Water and Health* 4(Suppl 2):89-99.

Chu, C., C. Lu, and C. Lee. 2005. Effects of inorganic nutrients on the regrowth of heterotrophic bacteria in drinking water distribution systems. *Journal of Environmental Management* 74(3):255-263.

Colford, J. M., Jr., J. R. Rees, T. J. Wade, A. Khalakdina, J. F. Hilton, I. J. Ergas, S. Burns, A. Benker, C. Ma, C. Bowen, D. Mills, D. Vugia, D. Juranek, and D. Levy. 2002. Participant blinding and gastrointestinal illness in a randomized, controlled trial of an in-home drinking water intervention. *Emerging Infectious Diseases* 8(1):29-36.

Colford, J. M., Jr., T. J. Wade, S. K. Sandhu, C. C. Wright, S. Lee, S. Shaw, K. Fox, S. Burns, A. Benker, M. A. Brookhart, M. J. Van Der Laan, and D. A. Levy. 2005. A randomized controlled trial of in-home drinking water intervention to reduce gastrointestinal illness. *American Journal of Epidemiology* 161(5):472-482.

Colford, J. M., S. Roy, M. J. Beach, A. Hightower, S. E. Shaw, and T. J. Wade. 2006. A review of household drinking water trials and an approach to the estimation of endemic waterborne gastroenteritis in the United States. *Journal of Water and Health* 4(Suppl 2):71-88.

Craun, G. F., and R. L. Calderon. 2001. Waterborne disease outbreaks caused by distribution system deficiencies. *Journal of the American Water Works Association* 93(9):64-75.

Craun, G. F., N. Nwachuku, R. L. Calderon, and M. G. Craun. 2002. Outbreaks in drinking-water systems, 1991-1998. *Journal of Environmental Health* 65(1):16-23.

Deb, A. K., Y. J. Hasit, and F. M. Grablutz. 1995. *Distribution system performance evaluation*. Denver, CO: AWWARF.

Edwards, M., B. Marshall, Y. Zhang, and Y. Lee. 2005. Unintended consequences of chloramine hit home. In *Proceedings of the Water Environment Federation Disinfection Conference*. Mesa, AZ. Pp. 240-256.

Egorov, A., T. Ford, A. Tereschenko, N. Drizhd, I. Segedevich, and V. Fourman. 2002. Deterioration of drinking water quality in the distribution system and gastrointestinal morbidity in a Russian city. *International Journal of Environmental Health Research* 12(3):221-233.

Emde, K., J. Talbot, A. Barry, N. Fok, B. Reilley-Matthews, L. Gammie, E. Geldreich, and J. Mainiero. 2001. *Waterborne gastrointestinal disease outbreak detection*. Denver, CO: AWWARF.

Flannery, B., L. B. Gelling, D. J. Vugia, J. M. Weintraub, J. J.Salerno, M. J. Conroy, V. A. Stevens, C. E. Rose, M. R. Moore, B. S. Fields, R. E. Besser. 2006. Reducing *Legionella* colonization of water systems with monochloramine. *Emerging Infectious Diseases* 12(4):588-596.

Frost, F. J., T. R. Kunde, L. Harder, and T. I. Muller. Unpublished. Preliminary report–Northwest epidemiological enteric disease study. Lovelace Clinic Foundation, Albuquerque, NM.

Gadgil, A. J. 1998. Drinking water in developing countries. *Annual Review of Energy and the Environment* 23:253-286.

Goh, S., M. Reacher, D. P. Casemore, N. Q. Verlander, A. Charlett, R. M. Chalmers, M. Knowles, A. Pennington, J. Williams, K. Osborn, and S. Richards. 2005. Sporadic cryptosporidiosis decline after membrane filtration of public water supplies, England, 1996-2002. *Emerging Infectious Diseases* 11(2):251-259.

Grigg, N. S. 2005. Assessment and renewal of water distribution systems. *Journal of the American Water Works Association* 97(2):58-68.

Haas, C. N., and R. R. Trussell. 1998. Framework for assessing reliability of multiple, independent barriers in potable water reuse. *Water Science and Technology* 38(6):1-8.

Hellard, M. E., M. I. Sinclair, A. B. Forbes, and C. K. Fairley. 2001. A randomized, blinded, controlled trial investigating the gastrointestinal health effects of drinking water quality. *Environmental Health Perspectives* 109(8):773-778.

Hellard, M. E., M. I. Sinclair, S. C. Dharmage, J. F. Bailey, and C. K. Fairley. 2002. The rate of gastroenteritis in a large city before and after chlorination. *International Journal of Environmental Health Research* 12(4):355-360.

Karim, M., M. Abbaszadegan, and M. W. LeChevallier. 2003. Potential for pathogen intrusion during pressure transients. *Journal of the American Water Works Association* 95(5):134-146.

Kim, M., C. Y. Choi, and C. P. Gerba. 2008. Source tracking of microbial intrusion in water systems using artificial neural networks. *Water Research* 42(4-5):1308-1314.

Kirmeyer, G., W. Richards, and C. D. Smith. 1994. *An assessment of water distribution systems and associated research needs*. Denver, CO: AWWARF.

Kirmeyer, G. K., M. Freidman, K. Martel, D. Howie, M. LeChevallier, M. Abbaszadegan, M. Karim, J. Funk, and J. Harbour. 2001. *Pathogen intrusion into the distribution system*. Denver, CO: AWWARF.

Kunde, T. R., F. J. Frost, L. S. Nelson, L. Harter, M. F. Craun, G. F. Craun, and R. L. Calderon. Unpublished. *Estimates of endemic waterborne risks. Community intervention study: reduced gastrointestinal illness rates associated with improved water treatment*. Preliminary Report–A 98. Lovelace Clinic Foundation, Albuquerque, NM.

LeChevallier, M. W., R. W. Gullick, M. R. Karim, M. Friedman, and J. E. Funk. 2003a. The potential for health risks from intrusion of contaminants into the distribution system from pressure transients. *Journal of Water and Health* 1(1):3-14.

LeChevallier, M. W., R. W. Gullick, and M. Karim. 2003b. *The potential for health risks from intrusion of contaminants into the distribution system from pressure transients*. Voorhees, NJ: American Water Works Service Company.

Lee, E. J., and K. J. Schwab. 2005. Deficiencies in drinking water distribution systems in developing countries. *Journal of Water and Health* 3(2):109-127.

Lee, J. J., P. Schwartz, P. Sylvester, L. Crane, J. Haw, H. Chang, and H. J. Kwon. 2003. *Impacts of cross-connections in North American water supplies*. Denver, CO: AWWARF.

Levy, D. A., M. S. Bens, G. F. Craun, R. L. Calderon, and B. L. Herwaldt. 1998. Surveillance for waterborne-disease outbreaks—United States, 1995-1996. *Morbidity and Mortality Weekly Report Surveillance Summaries* 47(5):1-34.

Liang, J. L., E. J. Dziuban, G. F. Craun, V. Hill, M. R. Moore, R. J. Gelting, R. L. Calderon, M. J. Beach, and S. L. Roy. 2006. Surveillance for waterborne disease and outbreaks associated with drinking water and water not intended for drinking—United States, 2003-2004. *Morbidity and Mortality Weekly Report Surveillance Summaries* 55(12):31-58.

McConnell, S., M. Horrocks, M. I. Sinclair, and C. K. Fairley. 2001. Changes in the incidence of gastroenteritis and the implementation of public water treatment. *International Journal of Environmental Health Research* 11(4):299-303.

Messner, M., S. Shaw, S. Regli, K. Rotert, V. Blank, and J. Soller. 2006. An approach for developing a national estimate of waterborne disease due to drinking water and a national estimate model application. *Journal of Water and Health* 4(Suppl 2):201-240.

Miles, S., C. P. Gerba, I. L. Pepper, and K. A. Reynolds. 2009. Point-of-use drinking water devices for assessing microbial contamination in finished water and distribution systems. *Environmental Science and Technology* 43(5):1425-1429.

NRC (National Research Council). 1983. *Risk assessment in federal government: managing the process*. Washington, DC: National Academy Press.

Park, S. R., W. G. Mackay, and D. C. Reid. 2001. *Helicobacter sp.* recovered from drinking water biofilm sampled from a water distribution system. *Water Research* 35(6):1624-1626.

Parkin, R. T. 2000. *Issues in modeling rare health events such as chronic sequelae associated with microbial pathogens*. Annual Meeting of the Society for Risk Analysis. Center for Risk Science and Public Health, Arlington, VA, December.

Parkin, R., L. Ragain, R. Bruhl, H. Deutsch, and P. Wilborne-Davis. 2006. Advancing collaborations for water-related health risk communication. Denver, CO: AWWA Research Foundation and American Water Works Association.

Payment, P. 1999. Poor efficacy of residual chlorine disinfectant in drinking water to inactivate waterborne pathogens in distribution systems. *Canadian Journal of Microbiology* 45(8):709-715.

Payment, P., L. Richardson, J. Siemiatycki, R. Dewar, M. Edwardes, and E. Franco. 1991. A randomized trial to evaluate the risk of gastrointestinal disease due to consumption of drinking water meeting current microbiological standards. *American Journal of Public Health* 81(6):703-708.

Payment, P., J. Siemiatycki, L. Richardson, G. Renaud, E. Franco, and M. Prevost. 1997. A prospective epidemiological study of gastrointestinal health effects due to the consumption of drinking water. *International Journal of Health Research* 7(1):5-31.

Pepper, I. L., P. Rusin, D. R. Quintanar, C. Haney, K. L. Josephson, and C. P. Gerba. 2004. Tracking the concentration of heterotrophic plate count bacteria from the source to the consumer's tap. *International Journal of Food Microbiology* 92(3):289-295.

Propato, M., and J. G. Uber. 2004. Vulnerability of water distribution systems to pathogen intrusion: how effective is a disinfectant residual? *Environmental Science and Technology* 38(13):3713-3722.

Pryor, M., S. Springthorpe, S. Riffard, T. Brooks, Y. Huo, G. Davis, and S. A. Satter. 2004. Investigation of opportunistic pathogens in municipal drinking water under different supply and treatment regimes. *Water Science and Technology* 50(1):83-90.

Reynolds, K. A., C. P. Gerba, and I. L. Pepper. 1996. Detection of infectious enteroviruses by an integrated cell culture-PCR procedure. *Applied and Environmental Microbiology* 62(4):1424-1427.

Reynolds, K. A., K. Roll, R. S. Fujioka, C. P. Gerba, and I. L. Pepper. 1998. Incidence of enteroviruses in Mamala Bay, Hawaii using cell culture and direct polymerase chain reaction methodologies. *Canadian Journal of Microbiology* 44(6):598-604.

Reynolds, K. A., C. P. Gerba, M. Abbaszadegan, and I. L. Pepper. 2001. ICC/PCR detection of enteroviruses and hepatitis A virus in environmental samples. *Canadian Journal of Microbiology* 47(2):153-157.

Reynolds, K. A., K. D. Mena, and C. P. Gerba. 2008. Risk of waterborne illness via drinking water in the United States. *Reviews of Environmental Contamination and Toxicology* 192(4):117-158.

Romero-Gomez, P., C. K. Ho, and C. Y. Choi. 2008, Mixing at cross junctions in water distribution systems. I. A numerical study. *ASCE Journal of Water Resources Planning and Management* 134(3):284-294.

Schmeisser, C., C. Stöckigt, C. Raasch, J. Wingender, K. N. Timmis, D. F. Wenderoth, H. C. Flemming, H. Liesegang, R. A. Schmitz, K. E. Jaeger, and W. R. Streit. 2003. Metagenome survey of biofilms in drinking-water networks. *Applied and Environmental Microbiology* 69(12):7298-7309.

Snead, M. C., V. P. Olivieri, K. Kawata, and C. W. Kruse. 1980. The effectiveness of chlorine residuals in inactivation of bacteria and viruses introduced by post-treatment contamination. *Water Research* 14:403-408.

Thomas, K. M., D. F. Charron, D. Waltner-Toews, C. Schuster, A. R. Maarouf, and J. D. Holt. 2006. A role of high impact weather events in waterborne disease outbreaks in Canada, 1975-2001. *International Journal of Environmental Health Research* 16(3):167-180.

Tobin-D'Angelo, M. J., M. A. Blass, C. del Rio, J. S. Halvosa, H. M. Blumberg, and C. R. Horsburgh. 2004. Hospital water as a source of complex isolates in respiratory specimens. *Journal of Infectious Diseases* 189(1):98-104.

Trussell, R. R. 1999. Safeguarding distribution system integrity. *Journal of the American Water Works Association* 91(1):46-54.

Vacrewijck, M. J. M., G. Huys, J. C. Palomino, J. Swings, and F. Portaels. 2005. Mycobacteria in drinking water distribution systems: ecology and significance for human health. *FEMS Microbiology Reviews* 29(5):911-934.

Wilhelm, A. A., P. Lucas, K. Reynolds, and M. R. Riley. 2008. Integrated capture and spectroscopic detection of viruses in an aqueous environment. In *Optical fibers and sensors for medical diagnostics and treatment applications VIII*. Proceedings of SPIE 6852:1-8.

Yoder, J., V. Roberts, G. F. Craun, V. Hill, L. Hicks, V. Radke, R. L. Calderon, M. C. Hlavsa, M. J. Beach, and S. L. Roy. 2008. Surveillance for waterborne disease and outbreaks associated with drinking water and water not intended for drinking—United States, 2005-2006. *Morbidity and Mortality Weekly Report Surveillance Summaries* 57(9):39-62.

4

Addressing Risk for Waterborne Disease

OVERVIEW

Contributors to this chapter discuss a broad range of responses to the threat of waterborne disease, including drinking water disinfection, increasing access to water, improving sanitation, and investment in and implementation of public health interventions. Among these, the most seemingly straightforward approach—water treatment—is actually far from simple, as Philip Singer, of the University of North Carolina at Chapel Hill, demonstrates in the chapter's first paper. Singer provides a quantitative overview of water quality and disinfection, emphasizing the use of chlorine as a disinfectant. He describes water quality factors (e.g., reduced inorganic material, dissolved organic carbon, and microbial contents) that influence chlorine's effectiveness, and explains how sanitary engineers use the concept of "chlorine demand" to assess and address these factors in order to achieve water disinfection with chlorine. He also discusses parameters and limitations of various approaches to water treatment, including solar radiation, giving special attention to the significant barrier to disinfection posed by particulate matter and its removal by various filtration and flocculation methods.

In the developing world, the profound disease burden attributed to diarrhea makes it the most important target for waterborne disease prevention, according to workshop speaker Thomas Clasen of the London School of Hygiene and Tropical Medicine. Following a systematic review of interventions to improve water quality for preventing diarrheal disease (Clasen et al., 2007a), which compared interventions at the both the source (protected wells, bore holes, and distribution to public standpipes) and in the household (improved water storage, solar disinfection, filtration, and combined flocculation-disinfection), he and coauthors concluded that household-based interventions were nearly twice as effective as source-based measures. Clasen and coworkers subsequently conducted a cost-effectiveness analysis to determine the cost per disability-adjusted life year (DALY, a measure of disease burden) averted for a similar range of source and household interventions (Clasen et al., 2007b). The researchers found that upon reaching 50 percent of a country's population, interventions involving household chlorination and solar disinfection paid for themselves and that all interventions were cost-effective.

The most prevalent method of home water treatment worldwide, boiling, was not included in these analyses. Although highly effective in reducing microbiological contamination, boiled water can be readily recontaminated; moreover, Clasen noted, boiling is relatively costly, is associated with risk for burn accidents, and results in indoor air pollution as well as carbon emissions (Clasen, 2008). Because of boiling's prominence, Clasen's group has conducted assessments of its microbiological effectiveness and cost in several developing country settings in order to establish a benchmark against which other safe drinking water interventions can be compared. For example, in a recent study in semirural India, where more than 10 percent of households disinfect their drinking water by boiling, the researchers found that boiling, as practiced in these communities, significantly improves the microbiological quality of water (on a par with water filters), but does not fully remove the potential risk of waterborne pathogens (Clasen et al., 2008). They also calculated that while the entry costs of boiling are the least of any water treatment option in this setting, the cost of continuing the practice annually is greater than the ongoing out-of-pocket cost of treating the same volume of water with sodium hypochlorite, or solar disinfection, and the five-year cost of boiling would also exceed most filtration options.

Efforts to increase the availability, uptake, and correct, consistent use of household water treatment and safe storage systems are spearheaded by the International Network to Promote Household Water and Safe Storage, a consortium of interested UN agencies, bilateral development agencies, international nongovernmental organizations (NGOs), research institutions, international professional associations, and private sector and industry associations (Clasen, 2008; WHO, 2008). The Network now claims more than 100 members from government, UN agencies, international organizations, research institutions, NGOs, and the private sector and has accomplished much in terms of advocacy, communication, research, and implementation. However, despite these achievements, the mission of the Network to "achieve a significant reduction of waterborne disease, especially among children and the poor" is far from realization. Presently, only a tiny fraction of the millions of people who could benefit from household water treatment and safe storage (HWTS) interventions—far more than the one billion who use "unimproved" water sources, as previously noted—are being served, and those who need them most are the most difficult to reach.

In a recent report authored for the World Health Organization (WHO), Clasen (2008) examined efforts to scale up other important household-based interventions (e.g., oral rehydration salts, treated mosquito nets) for lessons of potential value to scaling up HWTS. He found several important recurring themes applicable to scaling up HWTS. These include the need to

- focus on the user's attitudes and aspirations;
- take advantage of simple technologies (minimize behavior change);
- promote nonhealth benefits, such as cost savings, convenience, and aesthetic appeal;
- use schools, clinics, and women's groups to gain access to more vulnerable population segments;
- take advantage of existing manufacturers and supply channels to extend coverage;
- provide performance-based financial incentives to drive distribution;
- align international support and cooperation to encourage large-scale donor funding;
- use free distribution to achieve rapid scale-up and improve equity;
- use targeted subsidies, where possible, to leverage donor funding; and
- encourage internationally-accepted standards to ensure product quality.

In his workshop presentation, Clasen noted that all introductions of novel health interventions to low-income populations face similar challenges—creating awareness, securing acceptance, ensuring access and affordability, establishing political commitment, addressing sustainability—but several additional barriers exist that must be overcome to scale up HWTS. These include

- the widely held belief that diarrhea is not a disease;
- skepticism about the effectiveness of water quality interventions;
- technology shortcomings with the available interventions;
- need for correct, consistent, sustained use (as compared with one-time interventions, such as vaccines);
- the existence of several transmission pathways for waterborne disease;
- suspicion on the part of the public health sector regarding the commercial agenda and lack of standards governing HWTS products;
- the orphan status of HWTS within governmental ministries; and
- the lack of focused international commitment and funding for diarrheal diseases.

"The goal of scaling up HWTS will not be achieved simply by putting more resources into existing programmes or transitioning current pilot projects to scale," Clasen (2008) concludes.

> The gap between where we are and where we need to be is to great given the urgency of the need. What is needed is a breakthrough. The largely public health orientation that has brought HWTS to its present point now need to enlist the help of another group of experts: consumer researchers, product designers, educators, social entrepreneurs, micro-financiers, business strategists and policy advocates. The private sector is an obvious partner; they not only possess much of this expertise but also the incentive and resources to develop the products, campaigns and delivery models for creating and meeting demand on a large scale. At the same time, market-driven, cost-recovery models are not likely to reach vast populations at the bottom of the economic pyramid where the disease burden associated with unsafe drinking water is heaviest . . . mass coverage among the most vulnerable populations may be impossible without free or heavily subsidized distribution. For this population segment, the public sector, UN organizations and NGOs who have special access to these population segments must engage donors to provide the necessary funding and then demonstrate their capacity to achieve both scale and uptake. Governments and international organizations can also help encourage responsible action by the private sector by implementing performance and safety standards and certification for HWTS products; reducing barriers to importation, production and distribution of proven products; and providing incentives for reaching marginalized populations. (Clasen, 2008)

Many of the ideas raised by Clasen regarding appropriately scaled water and sanitation infrastructure for developing countries are expanded upon by workshop speaker Joseph Hughes and coauthors, who offer an engineer's perspective on water infrastructure in the developing world in the chapter's second paper. Caravati et al. envision a new model for water and sanitation infrastructure that addresses global complexities, rather than a "one size fits all" approach based on developed-world systems. The authors describe several promising technologies that may help to address water and sanitation challenges in developing countries. First, however, they provide comprehensive background information on the dynamics of natural water movement, as well as the passage of water through the "engineered hydrologic cycle" of water and wastewater collection, treatment, and distribution.

Conventional, developed-world water and sanitation technologies "are often chemical-, energy- and operational-intensive, are based on heavy infrastructure systems (i.e., dams, pumps, distribution grids, etc.), and require considerable capital and maintenance, all of which hinder their use in much of the world," the authors note. "If safe water and appropriate sanitation are to become accessible to those who are not currently served, new approaches and modern technologies must be employed. This will require a significant change in the way water and wastewater treatment systems are conceived and how they interact with other infrastructures systems (i.e., energy)." They outline a "new paradigm" for water and sanitation infrastructure and describe how progress under way in three vital areas—increased energy efficiency, availability of capital for business creation, and technology development—can advance this paradigm. Their contribution concludes with a review of research needed to fully develop a new, globally-appropriate model for water and sanitation infrastructure.

Given the global trend toward urbanization, particular attention must be paid to water and sanitation challenges for humans—tens of millions of them in megacities—living in close proximity to each other. The chapter's third contribution, by workshop speaker Pete Kolsky of the World Bank and coauthors Kristof Bostoen and Caroline Hunt, focuses on the complex relationships that must be understood in order to recognize and address the threat of waterborne disease in urban settings, particularly in low-income communities. This essay originally appeared as a chapter in the book *Scaling Urban Environmental Challenges: From Local to Global and Back* (Marcotullio and McGranahan, 2007).

Bostoen et al. begin by reviewing the effects of water supply, sanitation, and hygiene on health as viewed through two common models that clarify the complex interrelationships among these elements: classifications of water-related infections (see also Bradley in Chapter 1) and the F-diagram (depicted in Figure WO-13), a model of fecal-oral disease transmission. They then examine goals set by the international community for water and sanitation, along with obstacles that must be overcome in order to meet these goals, including the need to develop reliable measures of progress toward these goals. Following an examination of the significance of boundaries—"limits beyond which and individual or group feels no responsibility"—to urban water and sanitation issues, the authors conclude that institutional boundaries (which are central to many enviromental problems) must be identified and acknowledged. Improvements in water and sanitation services are significant only if they lead to change at the household level, they contend; therefore, household access to these services must be monitored and evaluated.

Ultimately, the threat of waterborne disease must be addressed through investment in safe water and sanitation interventions. Such investments are drastically underfunded, according to workshop speaker Vahid Alavian of the World Bank, who noted that the annual investment in water and sanitation needed to meet the MDGs exceeds $25 billion; only about half that sum is currently being spent. The World Bank is the largest global investor in water/sanitation investment, he added, but its portfolio of about $11 billion cannot begin to meet demand. His colleague Kolsky pointed out that the World Bank's water and sanitation program at the Bank has received a grant of $20 million from the Bill and Melinda Gates Foundation to support sanitation scale-up and hygiene promotion projects, of which a significant fraction (15 to 20 percent) will be spent to evaluate the effectiveness of scaled-up interventions.

The chapter's final essay, by speaker Sharon Hrynkow of the National Institute of Environmental Health Sciences (NIEHS) introduces a potential engine to drive the improvement of water quality and access in low-income settings: the phenomenon of social entrepreneurship. Using illustrative examples, she contrasts the social entrepreneur's approach to solving these problems by focusing on delivering interventions or gathering information for policy purposes with that of medical researchers, who attempt to identify connections between toxins or microbes and illness, and then to reduce human exposure to disease agents.

"Increasing the dialogue between the medical research community and the social entrepreneur community would likely enhance operations on both sides," Hrynkow concludes. In particular, she envisions an alternative to traditional medical grants, which rarely support policy development, that could support both medical research and social entrepreneurship and thereby encourage the transition of solutions for safe water and sanitation from basic science into practice.

MEASURES OF WATER QUALITY IMPACTING DISINFECTION

Philip C. Singer, Ph.D.[1]
University of North Carolina at Chapel Hill

This paper provides a discussion of important water quality factors impacting disinfection, with an emphasis on the use of chlorine as a disinfectant. It has been prepared to be somewhat tutorial in nature in an attempt to educate those unfamiliar with the complexities of water disinfection by chlorine. There are numerous textbooks with chapters on this subject (Letterman, 1999; MWH, 2005).

Drinking Water Disinfectants

Several different types of disinfectants are used to treat drinking water:

- free chlorine ($HOCl/OCl^-$)
- combined chlorine (i.e., monochloramine [NH_2Cl])
- ozone (O_3)
- chlorine dioxide (ClO_2)
- ultraviolet (UV) irradiation

When chlorine is added to water, it hydrolyzes to form hypochlorous acid (HOCl) and the hypochlorite ion (OCl^-). Hence, free chlorine in water is a combination of HOCl and OCl^-. Chlorine is the most widely used disinfectant for the purification of drinking water in the world. Ozone and chlorine dioxide are also used to disinfect drinking water in the United States, western Europe, and in some of the advanced Pacific Rim nations, but not in the developing world. UV

[1] Dan Okun Distinguished Professor of Environmental Engineering, Department of Environmental Sciences and Engineering, Gillings School of Global Public Health.

irradiation—including simple solar irradiation methods employed in the developing world—is a growing technology to disinfect drinking water.

Disinfection Kinetics

Free chlorine is an effective disinfectant for inactivating waterborne bacteria, viruses, and a variety of protozoan cysts (e.g., *Giardia*), but it is not effective against *Cryptosporidium*. Its effectiveness for inactivating microorganisms can be quantified under various conditions by a measure known as CT.[2] CT values are derived from the CT term in the Chick-Watson expression

$$dN/dT = -kCN \quad (1)$$

in which N is the number concentration of microorganisms, k is a rate constant, C is the concentration of the disinfectant, and T is time. Integration of Equation (1) yields the log of inactivation as a function of the concentration of disinfectant multiplied by the contact time, expressed in units of milligram-minutes per liter.

$$\mathrm{Log}_{10}\, No/N = kCT/2.3 \quad (2)$$

The rate constant, k, depends on the specific disinfectant, the type of organism, and temperature. N_0 is the initial concentration of organisms. Requisite CT values to achieve various degrees of inactivation are temperature-dependent.

Table 4-1 shows CT values for the inactivation of *Giardia* and viruses by chlorine over a range of temperatures. In water at 5°C, at a concentration of 1 milligram per liter (mg/L) of chlorine, it will take 149 minutes to achieve 3-log inactivation of *Giardia*. For colder waters, more chlorine and/or longer contact times are needed to achieve the same degree of inactivation. Table 4-1 also shows that the CT values for virus inactivation are smaller than those for *Giardia*, reflecting the fact that viruses are easier to inactivate with chlorine than *Giardia*. At residual chlorine levels of 0.2 to 0.3 mg/L under the same conditions, 3-log inactivation of *Giardia* will require on the order of 12.5 hours (not shown).

Factors Affecting Disinfection with Chlorine

pH

Several factors, in addition to temperature, influence the disinfectant potency of chlorine. The pH is an important consideration because it determines the form of chlorine present (HOCl or OCl^-). Hypochlorous acid is a more potent disinfec-

[2]CT = product of free chlorine residual (C) and contact time (T) required for disinfection.

TABLE 4-1 CT Values (mg-min/L) for Microbial Inactivation by Free Chlorine (pH 7.0, 1.0 mg/L Cl_2 residual)

Temperature (°C)	< = 0.5	5	10	15	20	25
4-log virus inactivation	12	8.0	6.0	4.0	3.0	2.0
3-log *Giardia* inactivation	210	149	112	75	56	37

SOURCE: Based on data in AWWA (2006).

tant than the hypochlorite ion; therefore, disinfection tends to be more effective with decreasing pH.

Chlorine Demand/Reducing Agents

Because chlorine is also a good oxidant, its stability in water is influenced by the presence of reduced inorganic and organic materials in the water, which exert a chlorine demand and chemically reduce the chlorine concentration. Additionally, the *type and state of the microbial agents* (i.e., whether the organisms occur as single cells or are associated with particles suspended in the water) affect the ability of chlorine to disinfect the water.

All of these factors determine the dose of chlorine that must be applied to a given water so that the target residual chlorine concentration (C) remaining at the end of a given contact time (T) can be achieved in order to meet the requisite CT value to ensure the desired degree of inactivation. The dose of chlorine applied, minus the chlorine residual, is known as the "chlorine demand" associated with a particular water supply. In a municipal water treatment facility, chlorine is usually applied to the raw water at the head of the treatment plant or after sedimentation or filtration, and the residual is measured at the point of entry to the distribution system. The difference between the dose and the residual is the chlorine demand and is due to consumption of chlorine by reduced organic and inorganic substances in the water. The higher the concentration of reduced organic or inorganic material, the greater the requisite chlorine dose needed to achieve a target residual and, hence, the greater the chlorine demand. In a village in which a woman collects water and carries it to her home where she adds chlorine to it, the chlorine demand reflects the amount of chlorine that must be added to the water in the container in order achieve the desired degree of inactivation in a specified time period, after which the water is presumed to be safe to drink.

To achieve the desired residual chlorine concentration to meet a target degree of inactivation as characterized by CT, one needs to calculate the dose of chlorine that must be added to any given water. To do this properly, one needs to know the degree to which reduced substances present in the water can lower the concentration of chlorine. This relationship is depicted in Figure 4-1, which compares chlorine dose and residual free chlorine concentrations for several raw

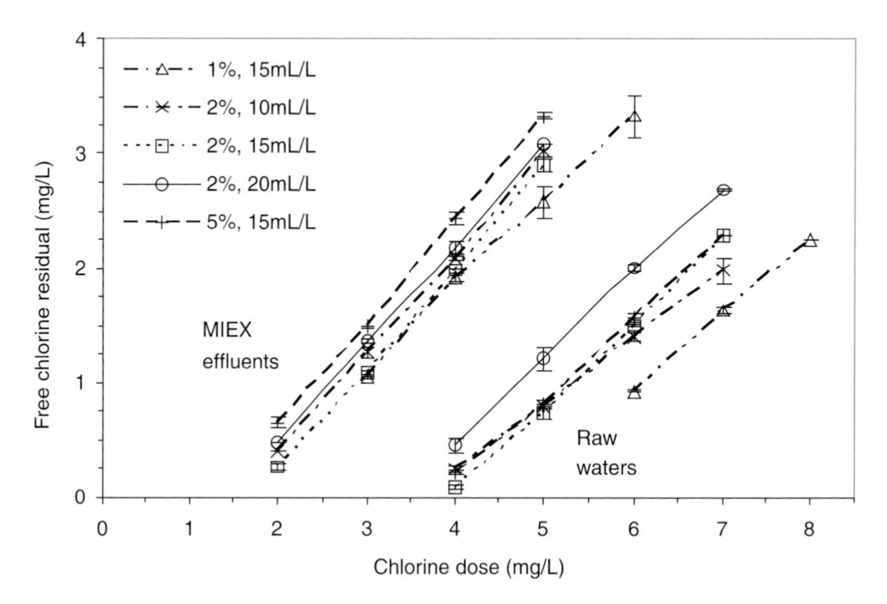

FIGURE 4-1 Chlorine demand of several raw waters and partially treated waters (MIEX® effluents).
SOURCE: Reprinted from Boyer and Singer (2006) with permission from Elsevier.

and partially treated waters (labeled here as MIEX® effluents). The figure shows that, for the raw waters, 5-6 mg/L of chlorine must be applied in order to achieve a free chlorine residual of 1.0 mg/L. In this figure, the contact time is 24 hours. Hence, the chlorine demand of the raw water is 4-5 mg/L. For the treated waters, because a significant amount of dissolved organic material has been removed by treatment, the chlorine doses needed to achieve the same 1.0 mg/L free chlorine residual is 2-3 mg/L, reflecting a chlorine demand of 1-2 mg/L over 24 hours. Hence, in this case, treatment removed approximately 50 percent of the chlorine-demand associated with the dissolved organic material in the raw water.

Table 4-2 presents some examples of chlorine-demanding reactions with four inorganic reducing agents commonly found in raw water supplies: reduced (ferrous) iron (Fe(II)), (manganous) manganese (Mn(II)), sulfide (S(–II)), and ammonia (N(–III)). These balanced stoichiometric reactions illustrate the amount of chlorine that must be added to water to overcome the chlorine demand of these reducing agents. Iron, manganese, and sulfide typically derive from natural sources, whereas ammonia is often associated with municipal and agricultural discharges.

Drinking water sources contaminated by sewage contain not only fecal bacteria and potentially pathogenic microorganisms but also organic material and ammonia, both of which exert substantial chlorine demands. As shown in

TABLE 4-2 Chlorine Demand of Various Inorganic Reducing Agents

Reaction	Chlorine Demand
$2Fe^{2+} + HOCl + 5H_2O \rightarrow 2Fe(OH)_3(s) + Cl^- + 5H^+$	0.64 mg/L of chlorine per mg/L Fe(II)
$Mn^{2+} + HOCl + H_2O \rightarrow MnO_2(s) + Cl^- + 3H^+$	0.93 mg/L of chlorine per mg/L Mn(II)
$H_2S + 4HOCl \rightarrow SO_4^{2-} + 4Cl^- + 6H^+$	8.86 mg/L of chlorine per mg/L S(–II)
$2NH_3 + 3HOCl \rightarrow N_2(g) + 3H^+ + 3Cl^- + H_2O$	7.61 mg/L of chlorine per mg/L NH_3–N

Table 4-2, the chlorine demand associated with ammonia is significant. Figure 4-2 depicts the progression of reactions that occur when increasing amounts of chlorine are added to water containing ammonia at a concentration of 0.5 mg/L as N. The first 2.5 mg/L of chlorine is converted to monochloramine; the next 2.5 mg/L of chlorine destroys the monochloramine. After this breakpoint is reached, free chlorine concentrations increase at essentially a 1:1 ratio as more chlorine is added. Thus, in order to get a free chlorine residual (the concentration of free chlorine beyond the breakpoint) necessary to meet the CT requirements for disinfection in water containing 0.5 mg/L of ammonia, at least 5 mg/L of chlorine must be added.

Natural organic material contains aromatic structures, unsaturated double bonds, and organic nitrogen, all of which react with chlorine. In addition to these oxidation reactions, chlorine participates in substitution and addition reactions to produce potentially carcinogenic halogenated byproducts. These include trihalomethanes, which are regulated in the United States by the Environmental Protection Agency and elsewhere in accordance with World Health Organization guidelines. On average, 1 to 1.5 mg/L of chlorine is consumed per mg/L of dissolved organic carbon (DOC) over the course of 24 hours, at pH 8 and 25°C.

Raw drinking waters generally contain a combination of chlorine-demanding impurities. A poor-quality surface water, for example, might contain 0.5 mg/L of ammonia and 6 mg/L of dissolved organic carbon, giving a total chlorine demand of 11-14 mg/L (5 mg/L to oxidize the ammonia in accordance with the breakpoint curve in Figure 4-2 and 6 to 9 mg/L for the 6 mg/L of dissolved organic carbon). For a better-quality surface water with 0.2 mg/L ammonia and 2 mg/L DOC, the chlorine demand would be 4-5 mg/L. For groundwater containing 1 mg/L iron, 0.5mg/L manganese, and 1 mg/L DOC, the chlorine demand would be on the order of 2.4 mg/L. Thus, different amounts of chlorine must be added in each case to achieve the same residual free chlorine levels needed for effective disinfection.

Measurement of Chlorine Residual

The most common method for measuring chlorine residual in treated water, the *N,N*-diethyl-*p*-phenylenediamine (DPD) colorimetric/spectrophotometric method,

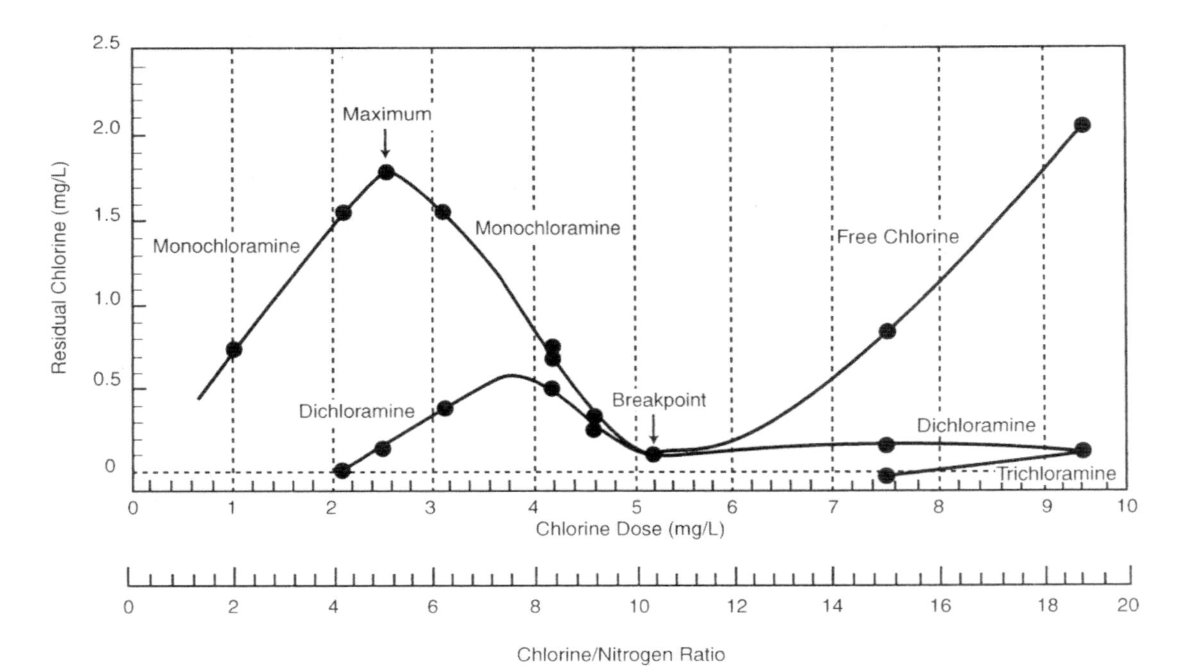

FIGURE 4-2 Breakpoint chlorination curve when chlorine is added to an ammonia-containing water.
SOURCE: Reprinted from *Water Chlorination/Chloramination Practices and Principles (M20)*, with permission. Copyright © 2006 American Water Works Association.

can distinguish between free and combined (monochloramine) chlorine species. However, because this method is subject to certain interferences, it must be performed and interpreted carefully, especially at low free chlorine concentrations and in waters containing dissolved organic nitrogen. Organic chloramines that are formed when free chlorine is added to water containing dissolved organic nitrogen, a component of the breakdown of proteinaceous material, also tend to react like free chlorine in the DPD colorimetric analysis. Because these organic chloramines do not have the disinfecting power of free chlorine, their presence gives an artificially high apparent free chlorine residual and a false sense of disinfection effectiveness. Additionally, while the minimum "detectable residual" with the DPD test is 0.2 mg/L chlorine, there are many instances where such detectable residuals have been measured by the DPD test but these waters have also been found to contain coliform bacteria that should not survive in the presence of chlorine at that concentration. Thus, measurements of free chlorine residual are subject to some uncertainty and must be interpreted carefully depending on water quality conditions.

Turbidity and Particle Content

The effectiveness of disinfection is impacted by the presence of particulate material in the water. Particles tend to protect the microorganisms from exposure to the disinfectant, especially if the microorganisms are present in an aggregated state. In the latter case, the organisms at the surface of the particle are exposed to the disinfectant but the organisms inside the aggregate are protected from exposure.

The turbidity of water is used as a surrogate for particle content. In the United States, water turbidity is monitored continually using a simple, relatively inexpensive device called a turbidimeter or nephelometer, which measures the light-scattering properties of particles at 90° to the incident light (see Figure 4-3). The intensity of scattered light is a function of the number, size, and shape of the particles present in the water (as well as of the wavelength of incident light, geometry and detection characteristics of the instrument, and its method of standardization and calibration). Water turbidity is therefore a collective reflection of a property of the particles (their light scattering characteristics) rather than a specific measure of particle size, number, or morphology.

In general, small particles scatter light more than larger particles, with the greatest degree of scattering resulting from particles that are about 0.5 microns (μm) in diameter, which is equivalent to the wavelength of the incident light. Viruses, which are much smaller in size (on the order of 0.03 μm), do not scatter light, whereas bacteria (0.5 to 1 μm in size), *Cryptosporidium* oocysts (3 to 5 μm), and *Giardia* cysts (10 to13 μm) do scatter light. Hence, the absence of a measurable turbidity does not guarantee that the water is free of harmful microorganisms.

A better measure of particle content can be achieved with particle size analyzers. A variety of particle size analyzers are commercially available for

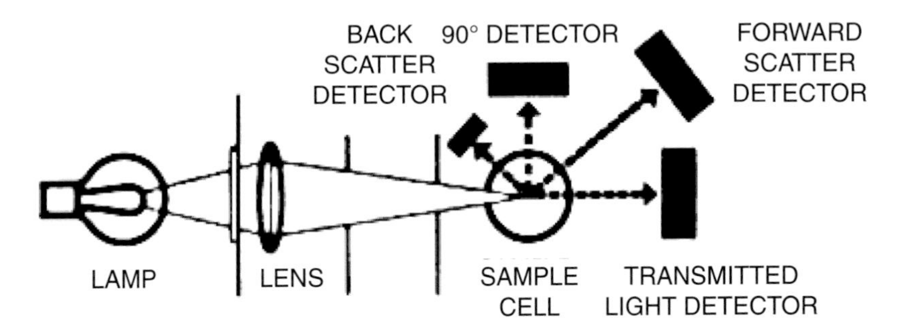

FIGURE 4-3 Schematic of a nephelometer used to measure turbidity.
SOURCE: Reprinted from http://www.eoc.csiro.au/instrument/html/marine/marine_images/hach_diagram.gif (accessed April 16, 2009) with permission from Hach Company.

characterizing particles in water. Particle counters can measure particle size and concentration, providing information about size distribution (i.e., the number concentration of particles of various sizes) in the water. Optical methods (see Figure 4-4) compare light blockage by the different particles as a known volume of water is drawn through an orifice. The degree of blockage is proportional to the cross-sectional area of the particle. As particles of different size are drawn through the sensing zone of the instrument, the extent of blockage of the incident light is sorted into different channels according to the amount of light blocked, giving rise to a particle size distribution based on the diameter of an equivalent sphere with the same cross-sectional area. Resistivity-based methods, which measure the volume displacement of water by particles of different sizes in a salt solution, provide information on the size distribution of particles according to their volume-equivalent particle diameter.

Image analyzers are recent additions to particle characterization techniques in water quality analysis. With these instruments, water samples can be continuously examined under a microscope, photographed, and the images stored in a computer file for subsequent review and analysis. These instruments permit identification of particle type and morphology. As previously noted, microorganisms in a sample may be present as single cells or as aggregates, and it is important to determine their degree of aggregation in order to assess their susceptibility to be inactivated by chlorine or any other disinfectant.

Particle Removal

Because microorganisms are often found in an aggregated state, and because particle-associated microorganisms are difficult to inactivate, the first line of defense against microorganisms of potential public health concern is filtration.

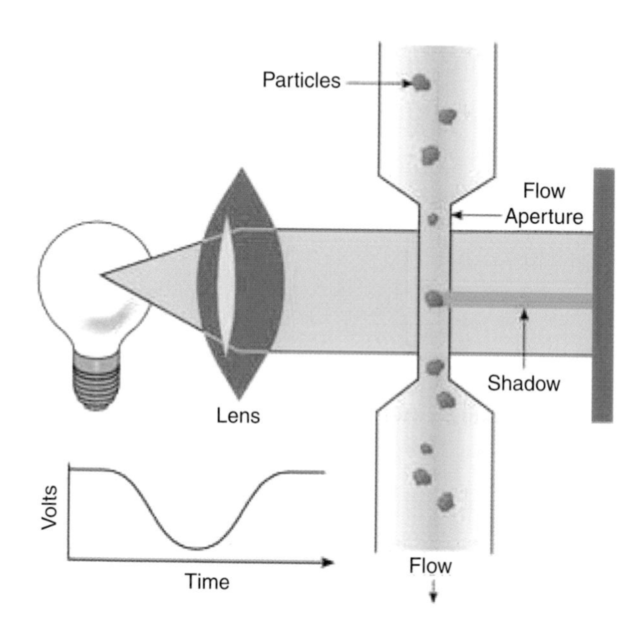

FIGURE 4-4 Schematic of an optical particle counter.
SOURCE: http://www.oilanalysis.com/backup/200207/PartCount-Fig1.jpg (accessed April 16, 2009) © Noria Corporation.

Filtration removes particles, both particles that are present in an aggregate state and free, single-cell organisms which are themselves particulate in nature. Disinfection with chemicals, such as chlorine, and physical inactivation methods, such as UV light, cannot be relied upon to safely disinfect water that has not undergone filtration. Effective filtration can occur via natural means (e.g., filtration that occurs by flow through porous media in a groundwater aquifer), engineered filtration (e.g., granular media filtration in a water treatment plant), or household filtration (e.g., filtration in a biosand or ceramic pot filter, both of which are being widely promoted for use in rural villages in developing countries). Once the raw water is relatively free of particulate material, chemical disinfectants or UV light can provide effective disinfection, provided the dose and contact time are sufficient.

CIVIL INFRASTRUCTURE FOR WATER, SANITATION, AND IMPROVED HEALTH: EXISTING TECHNOLOGY, BARRIERS, AND NEED FOR INNOVATION

Kevin C. Caravati
Georgia Tech Research Institute

Zakiya A. Seymour[3]
Georgia Institute of Technology

Joseph B. Hughes, Ph.D., P.E., BCEE[4]
Georgia Institute of Technology

Introduction

Civil infrastructure can be broadly described as the systems, services, and facilities needed for a functioning community or society. Examples of easily recognized civil infrastructure include dams, bridges, buildings, roads, transmission and distribution lines, and communication technologies. These and other infrastructure systems, collectively, are central to the improvement of health, quality of life, and prosperity of communities. Among the most basic of all civil infrastructure systems are those that store, convey, treat, and provide potable water, as well as collect, treat, and safely discharge wastewater. Together, these systems have a dramatic impact on human health and the health of the environment in which people live.

Engineers and scientists have studied methodologies to purify water contaminated with biological, chemical, and physical contaminants for centuries, and well-accepted techniques for water and wastewater treatment have been employed in the United States and other parts of the developed world for many decades. Yet access to water and sanitation remains one of the largest challenges for societies around the globe. Of the world's 6.5 billion-plus inhabitants, an estimated 1.2 billion people lack access to safe drinking water and 2.6 billion, or 42 percent of the world's population, lack access to basic sanitation (World Water Assessment Programme, 2006). Concerns of water access and improved sanitation are compounded by threats associated with climate change and a projected growth in population to 8 billion people by the year 2030. According to the Organisation for Economic Co-operation and Development (OECD), the number of persons

[3]School of Civil and Environmental Engineering.

[4]Schools of Civil and Environmental Engineering, and Material Science and Engineering. Corresponding author may be contacted at Georgia Institute of Technology School of Civil and Environmental Engineering, 790 Atlantic Dr. N.W., Atlanta, GA 30332-0355, Phone: (404) 894-2201, Fax: (404) 894-2278, E-mail: joseph.hughes@ce.gatech.edu.

in water-stressed countries is expected to increase to nearly four billion (OECD, 2008). By 2025, more than half the nations in the world will face freshwater stress or shortages, and by 2050 up to 75 percent of an estimated 9.1 billion people could face freshwater scarcity (Hightower and Pierce, 2008).

The future health, prosperity, and security of the human race will be strongly influenced by our ability to access clean water. Growing populations, rapid urbanization, expanding industrialization, changes in climate, growth in irrigated agriculture, and the globalization of corporations further contribute to our global water challenge. Infrastructure models deployed in the developed world over the past century are often ill suited to provide sustainable, scalable solutions. A new model is needed that addresses the complexities around global water and sanitation infrastructure.

Presented herein is an analysis of impediments to the solution of the world's water and sanitation needs through existing technology. First, a background section is prepared to provide brief reviews of (1) the cycle of water in nature and (2) the cycle of water through engineered systems. Second, an examination of well-established water and wastewater infrastructure models detail a range of factors that exist as barriers to the translation of models globally. Finally, a discussion of the potential for technological innovations to provide advances in improving water and sanitation challenges in developing countries is presented.

Background

Water is a widely occurring compound on Earth that is found in liquid, gaseous, and solid forms. Oceans cover approximately three-quarters of the Earth's surface and contain over 97 percent of the world's water supply. Oceanic water contains high salt concentrations and is unfit for human consumption without extensive treatment. Less than 3 percent of the Earth's water is "fresh water," and nearly 70 percent is in glaciers and icecaps, providing a mere 0.3 percent accessible for human consumption in lakes, reservoirs, rivers, and aquifers.

To understand water as a resource for human consumption, it is important to understand the dynamics of natural water movement among its three phases and the reservoirs for water that provide storage. Correspondingly, it is necessary to understand the flow, storage, and treatment of water through engineered systems used for the collection, treatment, and distribution of potable water and wastewater. For the purposes of clarity and consistency, the process of water movement and storage on Earth without human intervention is referred to as the "natural hydrologic cycle." Water movement through infrastructure systems for the delivery of clean drinking water to the public is referred to as the "engineered hydrologic cycle."

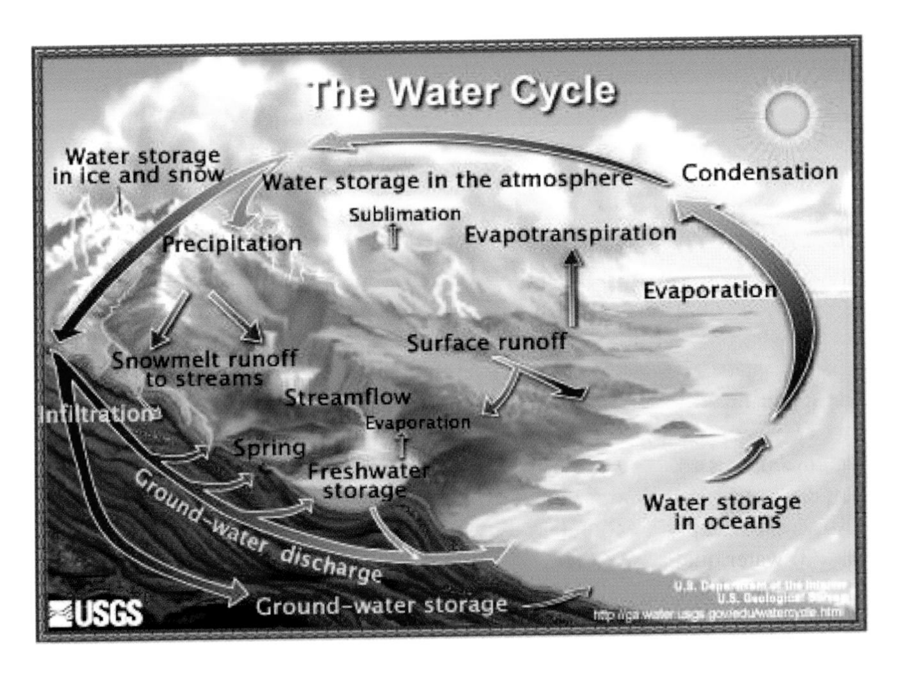

FIGURE 4-5 The hydrologic cycle.
SOURCE: USGS (2008).

Natural Hydrologic Cycle

The largest reservoir for water is in the oceans. Evaporation[5] results in the transfer of water to the atmosphere. Condensation of gaseous water to the liquid state occurs within the atmosphere forming clouds and precipitation. A diagram of the natural hydrologic cycle is presented in Figure 4-5. Roughly three-quarters of global precipitation results in water returning to the oceans. The fraction that falls on land initiates the terrestrial component of the water cycle. On land, water will be present as a solid (ice and snow) and as a liquid. The inflow, or precipitation that falls on land, is the predominant source of water required for human consumption, agriculture and food production, industrial waste disposal processes, heat dissipation in energy production, and for support of natural and seminatural ecosystems (World Water Assessment Programme, 2003). Outflow of the natural water cycle includes drainage to the oceans via rivers, lakes, and wetlands; evapotranspiration of water to the atmosphere from soils by plants; and evaporation from fresh water reservoirs. Water that collects into flowing bodies of water is categorized as surface water (streamflow). Water that seeps through soils

[5]The conversion of liquid to gas.

into underlying rock layers is contained as groundwater. Surface water systems are more rapidly recharged than groundwater. In fact, the flow of groundwater systems is very slow and the volume of water they can produce is often finite.

Engineered Hydrologic Cycle

Engineered water systems are processes that convey, store, and alter water quality. Many of these processes are designed to enhance what characteristically occurs in the natural environment (i.e., the breakdown of organic contaminants in wastewater by bacterial communities). A vast network interlinking water, sanitation, and energy infrastructure systems has been built to ensure clean drinking water and mitigate the impacts of waste on human and environmental health. The development and reliability of this network, collectively referred to as the "engineered hydrologic cycle," are critical factors in the growth and health of populations. Major components to describe the details of this network consist of (1) water supply creation and protection, (2) water and wastewater treatment, and (3) water quality.

Creation and Protection of Water Supply

An estimated 48 percent of the freshwater supply in the United States is used for energy production. The remaining percentage is classified into irrigation (34 percent), public use (11 percent), industrial use (5 percent), and mining, livestock, and aquaculture (less than 2 percent; USGS, 2005). While the focus of this discussion is public use, the largest supply issues are driven by the energy and agriculture sectors.

Due to the scarcity of freshwater throughout the world, careful consideration must be given to create and protect water supplies. Over the past 200 years, human activities have developed to such an extent that only a few natural water bodies remain (World Water Assessment Programme, 2003). Water management over the past century has focused on large-scale diversions of water out of natural systems; more than 60 percent of the world's rivers have undergone major hydrological alterations (Revenga et al., 1998). Throughout much of the developed world, large reservoirs with intake structures, dams, and distribution facilities provide clean water to millions over vast areas. Well fields that tap highly productive aquifers can do the same. Additional infrastructure changes also include extensive channelization of river systems, massive pumping of aquifers, and long-distance water conveyance systems. These engineered systems, representing the conventional approach for creating water supplies, have significantly altered lifestyles and the environment by ensuring the sustainability of water resources (Gleick, 2006).

In conjunction with creating water supplies, the development of resource protection programs ensures water quality. Resource protection involves several stakeholders, including regulatory agents, governments, commercial users, and residential consumers. It comprises a variety of watershed protection practices, including restricting land use in sensitive areas, managing solid wastes, and preventing saltwater intrusion. Collectively, through the application of these methods and the cooperative assistance of involved stakeholders, water supplies are safeguarded and public health is protected. The New York City Watershed Partnership project, described in Box 4-1, is an example of a successful large-scale water supply protection and infrastructure project. This partnership provides unfiltered drinking water to nine million people while preserving the economic viability and social character of the communities located in the upstate watershed (EPA, 2006).

Water and Wastewater Treatment

To ensure clean water supplies in the developed world, engineers have created treatment systems to provide safe drinking water from both surface water and groundwater sources, and to return the resulting wastewater to surface waters (on rare occasions water is returned to aquifers). These systems, which include drinking water and wastewater treatment facilities, conveyance systems, above-ground

BOX 4-1
Unfiltered Drinking Water for Millions: The New York City Watershed Partnership

New York City's water managers face two challenges: protecting the public health of both the city's population and the users of source water areas in upstate New York, and maintaining the economic viability of the upstate livestock and dairy communities. The agreement brokered by the New York State Governor and the EPA allows the Catskill Mountain/Delaware Watershed to remain unfiltered because of the high quality of the water supplies. New York City is required to implement a wide range of watershed protection programs at a cost of approximately $1.4 billion. Filtration of the Catskill Mountain/Delaware Watershed system would have cost the city $6 to $8 billion.

An effective watershed management program is the underlying success factor. The Safe Drinking Water Act requires that all drinking water taken from surface water be filtered to remove microbial contaminants. The law allows EPA to waive this requirement for water suppliers if they have an effective watershed control program and if their water meets strict quality standards. This initiative demonstrates that by protecting reservoirs and areas surrounding source waters it is possible to supply water for a massive urban population without the need for expensive filtration before chemical treatment.

SOURCE: EPA (2006).

storage facilities, and residuals management sites, are necessary for providing water for public use as well as for disposing of wastewater and wastes in a safe and efficient manner. A graphic depiction of these water and wastewater treatment systems is shown in Figure 4-6.

Initially, surface water and groundwater are directed to a drinking water treatment plant. The water is then treated to remove particulate matter (i.e., solids) and dissolved contaminants and is disinfected. Additional treatment for taste, appearance, hardness, and odor occurs before it is stored and/or piped to a distribution system. This distribution system delivers drinking water for consumption. After the public use, wastewater, through a collection system, is sent to a wastewater treatment plant for solids removal and biological and chemical treatment. Afterward, treated wastewater is discharged to a river, lake, reservoir, ocean, or aquifer system.

Water and wastewater treatment methodologies involve physical, biological, and chemical processes; these treatments are usually completed in series, and are, subsequently, also known as primary, secondary, and tertiary treatments. In potable water treatment, physical (i.e., filtration) and chemical (i.e., chlorination) are most common. In wastewater treatment, biological treatment is used to remove organic matter (i.e., biochemical oxygen demand or BOD) and treat residuals (i.e., anaerobic digestion). Physical treatment (i.e., gravity separations) and chemical treatment (i.e., chlorination) are also essential in wastewater treatment facilities. If the wastewater has been created in an industrial facility, additional treatment(s) may be required prior to its discharge. Listed in Table 4-3, as modified from Tchobanoglous et al. (2003) and in Viessman and Hammer (1985), are various methodologies that treat water and wastewater.

Treating water before and after public use is essential; often, surface waters that receive wastewater discharges are typically a drinking water source for communities downstream. Practices implemented in one part of a watershed have the capacity to impact its potential use as a water source. Consequently, the conventional engineered solution for drinking water and wastewater treatment, as seen in the developed world, imparts its own engineered hydrologic cycle in concert with the natural water cycle, providing integrated resource protection to reduce vulnerability and maintain the quality of water sources.

Water Quality

The quality of water found in rivers, lakes, reservoirs, and groundwater depends on several linked factors. These factors include geology, climate topography, biological processes, land use, and water residence time. Nonetheless, the degree of sanitation practiced is the most critical determinant of contamination of drinking water with pathogens.

Programs dedicated to water and sanitation typically spend 95 percent of their resources on water (Black, 2008), yet the global return on investments in

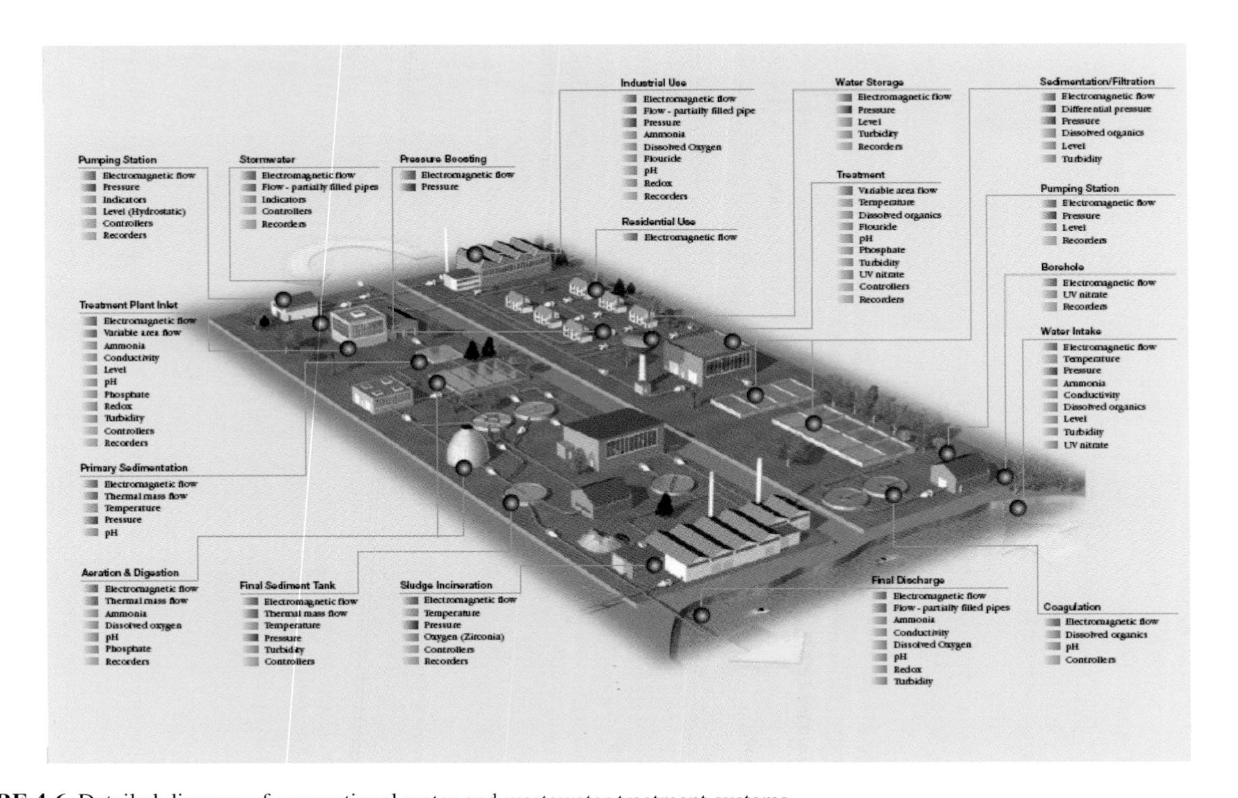

FIGURE 4-6 Detailed diagram of conventional water and wastewater treatment systems.
SOURCE: Reprinted from ABB (2007) with permission, www.abb.com/water.

TABLE 4-3 Conventional Water and Wastewater Treatment Methodologies

Treatment Level	Description
Water Treatment Level*	
Preliminary	From a shallow well or surface water intake, chlorine (for disinfection), lime (for softening), or alum (for coagulation), or combinations of these or other compounds, are added prior to mixing and flocculation. Activated carbon may be used for taste and odor control.
Presedimentation or mixing/flocculation	Mixing, flocculation, aeration, and sedimentation processes are used to remove excess lime, reduce hardness, increase dissolved oxygen, and reduce carbon dioxide.
Detention	Detention or contact tanks are used to complete oxidation reactions; activated carbon treatment also may be used for taste and odor control.
Sand filtration	Sand filters are used to remove unsettled floc and for iron and manganese removal.
Chlorination/fluoridation	Chlorine is added to establish a free chlorine residual in the distribution system; fluorosilicic acid may be added for fluoridation.
Waste removal	Waste removal may include chemical additions, flocculation, clarification/thickening, and filtration.
Water discharge and waste storage/disposal	Water discharge to a holding tank or gravity system; wastes are processed, collected, stored, and disposed.
Wastewater Treatment Level†	
Preliminary	Removal of wastewater constituents such as rags, sticks, floatables, grit, and grease that may cause maintenance or operational problems with the treatment operations, processes, and ancillary systems.
Primary	Removal of a portion of the suspended solids and organic matter from the wastewater.
Advanced primary	Enhanced removal of suspended solids and organic matter from the wastewater. Typically accomplished by chemical addition or filtration.
Secondary	Removal of biodegradable organic matter (in solution or suspension) and suspended solids. Disinfection is also typically included in the disinfection of conventional secondary treatment.
Secondary with nutrient removal	Removal of biodegradable organics, suspended solids, and nutrients (nitrogen, phosphorus, or both nitrogen and phosphorous).
Tertiary	Removal of residual suspended solids (after secondary treatment) usually by granular medium filtration or microscreens. Disinfection is also typically a part of tertiary treatment. Nutrient removal is often included in this definition.
Advanced	Removal of dissolved and suspended materials remaining after normal biological treatment when required for water reuse applications.

SOURCE: *Based on Viessman and Hammer (1985). †Reprinted from Tchobanoglous et al. (2003) with permission from McGraw-Hill and adapted, in part, from Crites and Tchobanoglous (1998).

low-cost sanitation provision may be in the range of $9 for each $1 spent (Hutton et al., 2006). Human and agricultural wastes pose a remarkably high risk to waterborne disease. The principal bacteria that cause intestinal disease include several genera of *Salmonella, Shigella, Vibrio cholerae, Leptospira*, and *Yersinia enterocolitica*. An average bowel movement weighs 250 grams and an average human produces 77 pounds of excrement per year (George, 2008); one gram of feces can contain 10 million viruses, 1 million bacteria, 1,000 parasite cysts, and 100 worm eggs (George, 2008), carrying the potential to threaten any water source. In nature, there is no pure water, but water that has been tainted due to poor sanitation is particularly poor quality.

The measurements of certain physical, chemical, and biological water quality parameters assist in determining the suitability of water for various purposes, the effectiveness of water and wastewater systems, and its potential impacts on public health. Water quality regulations in the United States and other developed countries require that water be properly managed to improve human health and minimize environmental degradation. In the United States, the number of water quality parameters evaluated on a routine basis at water treatment and wastewater treatment plants is considerable. A short list of examples includes pH, dissolved oxygen, hardness, turbidity, BOD, nitrate, ammonia, color, fecal coliform bacteria, heavy metals, and toxic organic pollutants, among others. Table 4-4, as provided by the American Water Works Association (AWWA, 1999), summarizes the source and significance of certain water quality parameters.

TABLE 4-4 Water Quality Parameters

Parameter	Sources	Effects of Water Supply
Solids, turbidity	Domestic sewage, urban and agricultural runoff, construction	Hinder water treatment process. Reduce treatment effectiveness. Shield microorganism against disinfectants. Reduce reservoir capacity.
Nutrients	Septic system leachate, wastewater plant discharge, lawn and road runoff, animal feedlots, agricultural lands, eroded landscapes, landfill leachate, rainfall (especially nitrogen)	Nitrates that may be toxic to infants and unborn fetuses; accelerates eutrophication; high levels of algae; dissolved oxygen deficiencies; increase algae activity; high color and turbidity; disinfection by-product formation; taste and odor problems.
Natural organic matter	Naturally occurring; wetlands in the watershed tend to increase concentrations	Influence nutrient availability; mobilize hydrophobic organics; disinfection by-product formation.

continued

TABLE 4-4 Continued

Parameter	Sources	Effects of Water Supply
Synthetic organic contaminants	Domestic and industrial activities, spills and leaks, wastewater discharges, agricultural and urban runoff, leachate, wastewater treatment and transmission	Adverse impacts on human health and aquatic life.
Coliform bacteria	Domestic sewage from wastewater discharges, sewers, septic systems, urban runoff, animal farms and grazing, waterfowl droppings, land application of animal wastes	Fecal coliform are indicators of warm-blooded animal fecal contamination that pose a threat to human health with microbial pathogens like *Giardia* and *Cryptosporidium*.
Metals	Industrial activities and wastewater, runoff	Adverse effect on aquatic life and public health.
Oil and grease	Runoff containing kerosene, lubricating and road oils from gas stations, industries, domestic, commercial and institutional sewage, food waste and cooking oil	Interfere with biological waste treatment causing maintenance problems; interfere with aquatic life; aesthetic impacts; human health impacts associated with selected hydrocarbons.
Sodium	Road deicing and salt storage	High blood pressure and heart disease.
Toxics	Agriculture, lawn care, industrial sites, roads and parking lots, wastewater	Toxic to humans and aquatic life.
Aesthetics	Taste and odor: industrial chemicals, algae metabolites, natural organic matter (NOM), urea; Color: metals, NOM, algae; Turbidity: solids and algae; Staining: Metals	Aesthetic problems; reduce public confidence in water supply safety.
Algae	Wastewater plant discharges, septic systems, landfill leachate, urban and agricultural runoff, precipitation	Taste and odor; filter clogging; some algae species toxic to aquatic life.
Dissolved oxygen	Organic matter, wastewater discharges, runoff, consumption by aerobic aquatic life and chemical substances	Water treatment problems; release of iron and manganese; taste and odor problems; ammonia.

SOURCE: Reprinted from AWWA (1999) with permission from McGraw-Hill.

Established Water/Wastewater Infrastructure Models

The development of the engineered hydrologic cycle is the basis for establishing water/wastewater infrastructure models. For discussion purposes, currently established water/wastewater infrastructure models common in the United States are referred to as the "conventional infrastructure model." Where traditional infrastructure models are not used, the term "household water treatment and storage model" is employed. Components of each model are discussed in the following section along with the barriers for increased implementation of both.

Conventional Water/Wastewater Infrastructure Model

The overall success of conventional water infrastructure in the developed world is attributed to the complementary advancements in engineered, economic, and societal systems since the mid-1800s. Conventional water and sanitation technologies, as seen in most developed regions, were mature decades ago—although through research engineers have continued to improve the performance of treatment systems. These technologies are often chemical-, energy-, and operational-intensive, are based on heavy infrastructure systems (i.e., dams, pumps, distribution grids, etc.), and require considerable capital and maintenance, all of which hinder their use in much of the world (Shannon et al., 2008). The success of the U.S. water infrastructure improvements relies on several interrelated and critical elements. Examples are provided below:

- *Energy Use.* Energy consumption represents the largest operational cost in water and wastewater treatment. It is essential for source collection and conveyance systems, distribution systems, and treatment. It is also needed for pumps (within wells, at surface water intakes, and for distributing water), aerators, chemical feed systems, and biological treatment systems.
- *Chemical Use.* Water and wastewater treatment require large quantities of chemicals to treat water efficiently and to remove elements that may inhibit water quality. Chemicals are also used for corrosion control and disinfection within the distribution systems. Chemical use can include ammonia, chlorine compounds, antiscalants, ozone, permanganate, alum, ferric salts, sodium hydroxide, hydrochloric acid, and ion exchange resins.
- *Subsidies.* Subsidies are often required to ensure widespread acceptance and adoption of water supply and wastewater treatment to the public. Additionally, certain subsidies, such as those provided to the agricultural industry within the United States, can deeply impact the availability of water. Regulatory agencies and governments must consider the interests of the public.

- *Regulatory Frameworks.* Regulatory frameworks are needed to establish ownership rights, to set standards for drinking water and industrial wastewater treatment, to develop water permitting requirements, and to determine resource allocations (power generation, recreation, agriculture).
- *Available Capital.* Determining the upfront capital available for the design and construction costs of a water infrastructure system is essential. Furthermore, careful consideration must be given to the long-term operation and maintenance needs of the facilities. In most developed countries, these expenditures are typically financed by the public sector.
- *Property Ownership.* Obtaining the appropriate property rights-of-way is necessary to protect the components of the water infrastructure system, including the watersheds, reservoirs, piping, and distribution sites. Additionally, determining property ownership will provide useful knowledge for the system users and the ability to establish a payment system for water usage.
- *Social Acceptance.* While regulatory frameworks establish the legality frameworks, the public must also accept components of a water infrastructure system. There are several decisions made when developing a water infrastructure system, such as determining financing options, permitting restrictions, and treatment and sludge disposal methods. These alternatives highlight the complexity of decision matrices and stress the need for collaboration with the public.

Barriers for Expanded Application of Conventional Infrastructure Models

The components essential for building conventional water infrastructure systems often do not exist in developing countries. Reliable, consistent supplies of energy are required, and chemicals for treatment may not be obtainable. Governments in developing nations often lack the authority and resources to implement large-scale programs, or they may be plagued by corruption or governance issues. Organizations may be able to obtain the initial funding but may not have sufficient revenue collection systems needed for operational costs. Additionally, concerns exist about the development of water resources through major dam-building programs. This lack of resources, revenue, and regulation impedes the successful development of the conventional water infrastructure model in developing countries. Specific examples of barriers that effectively prohibit the reliable and effective use of conventional water and wastewater infrastructure systems are illustrated below.

Water supply stress Worldwide water supplies and the quality of freshwater are being impacted by climate change, demographic shifts, and population growth, creating regions experiencing water stress. The traditional solution to water stress has been to enhance the water supply through conveyance from increasingly distant sources. Often, these systems include well-field and water distribution networks, household connections for wastewater conveyance and treatment, and the installation of dams for combined hydroelectric power generation and water supply reservoirs. Yet, this approach is frequently found unreliable in developing countries due to environmental, social, and economic reasons (Gleick, 2000). Traditional water management methods to divert water out of natural systems, if built, have not been maintained effectively; thus, much of the water supply is restricted to nearby surface water or shallow groundwater, making it more difficult to find and retain water supplies for croplands and urban centers. Both surface and groundwater supplies can become contaminated from human, industrial, and animal waste; depending on water demands, groundwater water tables can fall below acceptable recharge levels.

Concentrated population growth places a particular stress on water supplies. Over the next century, an additional three billion people will live in urban areas (Zimmerman et al., 2008), further focusing water demand and waste production. Outside of urban areas, the loss of forests and vegetation to support urbanization causes increased sedimentation, loss of wetlands, and eutrophication (Zimmerman et al., 2008). By the year 2025, water withdrawals are expected to increase from current levels by 50 percent in developing countries and 18 percent in developed countries (Zimmerman et al., 2008). Demands from the agricultural, industrial, and energy sectors will compete with the needs of coastal developments; economic development will likely take precedence over resource protection. Lacking strong governance systems, agencies are subject to corruption and an inability to create or enforce environmental protection practices to ensure the sustainability of water resources.

Energy The lack of reliable energy is problematic for establishing conventional solutions in developing regions. Energy requirements for water and wastewater treatment are staggering. Without reliable energy systems, conventional treatment processes simply fail. Equally important to having energy to treat water will be the impact of energy systems on water stress. Growth in future energy production is projected to be highest in water poor regions. Furthermore, the regions with the highest increased projected energy demand will include coastal North and South America, the Middle East, India, and China. Increasing energy production will require a reevaluation of water resources and reduce freshwater availability.

Poor sanitation The effect of poor or nonexistent sanitation on water resources can overwhelm the ability of conventional water treatment systems to provide safe water. Without regulations, surface and groundwater systems are subject to contamination with fecal matter, rendering them unfit for use as a drinking water source without advanced treatment systems. For millions of people, water is simply too precious to be used for disposal of waste, therefore the waste will not enter or be conveyed through a sewer system even if one existed. Conventional wastewater treatment requires that waste be conveyed by water. In many locations that simply will not occur and the engineered water cycle is "short circuited."

Economics and regulatory needs The price one actually pays for water is but a small fraction of what it truly costs to extract it, deliver it to users, and treat it after its use (Revenga, 2009). Subsidies that hide the true costs of water or subsidies for programs that can pay for themselves have proven to be ineffective. Furthermore, governments are less willing to subsidize large dam or water infrastructure projects and are shifting more responsibility to regional and local governments (Gleick, 2000).

Privatization of water services (and energy services) has been a "widespread failure" according to a recent United Nations Development Program study (Bayliss and McKinley, 2007). Private investors have shied away from investing in public utilities, and the private enterprises have focused primarily on cost recovery and not the provision of services at equitable prices to the poor. Current efforts tend to focus on building the capacity of the public sector, but funding sources for such projects are scarce. Energy subsidies in India encourage groundwater users to pump more than they actually need, leading to massive declines in local and regional water tables (Gleick, 2000). Agricultural subsidies in Asia encourage the consumption of inflows to the Aral Sea, resulting in shrinkage of the Sea and loss of species and livelihoods (Gleick, 2000).

Policies and pricing decisions that do not effectively price water lead to misuse. The poor pay more for water that is delivered from private sources compared to water provided by a municipality (World Water Assessment Programme, 2006). The World Health Organization defines reasonable access as the availability of at least 20 liters per capita per day from a source within one kilometer of the user's dwelling (WHO, 2008).

Regulatory frameworks are particularly important for sanitation services, but sanitation is typically underfunded. Waste disposal services are often nonexistent in poor rural areas or crowded urban centers or are unaffordable. Subsidies often are needed to promote access to utility services (Bayliss and McKinley, 2007), and some utilities have introduced "lifeline tariffs" in which minimal levels of utility services are provided free or at low cost. Oftentimes, utility connections to poor neighborhoods are subsidized; water delivery from these connections needs to be safe and reliable; otherwise public trust in the local authorities is lost.

Issues of property ownership plague megaslums and rural areas, which complicates the delivery of water and energy services. The ability of dwellers or residents to pay for services and the ability to collect connection fees and monthly user fees can be barriers to improvements.

The adoption of a traditional water infrastructure system by many institutions has reached a plateau as environmental and social concerns have increased. Population growth and changing demographics have limited unbridled expansion opportunities that existed in the nineteenth and twentieth centuries. Financing of major projects has declined due to rising material costs, legal opposition from stakeholders, and now a tightened credit market.

Household Water Treatment and Storage Water Treatment Model

The most common household water treatment and safe storage systems include chlorination, filtration (biosand and ceramic), solar disinfection, combined filtration/chlorination, and combined flocculation/chlorination. Lantagne et al. (2006) provide a thorough review of these systems with a summary of performance criteria for each option. These systems focus on point-of-use or point-of-entry drinking water treatment and storage. Chlorination and flocculation/chlorination methods are scalable at the village and national level; scalability issues exist for biosand, ceramic, solar disinfection, and filtration/chlorination options. The adoption of household water treatment and storage systems has increased in developing nations in recent years. These systems are small-scale applications of certain basic processes in the engineered hydrologic cycle discussed previously, but the treatment is not as comprehensive as is typically done in conventional systems.

This model has been most successful when local organizations have the capacity to provide materials and replacement parts, can provide technical assistance, and can facilitate behavior change communications (Lantagne et al., 2006). Significant challenges for these systems include evaluating the health impacts of these interventions in "real-world" settings, sustainability of the projects, and scalability in terms of reaching people without access to improved water sources (Lantagne et al., 2006).

Barriers for the Household Water Treatment System Model

Community education is often necessary to ensure adoption of the intervention and to ensure community acceptance. These efforts are labor- and time-intensive, and success is measured in small steps. Cultural traditions can impact the adoption rates of practices that are generally accepted in the developed world. For example, in Western Cameroon, it is considered "taboo" by some to use household chlorination (locally known as "poison") in drinking water (personal communication with P. Njodzeka of the Life and Water Development Group, Cameroon, December 30, 2008).

In rural farming villages, household water treatment systems are simply unaffordable to many where household income can be $0.25 per day or less, and residents collect water from springs or often drink from contaminated streams.

The adaptability rate of a household intervention is an area for further research. While knowledge of the intervention may be widespread, actual adoption rates may lag due to economic factors, a perceived lack of personal benefit from the intervention, and cultural or political issues.

A Paradigm Shift for Water and Sanitation Infrastructure

Variability exists in the designs and operation of conventional water infrastructure systems, but these generally follow a "prescriptive approach" based on Victorian age (or older) methodologies. Even if the capital and human will existed to create conventional systems for all people, it would take decades to build and it is uncertain that these systems are well suited to meet the demands of population growth and urbanization. If safe water and appropriate sanitation are to become accessible to those who are not currently served, new approaches and modern technologies must be employed. This will require a significant change in the way water and wastewater treatment systems are conceived and how they interact with other infrastructure systems (i.e., energy). Proposed features of a paradigm shift to address the complex challenges around global water and sanitation are shown in Table 4-5.

The Innovation Challenge

The benefits of conventional water infrastructure systems are undeniable. Human health has been improved as has life expectancy. Water supplies in most developed nations are relatively clean and reliable. Many of the water-related diseases rampant in Europe and North America in the late 1800s are no longer a concern in those regions (Gleick, 2000). However, the implementation of these systems carried a tremendous economic cost, and they require a continuous investment in operations and maintenance. In addition, they have greatly disturbed many ecosystems, displaced populations, and created new health concerns such as the formation of trihalomethanes, a disinfection byproduct linked to cancer.

Research activities in the United States, and other parts of the developed world, have focused on advancing the water and sanitation approaches within the conventional water and wastewater treatment paradigm. By comparison,

TABLE 4-5 Paradigm Shifts Addressing Water and Sanitation Infrastructure

Old Paradigm	New Paradigm
Slow implementation	Rapid implementation
Prescriptive technologies	Adaptive solutions
Low social acceptance criteria	High social acceptance criteria
One water quality type fits all	Provision of water quality based on use
Low priority on energy efficiency	High priority for energy efficiency
"Siloed" health, economic, engineering	Integrated systems approach
Financing via taxes, subsidies, tariffs	Innovative financing and business models
Centralized energy provider	Distributed energy systems
Less priority on resource conservation	High priority on resource conservation

research and technology for the development of nonconventional water and sanitation has been very limited, and market penetration of many of the proposed solutions has been inadequate and difficult to sustain. In order to make rapid progress in solving the world's water challenge, it is essential that research be directed at approaches that can be developed within the "new paradigm" presented earlier. In addition, research must be focused on water efficiency in the major water usage areas of energy and agriculture. Whether interest is directed to new treatment systems or to increased efficiency, there is reason to believe that new solutions are possible and with focused and sustained efforts, significant improvements in water access and sanitation can be realized. In this section, three specific areas where innovation yields interest today are presented. First, are changes in the water-energy nexus. Second, is the availability of capital for business creation in developing countries. Third, are examples of technology development that suggest that nonconventional approaches to treat water and wastewater are in existence and could be refined to meet the conditions of a "new paradigm."

The Water-Energy Nexus

Energy and water use are intimately linked. Domestic water use requires significant energy for pumping and treatment. American public water supply and treatment facilities consume about 56 billion kilowatt-hours per year—enough electricity to power more than 5 million homes for an entire year (EPA, 2009). Thus, saving water saves energy and results in fewer greenhouse gas emissions. On the other hand, energy production uses and impacts domestic water supplies. As stated previously, cooling water represents nearly half of the volumetric water use in the United States annually. The power industry requires reliable supplies of water for cooling, for flue gas desulfurization and ash handling, and for hydroelectric power generation. As population grows, so does the demand for electric power and water for agricultural, municipal, residential, commercial, industrial, and power generation uses, potentially straining water supplies. Because the water supply is limited, growth in demand can only be met by developing technologies that reduce the volume of water required per kilowatt-hour of power generated. In short, the ideal way to reduce water consumption is to increase energy efficiency.

Rapidly developing nations must be able to realize the benefits of energy efficiency while building their infrastructure. A study by the McKinsey Global Institute (2007) reported that the global demand for energy is estimated to grow at a rate averaging 2.2 percent a year up to the year 2020, a rate that is the fastest since 1986. Developing countries will account for an overwhelming 85 percent of energy demand growth to the year 2020. However, the McKinsey study reported that it is possible to improve energy efficiency in a manner that could cut energy demand growth by at least half. The greatest productivity improvement opportunity is in the global residential sector (the world's largest consumer of energy). Bringing the preferred existing, or yet to be discovered, ways to increase energy efficiency in buildings, residences, and other operations that derive electricity from stationary power sources is essential to mitigate water stress and decrease water use.

The water-energy nexus extends beyond efficiency. Today, a transition in energy production is occurring for a model of distributed power systems (Platz and Schroeder, 2007). While most of the world's power produced today comes from centralized power plants using fossil fuel combustion or nuclear fission to drive steam turbines, these centralized power systems require significant capital investments and extensive distribution networks to reach consumers. The development of small-scale, distributed energy systems is an area of research and development today. This includes solar generation, wind and water turbine generators, and other technologies that create electricity at the point of demand. As markets increase, the cost of onsite power generation will decrease due to economies of scale and it should be possible to infuse developing countries with power generation without the 10 to 20 years of construction of conventional stationary power systems. With the advent of distributed energy comes the potential for distributed water and wastewater treatment, using technologies or approaches that differ significantly from conventional water and wastewater systems. In fact, it is possible to create wastewater treatment systems that generate electricity that would themselves be a distributed power generator.

Innovative Financing and Business Models for the Water, Sanitation, and Energy Sectors

Intermediate solutions that address the technical and financial gap between point-of-use technologies and the heavy infrastructure projects require innovative financing approaches. National governments often lack the financial means to extend water, sanitation, and energy coverage through infrastructure investments. Cardone and Fonseca (2006) describe innovative financing trends and case studies for small-town water and sanitation services.

In most developing countries, financial services such as bank loans, insurance, and pension funds are not readily accessible by the poor (NWP, 2007). Microfinancing services supply capital to poor people considered "unbankable" by the conventional financial sector, with loans for as little as $50 (Morgan Stanley, 2007). Since 1976, microfinancing mechanisms have been providing small loans to the poor and the microfinance sector has grown significantly despite the absence of specific financial sector policies.

Microfinancial institutions (MFIs) differ from traditional banking institutions in several ways (J. P. Morgan, 2009). MFIs emphasize both their financial profitability and their social impact (i.e., double bottom line). MFIs also have higher net interest rate margins compared to commercial banks. This is due to the relatively high interest rates charged to microfinance clients typically resulting from higher administrative expenses because of the location of clients, small transaction size, and frequent interaction with MFI staff. In 2006, the average worldwide microfinance lending rate stood at 24.8 percent (J. P. Morgan, 2009). Furthermore, MFIs have typically had a stronger asset quality than mainstream banks in emerging markets, due to a good knowledge of customers and strong incentives for clients to pay and establish a good credit history (J. P. Morgan, 2009).

Microfinancing has typically not been available for financing water supply and sanitation activities primarily because of a lack of awareness of the business case for water and sanitation projects. Water and sanitation projects become bankable when assets such as pumps, turbines, and solar panels are introduced. These assets provide collateral and the means for generating recurring income streams and fee-based services. Emphasized investment in distributed energy technologies for the poor combined with local currency lending mechanisms may flourish as the microfinancing markets mature and infrastructure spending increases.

Microfinancing mechanisms can serve those at the household and small community levels. Platz and Schroeder (2007) provided case studies in Africa and Latin America that describe how larger scale programs can be financed and implemented. The delivery of water and electricity to the poor is characterized by low levels of cost recovery and requires long-term financial investments. Full cost recovery in the least developed countries can be problematic, and targeted subsidies often are needed to finance large projects.

Although the current financial crisis is not expected to affect long-term investment in energy infrastructure projects, delays in bringing current projects to completion are expected (IEA, 2008). Privatization projects in the water and sanitation sector have largely failed (Bayliss and McKinley, 2007), and global private water investments in 2007 were low, at just $3.2 billion (World Bank, 2008). Substantially more funding is needed to strengthen public sector services, and clean renewable technologies have the potential to attract new significant investment while offering a return on investment and societal benefits. Bringing these opportunities to fruition requires strong collaboration between the engineering, financial, and health sectors to ensure community acceptance and economic sustainability.

New Water and Sanitation Technologies

As was discussed previously, most research in the United States and elsewhere has focused on improving conventional infrastructure. In some cases, the impediments to improve existing technology have resulted in the development of new approaches to replace older technology. Examples of this, which have applications in developing countries, are membrane separations and UV disinfection. Membrane processes are advanced filters that are capable of removing particulate and, in some cases, dissolved contaminants of water. They are effective in the removal of bacteria and viruses and can also remove other contaminants depending on the fabrication of the membrane itself. Much of the most recent research has focused on reducing the pressure needed to drive water through the membrane, and considerable progress has been made in this area. Decreasing pressure results in lower energy needs and should allow for membrane treatment using some of the distributed power systems discussed previously. Small-scale household or community production of water for drinking and body contact using coupled energy production-membrane filtration units may represent an area of study.

UV disinfection has been an area of research to replace existing disinfection strategies (e.g., chlorination) that result in the formation of unwanted disinfection byproducts. Classic UV disinfection uses mercury vapor lamps, which are not likely to be used in the developing world due to cost, poor durability, energy consumption, and the potential effects of mercury wastes. Considerable technology development has been occurring in the area of light emitting diodes and other "lamps" that are far less energy consuming than traditional light bulbs. They also offer increased durability and longevity. The development of low power lamps that produce UV in a spectrum similar to a mercury vapor lamp would be ideal for use in the developing world to disinfect water, again coupled with a distributed energy source.

Microbial Fuel Cells

One promising technology for the treatment of human and animal wastes is microbial fuel cells (MFCs). While simple in design, MFCs harness the natural ability of bacteria to break down organic matter and create electricity directly. MFCs function with bacteria to oxidize organic matter (i.e., the electron donor) on the anode under anoxic conditions and transfer the electrons to a cathode through a wire. Oxidation of these recently fixed sources of organic carbon does not contribute net carbon dioxide to the atmosphere and, unlike hydrogen fuel cells, there is no need for extensive preprocessing of the fuel or for expensive catalysts (Lovley, 2006). By converting biochemical to electrical energy, their most likely near-term application is as a method of producing energy from wastewater (Logan and Regan, 2006). What currently is not known is how best to integrate MFC approaches in systems such as latrines or small-scale waste facilities. Electricity production from waste is an interesting possibility that may create a market for sanitation that currently is not in existence.

Geographic Information Sciences for Decision Makers

Geographic information sciences (GIS) combine remote sensing, geographic information systems, cartography, and surveying, interrelating with mathematics, and the physical, biological, and social sciences. It empowers researchers and decision makers to evaluate complex environmental systems and the interconnections with human health. New tools for modeling, predicting, and forecasting the water resources sustainability, quantity, and quality are being used in developing countries. The use of novel sensors, wireless and broadband technologies, high-performance computing, and real-time data assimilation is being promoted with the objective of better understanding the Earth's water resources and related biogeochemical cycles; this goal could lead to better management of activities that impact human health (Schnoor, 2008). Multidisciplinary innovations such as these provide a foundation for better decision making on global issues of water quality, water resources, and sanitation.

Nanomaterials Applications of nanoscience for water treatment are in the market and expanding (Hillie et al., 2007). They include

- nanofiltration membranes for removal of salts and micropollutants, and for wastewater treatment;
- use of Attapulgite clay, zeolite, and polymer filters, which can be manipulated on the nanoscale for greater control over pore size of filter membranes;
- nanocatalysts and magnetic nanoparticles that will enable the use of heavily polluted water for drinking, sanitation, and irrigation; and
- nanosensors for detection of chemical and biochemical contaminants.

The impact of nanotechnologies on human health is a growing research need. Risks and social issues associated with promoting advanced filtration techniques in developing countries need to be understood and communicated to promote transparency and adoption of the technologies.

Research Needs for a New Water and Sanitation Infrastructure Model

The development of a new water and sanitation infrastructure model requires multidisciplinary, systems-based thinking and innovations that vary from successful approaches adopted by developed nations. Advances and improvements in the necessary disciplines are occurring on a global scale; integrating the health, engineering, economic, political, technological, and social aspects for sustainable solutions for the world's poor calls for research in the following subjects:

- Development of robust and appropriate water technologies to create safe drinking water;
- Development of robust and appropriate wastewater and dry sanitation technologies;
- Technology transfer and development of markets for these technologies;

- Development of best practices in watershed protection, water conservation, and sustainable energy systems for rural and urban populations;
- Tailoring water and sanitation for microfinanced enterprises and investment banking institutions;
- Scalability of innovative technologies for maximum impact;
- Health and epidemiological concerns of a successful implementation; and
- Outreach and education to improve transparency at all levels of government.

IMPROVING URBAN WATER AND SANITATION SERVICES: HEALTH, ACCESS, AND BOUNDARIES[6]

Kristof Bostoen, Ph.D., M.Sc.
London School of Hygiene and Tropical Medicine

Pete Kolsky, Ph.D.
The World Bank

Caroline Hunt, Ph.D.

Introduction

Those who live in cities depend upon resources from outside city boundaries. Use of external water resources is one of the important ways a city affects, and is affected by, its surroundings. During rapid urban growth, these interactions become increasingly important for both the city and its environment.

Water flows back and forth between the natural environment and the urban community (Figure 4-7). Water supply brings water from the broader environment into the community, while drainage and sewerage returns it to the 'natural' environment. Water in such transfers is never pure H_2O, but is always mixed with other matter, as illustrated in Figure 4-7. Often this 'other matter' includes pathogens (disease-causing organisms).

Whatever water comes into a community has to be returned to the natural environment. Even with recycling and storage, the outflow must more or less equal the inflow, or else flooding will occur. Despite this fairly obvious fact, efforts are frequently made to improve community water supply without improv-

[6]This paper is reprinted with permission from Bostoen, K., P. Kolsky, and C. Hunt. 2007. Improving urban water and sanitation services—health, access, and boundaries. In *Scaling urban environmental challenges: from local to global and back*, edited by P. J. Marcotullio and G. McGranahan. London: Earthscan Publications.

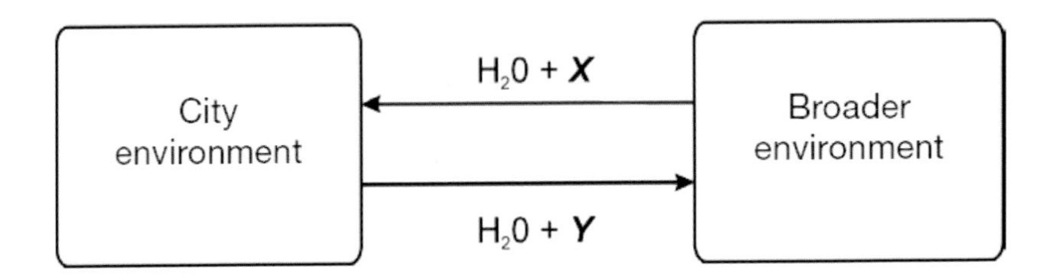

FIGURE 4-7 The water balance.

ing drainage. If water is returned to the natural environment with chemical or biological pollution, the contamination does not always disappear or die off, but can return to threaten the health of the polluting community or that of one downstream.

This chapter examines issues related to water supply and sanitation services, which are of particular relevance to low-income communities. The second section of this chapter looks at health issues relating to water, while the third section examines the targets set by the international community for water and sanitation and the challenges regarding the achievements of these targets.

As problems of water shortages and pollution are transferred from the local to the broader environment, the challenge shifts from one of maintaining human health to one of preserving the integrity of life-support systems for future generations (McGranahan et al, 2001). These transitions are well known and documented with regards to the water cycle.

The water cycle within the urban area, as illustrated in Figure 4-7, occurs at each spatial scale of the city; for a given neighbourhood, the rest of the city is seen as the broader environment. These different subdivisions, or boundaries, often create institutional issues which in turn have an impact upon service quality and health; these are explored in the fourth section of the chapter.

How Water Supply, Sanitation and Hygiene Affect Health

Below, in the following two sub-sections is a description of two common models to describe the relationships between water supply, sanitation, hygiene and human health, as they are understood at present. The third sub-section refers to recent and forthcoming research in this field.

Classifications of Water-Related Infections

The first model has evolved from earlier work, grouping *water-related* infectious diseases by broad routes of transmission (Feachem et al, 1977; White et al,

1972). The categories are defined by the types of intervention that can control morbidity and mortality, rather than by the biological taxonomy of the organisms that cause them. As such, this model has helped engineers and public health professionals to work together on practical control strategies (Kolsky, 1993). A similar classification exists for *excreta-related diseases* (Feachem et al, 1983a) but has been less widely used. There are four categories in the Bradley-Feachem classification of water-related disease:

- *Faecal-oral* (waterborne *and* waterwashed). These include infections that are transmitted by swallowing faecally contaminated matter (food and water) containing pathogens. They can be caused by lack of sufficient water to maintain personal and domestic hygiene *as well as* by drinking contaminated water. Diseases in this group include, among others, diarrhoeal diseases, typhoid, cholera and hepatitis A and E.
- *Strictly water-washed* (skin and eye infections). These are conditions that are exacerbated by lack of water for washing and hygiene, but are *not* faecal-oral. These diseases are largely related to skin and eyes, such as scabies, trachoma and conjunctivitis.
- *Water-based aquatic intermediate host.* Aquatic organisms such as snails act as hosts to parasites, which then infect humans either by being swallowed or through contact in water (e.g. by piercing the skin of those wading in the water). Diseases in this group include guinea worm and schistosomiasis.
- *Water-related insect vector.* These diseases depend on insect vectors, such as mosquitoes and flies, which breed in or near water. They transmit disease to humans, for example, through bites. The diseases involved include malaria, filariasis, yellow fever, dengue and onchocerciasis (river blindness).

From the four categories in the Bradley-Feachem classification it becomes clear that interventions focused on water quantity have broader impact than those focused on water quality. Water quality only affects faecal-oral diseases, whereas quantity affects both faecal-oral and water washed diseases. The relative importance of water quantity and its quality will be discussed later in this chapter.

Diarrhoeal diseases, which are faecal-oral, are responsible for the greatest number of episodes of illness (morbidity) and deaths (mortality) worldwide, compared to any other single classification of water and sanitation-related disease. This is shown in Table 4-6, based on data presented for World Health Organization (WHO) member states. It has been estimated that diarrhoeal disease represents 90 per cent of the health impact associated with water supply and sanitation (White et al, 1972). Diarrhoeal diseases are estimated to kill around 1.8 million people every year worldwide (WHO, 2004) of which the overwhelming majority is children. This toll is equivalent to 12 jumbo jet crashes every

TABLE 4-6 Health Impacts of Water- and Sanitation-Related Diseases

	Mortality Estimates for 2000 (thousands)	DALYs* Estimates for 2000 (thousands)
Faecal-oral		
Diarrhoeal disease	1,798	61,966
Poliomyelitis	1	151
Water-washed		
Trachoma		2329
Water-based		
Schistosomiasis	15	1702
Water-related vector		
Malaria	1,272	46,486
Lymphatic filariasis	19	5777
Dengue		616
Intestinal nematode infections	12	2951

NOTE: *DALYs or Disability-Adjusted Life Years is a indicator attempting to quantify 'time lived with a disability' and the 'time lost due to premature mortality' developed for the World Bank's 1993 World Development Report: Investing in Health.
SOURCE: Adapted from WHO (2004).

day or almost twice (1.9) the number of people who 'died in the World Trade Center on the 11th of September 2001' per day. There is some reason to believe that the number of deaths has fallen since the 1980s, possibly due to water and sanitation programmes and increased use of oral rehydration therapy (Bern et al, 1992). However, it appears that the number of episodes of diarrhoeal disease has remained constant.

Approximately 90 per cent of diarrhoeal disease cases are estimated to be attributable to environmental factors (Murray and Lopez, 1996). Apart from water supply, sanitation and hygiene, diarrhoeal disease is also associated with a number of other risk factors including age, malnutrition, lack of breastfeeding, and seasonality.

The F-Diagram

A second model is the F-diagram, depicted by Wagner and Lanoix (1958) (Figure 4-8), which has been widely used as a model of faecal-oral disease transmission. Unless faeces are isolated from potential contact with humans, animals and insects, pathogens may be carried on unwashed hands, in contaminated water or food, or via flies and other insects on to further human hosts. The first way to stop or reduce transmission is to ensure the safe disposal of faeces, through sanitation. Safe excreta disposal and washing hands following defecation is referred to as 'the first barrier' and considered the most important health intervention, as it keeps faecal pathogens out of the living environment. Children's faeces in par-

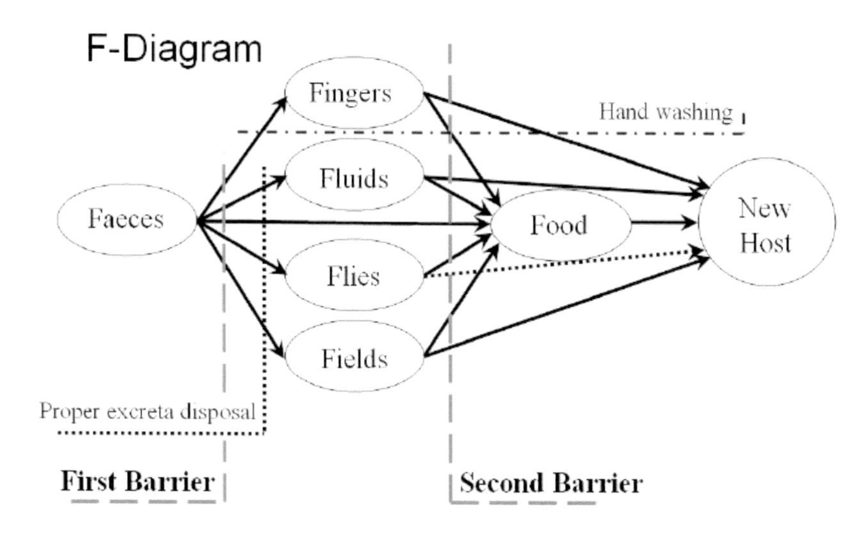

FIGURE 4-8 F-diagram.
SOURCE: After Wagner and Lanoix (1958). With permission from the World Health Organization.

ticular are known to contain a high load of pathogenic organisms, such as Ascaris and Trichuris, but are also least likely to be safely disposed of (Cairncross, 1989; Kolsky, 1993). The secondary barriers to faecal-oral disease transmission protect people from whatever faecal contamination of the environment is present. These are based on hygienic practices, such as washing hands before handling food, fly control, safe food storage and the use of footwear.

The F-diagram graphically presents multiple routes of transmission. A single type of pathogen may be transmitted by several of these routes, and the population at risk may be vulnerable to many different pathogens, which may favour different routes. Numerous commentators have advocated integrated measures to control diarrhoeal disease by combating multiple routes of transmission (Lewin et al, 1996; Curtis et al, 2000a). The greater the range of interventions, the greater the chance in successfully reducing diarrhoeal disease transmission. The F-diagram also shows that while water quality only affects one route, the quantity of water available for personal and domestic hygiene affects almost all routes.

Following the discovery in the 19th century of the undeniable role that water quality played in the epidemics of cholera and typhoid, there was a natural focus on the improvement of drinking water quality. This focus produced dramatic results in the reduction of waterborne epidemics. The F-diagram clearly shows that this would be the case where water contamination is the main route of transmission.

Where routes other than water consumption are more important for disease transmission, however, improving water quality will have far less effect. While waterborne epidemics are dramatic and alarming surprises, the sad truth is that the everyday endemic (non-epidemic) toll of faecal-oral disease is far, far higher, and most of the latter seems to be transmitted through routes other than water. While improving water quality does not necessarily affect endemic transmission, increasing the quantities of water available to improve personal and domestic hygiene *can* have a greater effect on this unacceptable toll (Cairncross, 1995). Most health benefits will be obtained from large amounts of water of a good quality. But if resources are scarce, public health professionals generally recognize the greater importance of access to water in quantity for hygiene, compared with the quality of that water (Esrey et al, 1985, 1991). Unfortunately, in practice, the main efforts in 'water and sanitation for low-income areas' are often still directed towards the improvement of the water quality of the public water supply, rather than improving access by poor households, and thus the quantity of water that those households can use. The beneficiaries of such efforts are more likely to be people who already have access to water than people with no access. Sanitation and hygiene promotion are even lower priorities in practice, although the principles of the F-diagram suggest that they should have equal or higher priority (Curtis et al, 2000b).

Recent and Forthcoming Research

Most studies of the impact of water quality have been based upon water quality measurements at the source or collection point. It is known however, that the degree of faecal contamination of water increases during transport to the household (Clasen and Bastable, 2003). There is also increasing evidence that improving water quality at the point of use has a positive health impact (Conroy et al, 2001; Iijima et al, 2001; Fewtrell et al, 2005; Clasen, 2006). This is regarded by some as an exception to the dominant paradigm (Clasen and Cairncross, 2004). While data support health benefits for people that have at least 15 litres of water per capita per day there are reasons to believe that benefits are reduced when access levels to water are lower (Clasen, 2006). However, at this time not enough data are available to substantiate this (Clasen, 2006). Systematic reviews and meta-analyses, such as those of the Cochrane Library infectious diseases group (Clasen et al, 2004) and field research will be needed to clarify these new findings and examine if these are in conflict with the current paradigm.

What Does All This Mean?

These models clarify the complex relationships between water, sanitation, hygiene behaviour and health. For example, good hygiene is more important in low-income areas where environmental exposure to pathogens is greater; residents of relatively clean areas can (and often do!) get away with lower standards of hygiene. Those who practise poor hygiene are certainly at greater risk than those who practise good hygiene, even in relatively clean environments. Water quantity is generally more important than water quality, because increased quantities of water promote good hygiene, and can prevent faecal-oral transmission by a number of different routes; increased quantities of water also reduce skin and eye infections. Only when drinking water is the main source of infection will water quality be more important than quantity. This is rarely the case where diarrhoea is endemic.

This means that in most cases within an urban setting, water distribution (access to water in quantity) is more important than public water treatment (its quality) until a certain relatively high level of environmental hygiene has been reached. Water treatment at the point of use seems to give health benefits, but it is not clear if a certain level of access to water is required to profit from such an approach. Finally, the quantity of water which people can actually use is vitally dependent upon access, as shown by Figure 4-10; the issue of access will be further explored in the following section.

Access to Improved Sanitation and Water Sources

'Water for life', the 2005 global water and sanitation assessment, contains the most up-to-date coverage data for most of the countries in the world (WHO/UNICEF, 2005). Since the Global Assessment 2000 Report (GA2000) (WHO/UNICEF, 2000) the United Nations (UN) Joint Monitoring Programme (JMP) does not report on 'safe' drinking water and 'adequate' sanitation. Instead, access to 'improved' water supply and sanitation technology types are now reported (see Table 4-7). This change in terminology reflects both past misrepresentation, and future uncertainty, in judging and defining services as *safe* in terms of human health. According to the report, over 2.6 billion people worldwide are without access to improved sanitation and over 1.1 billion do not have access to improved water supply. While many people have gained access since 2000, the number without access has remained the same (WHO/UNICEF, 2006).

Asia and Africa have the lowest levels of service coverage. In Asia, less than half the region's population have access to adequate sanitation. When comparing individual countries, the African region has the highest proportion of countries with less than 50 per cent water supply and sanitation coverage. In all regions, apart from North America, rural coverage is lower than urban coverage for both water supply and sanitation.

The Global Assessment 2000 presented the status of the sector using consumer-based data for the first time. These data were drawn from large nationally representative household sample surveys, such as the United States Agency for International Development's (USAID) Demographic and Health Survey (DHS) and United Nations Children's Fund's (UNICEF) Multiple Indicator Cluster Survey

TABLE 4-7 Water Supply and Sanitation Technologies Considered to Be Improved and Unimproved in WHO/UNICEF Global Assessment 2000

The following technologies were considered to be improved:	
Water supply	**Sanitation**
Household connection	Connection to a public sewer
Public standpipe	Connection to septic tank
Borehole	Pour-flush latrine
Protected dug well	Simple pit latrine
Protected spring	Ventilated improved pit latrine
Rainwater collection	
The following technologies were considered not improved:	
Water supply	**Sanitation**
Unprotected well	Service or bucket latrines (where excreta manually removed)
Unprotected spring	Public latrines
Vendor-provided water	Open latrine
Bottled water*	
Tanker truck provision of water	

NOTES: *Bottled water has been reclassified by the JMP as an '"improved" source of drinking only when there is a second source that is improved'.
SOURCE: Reprinted from WHO/UNICEF (2000, 2005) with permission from the World Health Organization.

(MICS) and national census data. The GA 2000 thus presented a better baseline for future targets than previous reports. The WHO/UNICEF Joint Monitoring Programme for Water Supply and Sanitation (JMP), which published the GA 2000, continues to update these data to monitor progress to the Millennium Development Goal Target for water and sanitation. The latest report to date is *Meeting the MDG Drinking Water and Sanitation Target* (WHO/UNICEF, 2006), which is available together with current data at the JMP website (www.wssinfo.org).

Targets For the Future

The UN Millennium Summit adopted the target of halving the proportion of people who are unable to reach, or to afford safe drinking water by the year 2015. The 2002 UN World Summit on Sustainable Development (WSSD) in Johannesburg has adopted the same target for access to sanitation facilities and the application of hygiene practices.

The compilation of the GA2000 has greatly improved data quality, using survey data. However, there is still a need to standardize survey outcomes to make results comparable. Figure 4-9 shows the typical scatter of results of the different surveys over the last 20 years for access to improved water in urban Niger. The variation in results is less for urban than for rural areas. Variations between different surveys are also less for household connections than for other

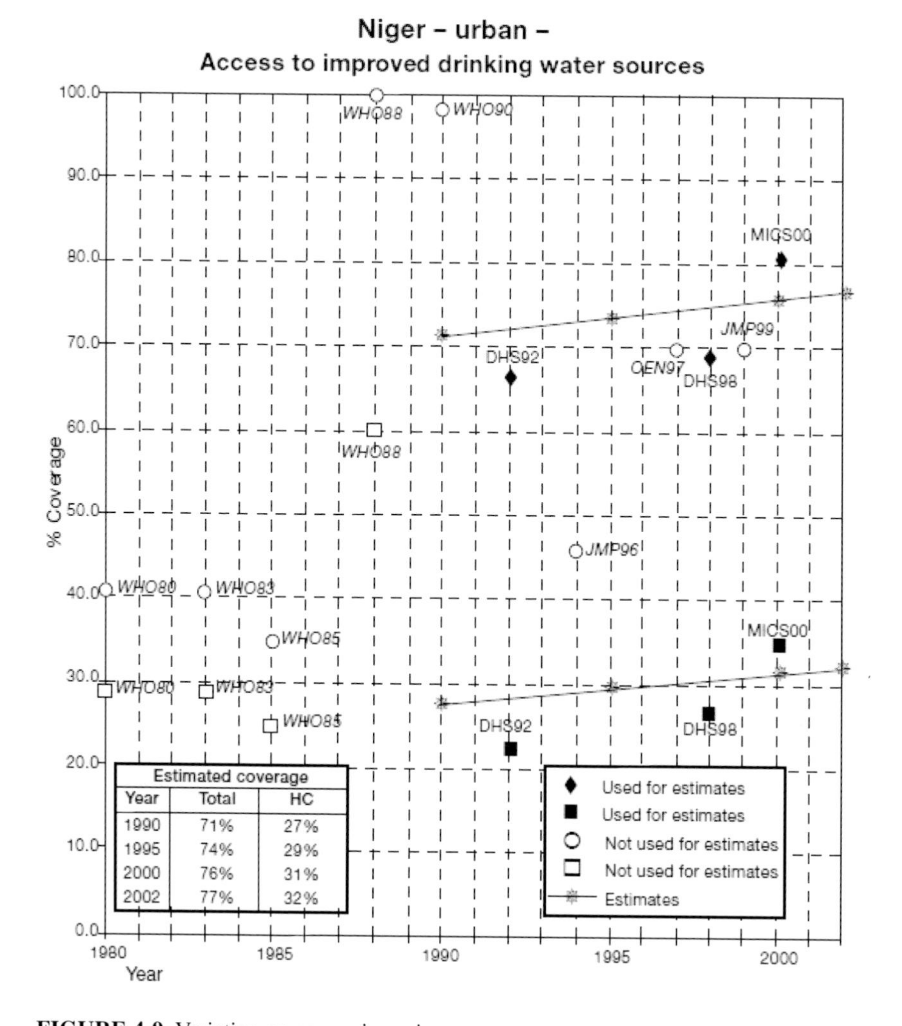

Estimated coverage		
Year	Total	HC
1990	71%	27%
1995	74%	29%
2000	76%	31%
2002	77%	32%

FIGURE 4-9 Variation on access in various surveys.
NOTE: HC = household connections.
SOURCE: UNICEF. NOTE: The original source of this figure is http://www.childinfo.org/files/NER_wat.pdf.

improved access, which is probably a reflection of the interest of the water utility in keeping track of its customers and the ease of defining this way of delivering water to households.

Urban populations in Asia and Africa are predicted to almost double over the next 30 years. Against this trend, meeting the International Development Target of halving the proportion of those unserved by water by 2015 would mean providing water services to more than 300,000 additional people every day over the next 15 years. Halving the number unserved by sanitation requires provision of services to over 400,000 additional people per day.

Limitations to the Use of Routine Data Sources

The use of existing surveys such as the DHS, MICS and national censuses, as in the GA 2000, has the advantage of being cost efficient, but it also has its drawbacks. These surveys have usually been designed to give a picture of 'the average household' on a national level and sometimes on a regional scale. Designed for other purposes, they cannot provide sector-specific (water, sanitation and hygiene) data at a local level that can be used for project implementation, evaluation and local decision making.

Large surveys such as the DHS require major administrative and organizational work, which means that they are unlikely to take place in countries or areas experiencing conflict or natural disasters.

To achieve better measurements in the field of water, sanitation and health practices there is a need for a simple standardized sample technique and a standardized set of indicators. Moreover, not only the collection techniques, but the interpretation and the type and extent of analyses need to be agreed upon if results are to be compared worldwide.

Access to Improved Services and Its Relation with Health

The level and type of service both have the potential to influence health. However, numerous other factors, which influence the use and nature of the service, also affect health risk, in some cases to a greater extent than the level or type of service itself. These factors include: access to, and use of services; system maintenance; treatment; seasonality; water sources; and pathogenspecific factors. Poverty is very often a key variable behind many of the factors listed above, most notably access to services.

The improvement of water supply and sanitation has attracted particular interest in reducing diarrhoeal disease (Feachem et al, 1983b; Esrey et al, 1985). These environmental improvements, together with improvements in living standards, played a major role in reducing diarrhoea rates and controlling endemic typhoid and cholera in Europe and North America between 1860 and 1940 (Esrey et al, 1985). Similar effects were anticipated from equivalent improvements in low income countries, and these expectations contributed to the declaration of the Water Decade. In 1977, the UN Water Conference at Mar del Plata set up an 'International Drinking Water Supply and Sanitation Decade' for 1981–1990. Its aim was to make access to clean drinking water available across the world.

Although improving access to water and sanitation projects improves health (Feachem et al, 1983b; Esrey et al, 1985), it is difficult to link these achieved health benefits back to *specific* improvements. Many attempts have been made to measure the health impact from water supply, sanitation and, more recently, hygiene practices. Even attempts under the supervision of eminent specialists to measure the health impacts of water supplies and sanitation have produced almost useless or meaningless results (Cairncross, 1999). Health impact studies are, for that reason, not an operational tool for project evaluation or 'fine tuning' of interventions (Cairncross, 1990). This led various organizations like the WHO and the World Bank to adopt the Minimum Evaluation Procedure (WHO, 1983), which concentrates on measuring functioning and use of services rather than measuring their health impacts (World Bank, 1976; Briscoe et al, 1985; Esrey et al, 1985; Cairncross, 1999).

While the health benefits of increased water quantity are known, they are difficult to measure and to attribute back to increases in supply. There is, for example, a clear but counterintuitive relation between the time needed to collect water and the amount of water collected (Figure 4-10). It is known that an increase in water consumption increases the water used for hygiene, which improves health. So, one might expect that reducing the time it takes to secure daily supplies to below 30 minutes would have a beneficial health impact. However, a reduction of the collection time of between 30 and 3 minutes will actually have little impact. Those who spend less than 3 minutes for water collection usually have a house-

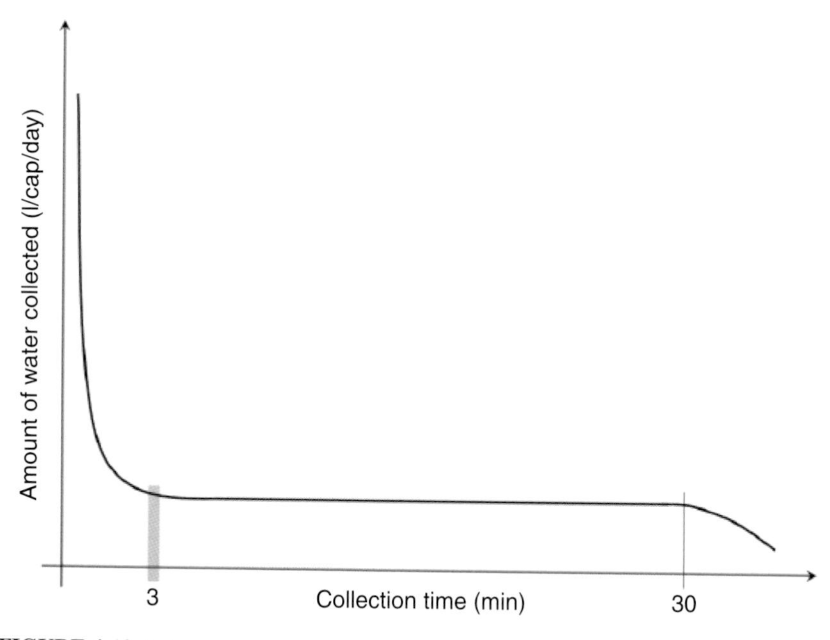

FIGURE 4-10 Relation between water consumption and time involved in water collection. SOURCE: Reprinted from Cairncross and Feachem (1993) with permission from John Wiley & Sons.

hold connection. Note also that collection time includes queuing time, which can be significant in areas with relatively closely spaced taps with intermittent service, or in areas that are serving large populations. In various parts of Africa, reports show that while distance to source diminished or stayed the same, collection times for water increased (Thompson et al, 2002; UN-Habitat, 2003).

What Does All This Mean?

While the main ways water, sanitation and health practices relate to health are broadly understood in theory, their real-world interactions are far more complex. However, increasing access to water and sanitation is clearly recognized as leading to health benefits, and new international targets have been set for that reason. Although improved access might improve health, it is methodologically extremely difficult to attribute improvements in health exclusively to specific interventions, on a project-by-project basis. This makes health impact an unsuitable outcome measure for project evaluation. The current indicators based on level and type of service also have their limitations. Better sector specific indicators and survey tools need to be developed.

Boundary Issues and the Urban Environment

Differing Perspectives

Figure 4-11 shows the urban environment from the point of view of the householder. The home is at the centre, and is the householder's first priority for environmental management. If the householder is able to maintain the home in a relatively clean and pleasant condition, the next priority becomes the surrounding street (peri-domestic). Local and informal lane arrangements may, for example, develop among neighbourhoods to ensure that rubbish does not pile up in the street, or clog drains. If the home and street are relatively clean, then some citizens will be concerned about the state of the environment in their larger neighbourhood (ward); when these problems are addressed, attention can then be focused on the rest of the city, and eventually, on the environment outside the city. It is natural that the focus of environmental concern spreads further outwards as the problems at each of the smaller (inner) scales are resolved.

It turns out that the householder's perspective and priorities are similar to those that emerge from a public health perspective. As most of the victims of poor environmental health are children under five, it makes sense to focus attention on where they spend the most time, which is at home. We have also seen that water access *at the household scale* is critical to increasing the quantity of water used to improve hygiene. The construction of public toilets half a kilometre from the house may offer limited improvement in convenience and dignity of some adults, but will not significantly improve the health of children in the community, who

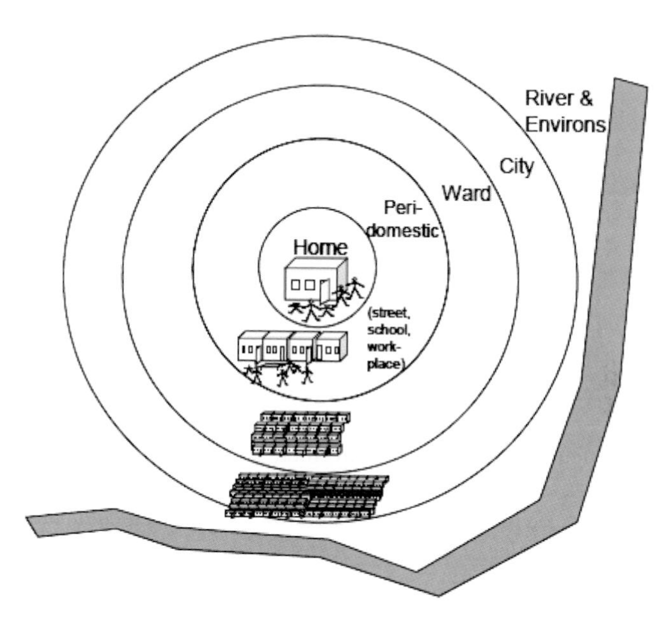

FIGURE 4-11 Scales of the urban environment, as seen by a householder.
SOURCE: Kolsky (1996) unpublished lecture notes, London School of Hygiene and Tropical Medicine.

will rarely if ever use such services. Improving the quality of river water by controlling the quality of the wastewater discharge may be of ecological benefit, but it makes no difference to a household's health *unless* that improvement is translated into improved drinking water quality at the household scale. The public health priority for environmental improvement thus becomes the household, followed by its immediate neighbourhood.

These differing scales of the urban environment are reflected in the structure of environmental service provision. Water supply, sewerage, storm drainage and solid waste management all involve the flow of mass between the individual household and the larger environment. Figure 4-12 shows the superimposition of a water supply system upon the scales of the environment shown in Figure 4-11, and similar sketches can be prepared for the other environmental services such as sewage disposal and solid waste management. Table 4-8 also shows the physical infrastructure associated with each level of aggregation.

The perspective of the service provider (e.g. the sewerage utility, the water supply engineer, and so on) is often different from that of the householder and public health specialist. They will look at the same system, but usually focus on different concerns, as shown in Figure 4-13. The highest priority of the water

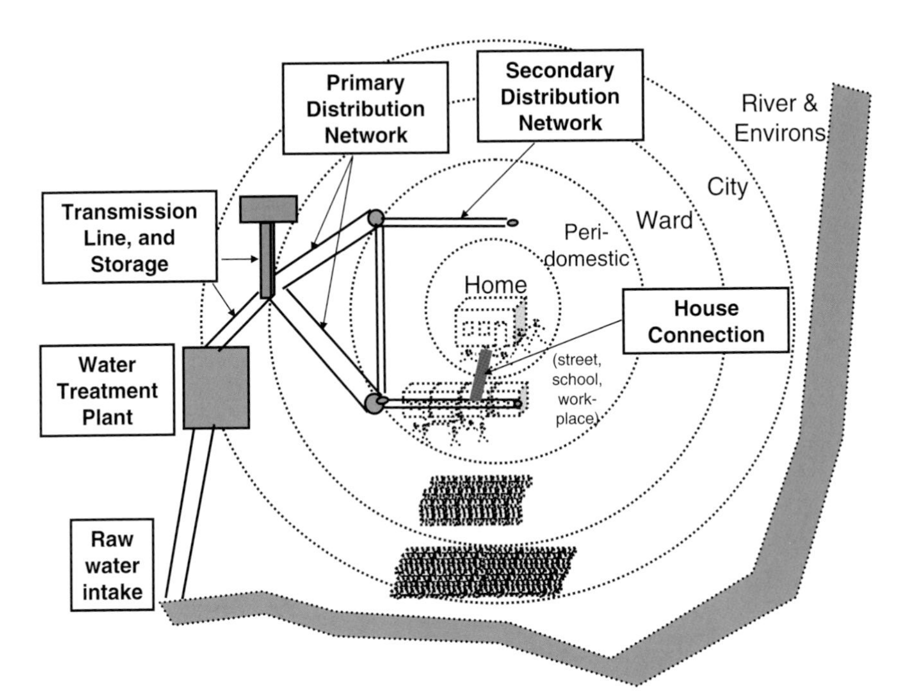

FIGURE 4-12 Scales of water supply infrastructure matched to the urban environment. SOURCE: Kolsky (2006) unpublished lecture notes, London School of Hygiene and Tropical Medicine.

engineer is often the intake and central treatment works. Their attitude is, if this link of the chain fails, all the rest will fail.

Construction of centralized works often involves substantial amounts of financial capital and technical sophistication, both of which contribute to the professional standing of the individuals involved. Primary mains are the second priority after the central works, because of the relatively large impact of the failure or inadequacy of these links in the chain. While central pipes are more expensive per metre, the majority of the cost in the distribution system is tied up in the large number of small outlying lines. Individual street mains and house connections are often at the periphery of the technical professional's vision. Indeed, in many cases the 'outermost rings' of the technical professional are virtually ignored; water is provided to a public tap to serve a neighbourhood, and what happens to the water after that is not the practical concern of the utility. This focus on centralized works is even less appropriate for sanitation infrastructure, where large amounts of resources may easily be spent upon centralized waste treatment works of marginal public health benefit in comparison with the provision of basic household access.

TABLE 4-8 Scales of the Urban Environment and Water, Sewerage and Drainage and Solid Waste-Related Infrastructure Issues

Service	Household	Street	Neighbourhood	Ward	City	Environs
Water	house connection or vendor	street main or public tap	secondary main	primary main	treatment works	intake
Sewerage and drainage	sewer connection	street sewer	secondary collector	primary collector	treatment works	outfall
Solid waste	collect or carry to bin	collective bin or skip	Recycling point	transfer station	Transfer/recycling centre	landfill

NOTE: Recycling often occurs at each stage of the solid waste chain.
SOURCE: Kolsky (1996) unpublished lecture notes, London School of Hygiene and Tropical Medicine.

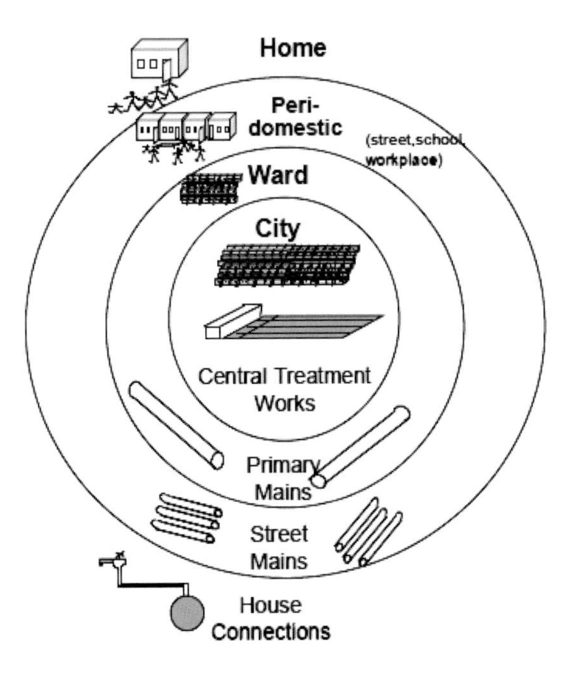

FIGURE 4-13 Water supply infrastructure and priorities, as seen by technical professionals.
SOURCE: Kolsky (2006) unpublished lecture notes, London School of Hygiene and Tropical Medicine.

Unfortunately, the poor live in the outermost 'marginal' rings. These 'rings' often represent very clearly definable neighbourhoods which are rarely laid out for service provision, have dubious tenure relationships, and are thus not a priority for service delivery. They are therefore easily excluded, yet it is here where the battle for urban public health is won or lost.

Boundaries and Their Impact on Services

We define a boundary as the limit beyond which an individual or group feels no responsibility. Boundaries can be marked by physical, legal, bureaucratic, psychological or customary limits. They can include formal district boundaries in rural areas, and ward boundaries in urban areas; they may also be as informal, but also as strong, as the sense of community around a courtyard or square.

The use of a subjective word like 'feels' in the definition seems odd, but the reason for its use is a pragmatic one. The world is filled with legal or administrative boundaries without practical meaning because those within the boundary do not feel, directly or indirectly, the consequences of their actions.

Boundaries permit the breakdown of complex problems into simpler parts, and the corresponding delegation of responsibilities and tasks. By limiting responsibilities, they become manageable, both for the person delegating and for the person performing the task. It is difficult to be responsible for everything, but we can accept responsibility within given boundaries.

Classification of Boundary Problems

While boundaries are a useful political, social and administrative device, they are not without drawbacks. There are, at least, four related categories of problems arising from boundaries.

Complete externalities When the damage of my actions to others does not affect me, I have no incentive to change my behaviour. This situation is known to economists as an 'externality', and in this case, explains why wastewater treatment always lags behind drinking water as a community priority. Drinking water affects the members of a community directly, so it is in their interest to take appropriate action to ensure its quality. Sewage effluent quality affects only downstream users; upstream community members causing the problem have no direct interest in resolving it for their downstream neighbours. Government action is often required to solve such problems. Although environmental pollution is the classic example of such a boundary problem, there are also administrative examples of externalities, as described below.

Partial externalities In some cases, I *do* care about the damage I cause outside my boundaries, but if most of that damage will occur in any event, why should I change my behaviour? Hardin describes this in his classic paper *The Tragedy of the Commons* (Hardin, 1968). The paper examines the behaviour of individual shepherds responsible for grazing on common land. Collectively, it is clear that they would be better off with fewer sheep grazing on the commons, because in the long term, overgrazing will destroy the resource. Individually, however, each shepherd is better off grazing as many of his sheep on the land as possible; after all, if he restrains his behaviour, the commons will still be destroyed by others, and he will have sacrificed in vain. Some form of social arrangement needs to be worked out, or else the commons will be finished. The difficulty of working out such social arrangements, complete with effective sanctions against those who violate them, lies at the heart of many environmental and social problems. Rubbish dumping in urban slums often falls within this category.

Badly drawn boundaries In some cases, things could work better if boundaries were simply redrawn. Engineers often refer to drainage networks and other urban infrastructure as trees. Such trees have large 'trunk' mains, and smaller 'branch' lines, and the image of a tree effectively conveys the notion of many smaller entities combining into a larger one. Most of the length of a drainage network, and most of the cost, lies in the many small branches rather than in the trunk line. While trunk lines are certainly more expensive per metre, the total cost of a network is dominated by the smaller branches at a smaller unit cost making up the outer branches. Such lines are often technically simple and individually not expensive; added up they represent the main cost of the network.

Traditionally, centralized authorities have taken responsibility for entire drainage networks on the grounds that all drains up to the individual property boundaries are public goods. This model was developed in the industrialized world and has served quite well there. Alternative approaches, however, have emerged in the cities of the developing world. Municipal authorities there are devolving the responsibility for these smaller lines to local community groups or non-governmental organizations (NGOs). This can be done because small branches are technically simple, and can be managed better by closer supervision within the community than by the distant municipal authorities. This does not mean that central authority should disengage totally in regards to compliance and appropriateness of these activities.

Redrawing boundaries in the water sector is not a new idea; the regional water authorities in Britain (Okun, 1978) are another practical demonstration of the benefits of more rational boundary definition. The environmental economics literature (Ruff, 1970) stresses the need to draw boundaries so that benefits and costs of activities are felt by those who take the action, thus reducing the boundary problem of externalities.

The category of badly drawn boundaries includes also cases where there are gaps between boundaries (so no group feels responsible), and cases where boundaries overlap (where more groups claim the same responsibility or authority).

Boundaries as barriers Movement of ideas and resources across boundaries is always more difficult than *within* boundaries. For example, at the start of the International Drinking Water Supply and Sanitation Decade, a need was recognized to integrate the services of water supply and 'sanitation', used here as a euphemism for excreta disposal. This made sense both in conceptual public health terms (as both are involved in the spread of faecal-oral disease) and in practical engineering terms where water-based sewerage is the main means of excreta disposal, and thus depends directly on the water supply. Communications between these services improved as intended, under the new boundaries.

Redrawing the boundaries around water supply and excreta disposal, however, caused the nominal separation of sanitary sewerage and surface water drainage even where these systems are physically interconnected. In the same street separate crews clean these 'separate' drains, requiring separate transport and equipment at different times. Because of the new administrative boundaries, these crews no longer communicate, and no longer share access to the same resources.

Other types of boundary create other communications problems in water, sanitation and other sectors. Because external support agencies do not wish to become entangled in an open-yended [*sic*] commitment to paying recurrent costs, they have traditionally limited their involvement to capital investment. This means that both international and local resources are drawn to the investment sector, to the neglect of the operational aspects. Recently created municipal development authorities created in recent decades are responsible for municipal investment, but not for the day-to-day maintenance of the infrastructure they develop. There is often a strong feeling among those running the infrastructure that those planning and designing it don't understand basic operational reality.

Boundaries define the problem considered by those with responsibilities within these same boundaries. There is, for example, an administrative boundary between street sweeping and drain maintenance. The street sweeper's boundary extends only to keeping the street clean, so sweeping sediment and debris into a drain is seen as perfectly acceptable. Similarly, the drain cleaner's boundary extends only to keeping the drain clean, so that emptying debris onto the street for subsequent pick-up by the solid waste department is also seen as acceptable. If the street sweeper returns before the solid waste department picks up the debris, then it will be swept back into the drain, thus achieving the environmental goal of perfect recycling! In this case, the real issue is 'solid waste management' which cuts across the pre-existing boundaries of street sweepers and drain cleaners.

Public and Private Domains as a Special Example of Boundary Problems

For a long time public health engineering has been focused on *public* domains to bring health improvements to deprived populations. This has involved construction of large-scale urban water and sanitation infrastructure. A spatial model of disease transmission in public and private domains illustrates the move away from this traditional, engineering approach to public health (Cairncross et al, 1996). Diseases transmitted on the household level have to be dealt with via interventions that reach the household level. This puts more emphasis on *private* health at household level and the need to understand decisions made and actions taken at household level and their relation to environmental health. The private domain is distinctly different from the public domain in which the intervention of a public authority is required to prevent disease transmission. Some of these spatial problems relating to this model have been discussed in the paragraphs above. This model acknowledges the importance of household practices and behaviour without ignoring the public domain.

Many studies have investigated the links between public water supplies and household contamination of stored water (Kirchhoff et al, 1985; Deb et al, 1986; Jonnalagadda and Bhat, 1995; Mintz et al, 1995; Jagals et al, 1997; Quick et al, 1999). Findings have been rather mixed. There is a growing consensus that diarrhoeal disease pathogens originating within the home, as found in household water storage vessels, are less of a threat to household health than pathogens found in source water supplies (e.g. from public wells) (VanDerslice and Briscoe, 1993). There appears to be degrees of immunity to pathogens commonly found within the household. These complexities are acknowledged and explored within the public-private domain model. The recent literature on point-of-use treatment of drinking water (Conroy et al, 2001; Hellard et al, 2001; Iijima 2001; Fewtrell et al, 2005; Clasen, 2006) seems to indicate that there are health benefits from improving drinking water quality at the household level. Further research will be required to determine if this is true for households with limited access to water.

Several studies have looked at the impact of interventions at household scale when the neighbourhood is contaminated with faeces (Cairncross et al, 1996). Others have logically argued that interventions should be targeted at those most in need. It has been suggested that children who are not breast-fed are more susceptible to diarrhoeal disease and as such may benefit more from water supply and sanitation interventions than breast-fed infants (Esrey et al, 1985, 1991). The logistical difficulties in carrying out this approach may, however, make it impracticable.

Water Stress at Global and Household Scales, a Boundary Point of View

At the International Conference on Freshwater in Bonn in 2001, water scarcity was attributed to growing demand and increasing pollution and waste of freshwater sources. There is confusion between water stress at the household and regional scales. Regional water stress is sometimes portrayed as the major determinant of households' access to adequate water and sanitation, as well as the prevalence of water-related diseases (UN-Habitat, 2003). The amounts of water required to meet basic needs are relatively modest. It is estimated that on a worldwide basis, agriculture accounts for about 69 per cent of annual water withdrawals; industry about 23 per cent, and domestic use about 8 percent (Hinrichsen et al, 1997). In Africa the percentage of domestic water use is estimated to be 7 per cent, while in Asia only 6 per cent (Hinrichsen et al, 1997).

The most common measure for water stress at a national level is the Falkenmark indicator (Falkenmark et al, 1989) which estimates the amount of freshwater available per capita per year. Benchmark values for the Falkenmark indicator are less than $1700m^3$ per capita per year, which indicates water stress, while less than $1000m^3$ per capita per year indicates severe water stress. In Figure 4-14 the relationship between urban water access, national water stress and national gross domestic product (GDP) per capita is shown based on data from the United Nations Environment Programme (UNEP) Data Compendium (UNEP, 2002). The figure shows the counter-intuitive relationship that water stressed nations have a larger proportion of their populations with access to water than those nations not considered 'water stressed'. Figure 4-14 shows that it is an erroneous over-simplification to extrapolate freshwater stress at a national level

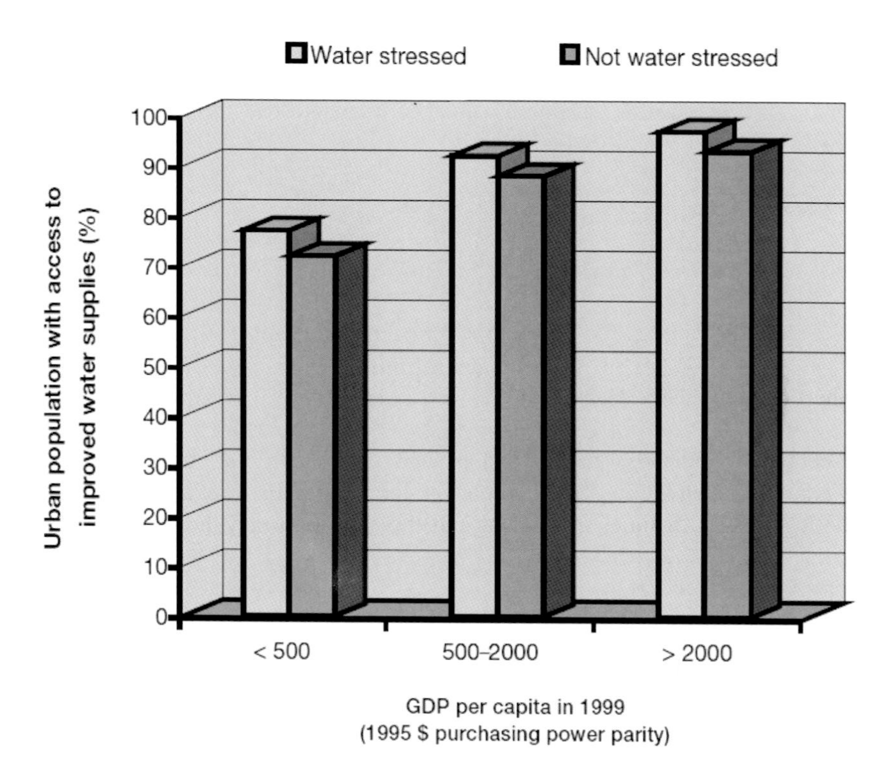

FIGURE 4-14 Relationship between urban water access, national water stress, and national GDP per capita.
SOURCE: UN-Habitat (2003).

to imply similar reduced access to water at the household level, regardless of the GDP per capita. Apart from a variety of other possible explanations, it is crucial to bear in mind that domestic drinking water supply and sanitation make up a very small fraction of any nation's demand for water resources.

Solutions to Boundary Problems?

Boundaries at different levels can be helpful in identifying problems within environmental services. Water provision by a utility to the neighbourhood level, but not to the individual household, invites the creation of an informal sector that may or may not operate as an extortionate cartel. The interface between differing levels can also be critical where the volumes in and out of the interface do not match. In the case of solid waste management, waste may be carried to a neighbourhood bin, but not picked up from that bin by the municipal service.

Local drains that serve one neighbourhood by flooding their neighbour can also reflect a failure to consider the whole system. So service provision must extend to the true 'end user' in the outer boundaries to ensure that the goals of access are really achieved.

Solutions to boundary problems are rarely obvious, or else sound trite. Nevertheless, two strategies emerge:

1. *Careful definition of boundaries, with flexible mechanisms for redefining them.* Redrawing boundaries along hydrological rather than political lines has worked well in some European watersheds (Okun, 1978). This has resulted in the advocacy for an integral holistic and transboundary management system of river basins (BHS, 1998). Unforeseen problems will, however, still arise, and there must be a reasonable way to refine or redraw the boundaries in the light of experience.
2. *Good mechanisms to identify and confront cross-boundary problems. Identification* of cross-boundary problems is not so difficult, especially if a group is given the task of finding out, for example, why government services don't perform properly. *Resolution* of inter-departmental conflicts, however, is a timeless management problem (Handy, 1985). 'Interdepartmental Task Forces' can be effective, or simply a sop offered in order to appear to be doing something about the problem.

Conclusion

The problems of the poor, are suffered by the poor, and dealt with by the poor, The problems of the rich, are suffered by the public, and dealt with by the Government.

Marianne Kjellen

Scale of the Problem

At present, an estimated 2.6 billion people lack access to improved sanitation and 1.1 billion people lack access to improved water supply. While there is substantial uncertainty about the definition or measurement of these terms, there is no doubt that a very large proportion of the human race does not have what public health authorities could accept as reasonable access to water and sanitation, and thus to the prerequisites for better hygiene. The toll of this inadequate access is high in terms of health, time, money, comfort and dignity. More than 2 million people, mainly children, die every year from diarrhoeal disease, equivalent to a jumbo jet crash every two hours. Unfortunately, this burden falls almost exclusively upon the poor, and most of it could be avoided with promotion of better water, sanitation and hygiene services.

Water, Sanitation, Hygiene and Health

It has been well established for nearly 30 years that the quantity of water that people use for personal and domestic hygiene is more important in maintaining health than the quality of the water they drink. While recent epidemiology on household water treatment justifies greater attention to this intervention, the implications for the public utility remain the same: that physical access to nearby waterpoints is a key aspect of service provision. The importance of excreta disposal and hygiene practices such as handwashing are obvious from the models mentioned and clearly visualized in the F-diagram.

Access to Services

The practical interrelationship between water, sanitation and hygiene in relation to health are far more complex than the theoretical links. This makes health impacts difficult to attribute to improved services, and the measurement of health impacts unsuitable for routine project evaluation. Increasing the access to water and sanitation is widely recognized as leading to health improvements, and new targets have been set by the international community to improve worldwide morbidity and mortality.

So far, the type of technology used at household level has been used as the main proxy for access and health impact. More careful consideration shows that this idea has severe limitations, and that better indicators of access to environmental health services need to be developed. Reasonable access to services such as 'improved' water is a complex idea with many facets, encompassing not just water quality, but also its quantity, cost, operational reliability, seasonal availability and collection time and effort. Measuring access to improved services, although essential, is not straightforward. To measure progress on these targets, more powerful and sector specific survey tools will need to be developed, along with the will to use them.

Domestic water consumption increases greatly with convenient and affordable water delivery to the household, reflecting the importance of collection time as a determinant in the amount of water used. These types of proxy indicators are easier to measure than the health impacts obtained by these improved services.

Boundaries, Scales of Environmental Challenges and Access

Before they can be expanded, services have to become more efficient. Institutional boundaries are central to many environmental problems, and are closely linked to the economist's idea of 'externalities', which occur when individuals or agencies are unaffected by the consequences of their actions. In many cases, these boundaries are regarded as immutable, because of institutional resistance to change. Identifying, acknowledging and addressing the various problems associated with institutional boundaries requires time, energy and goodwill. Successful models can evolve only after trial and error, careful monitoring and evaluation and honest documentation of both success and failure.

Environmental health operates on a variety of levels. One simple spatial division is between public and domestic domains of infection. The most vulnerable groups in society in terms of mortality and morbidity (children and the elderly) spend most of their time in the domestic domain. Recent epidemiology of disease transmission and hygiene has stressed the need to create conditions in which households can manage the domestic domain more effectively, through increased water use, household sanitation and improved personal hygiene.

Regional water stress is sometimes portrayed as the major determinant of households' access to adequate water and sanitation. Household consumption in the world is below 8 per cent of the global water use, and this chapter shows that household access to water in urban areas is, on average, higher in water-stressed nations than in countries where there is no water stress, according to the Falkenmark indicator.

This chapter has also presented a range of scales in which to examine environmental challenges, varying from that of the household up through street, ward and city levels. These scales often reflect their own boundary problems, as when the capacity to collect waste at one point is not equal to the capacity to remove it to the next stage. Householders, public health experts, and infrastructure professionals view these environmental scales differently. Both householders and public health experts see the household as the natural focus for environmental service provision, while infrastructure professionals tend to focus, for good and bad reasons, upon the centralized infrastructure (treatment works, centralized pumping, and so on). Environmental service provision is a chain, and, like all chains, is only as strong as its weakest link. Improvements in environmental health infrastructure will only be significant if they lead to changes at the household level. This public health reality underscores the need for improved access to environmental services at the household level. We need reliable measures of such access if we are to improve it, and the need for better practical indicators of access to environmental services is thus critical for the sector at this time.

Disclaimer

The findings, interpretations and conclusions expressed in this article are entirely those of the authors, and should not be attributed to the World Bank.

MEDICAL RESEARCH AND SOCIAL ENTREPRENEURSHIP COMMUNITIES: INCREASING THE DIALOGUE MAY LEAD TO NEW INSIGHTS FOR PUBLIC HEALTH

Sharon H. Hrynkow, Ph.D.[7]
National Institutes of Health

Introduction

A wave of enthusiasm for "social entrepreneurship" is sweeping across university campuses in the United States, the not-for-profit or citizen sector, and the private sector. While the concepts underpinning social entrepreneurship are not new, having been introduced as a discipline nearly 30 years ago, the vigor with which theses concepts are being applied to tackle tough societal problems is new and increasingly sophisticated. And, the numbers of individuals, non-governmental organizations (NGOs), and for-profit entities applying the principles are only increasing.

At the same time, the health and medical research community in the United States recognizes that new approaches in the research arena are needed in order to translate basic science knowledge into public health practice. As stated in the 2009 Budget Request Congressional Justification for the National Institutes of Health (NIH), "health care costs will not be tempered unless we accelerate the discovery of transformative ways of practicing medicine—which can only happen through research" (NIH, 2008). The principles and successes of social entrepreneurship, which introduce new and sometimes disruptive ideas and completely new delivery mechanisms, bear examination as nations work to lower health care costs.

This paper examines four strategies to provide clean water or sanitation in resource-poor settings. Each case is examined through the lens of social entrepreneurship and with a view toward identifying lessons learned to inform future efforts in the medical research arena. Examples are not delineated in depth but were chosen to offer illustrative examples. Clean water and sanitation was selected as the topic area in part because of the enormous toll in death, disability, and human suffering exacted due to poor water quality and sanitation: nearly 10 percent of the global burden of disease could be prevented by improving water, sanitation, and hygiene, and in the 32 worst-affected countries, this figure would be 15 percent (Prüss-Üstün, 2008).

[7] Associate director, National Institute of Environmental Health Sciences.

What Is a Social Entrepreneur?

Much has been written about social entrepreneurs and the movement in the past two decades, and there have admittedly been many champions. Bill Drayton, founder of Ashoka, the first and one of the best known organizations that invests in social entrepreneurs, states that "There is nothing more powerful than a new idea in the hands of a true social entrepreneur" (Ashoka, 2009a). But what is a social entrepreneur? Most agree that social entrepreneurs are "individuals with innovative solutions to society's most pressing social problems" (Ashoka, 2009b). Among the qualities that typify a social entrepreneur, they are determined, ambitious, and persistent; they persuade societies to take new leaps; and they are visionaries and ultimate realists at the same time. In his book *How to Change the World: Social Entrepreneurs and the Power of New Ideas*, David Bornstein examines the social entrepreneurship wave and offers perspectives on the qualities that typify these innovative leaders: willingness to self-correct, share credit, break free of established structures, cross disciplinary boundaries, work quietly, and with strong ethical impetus (Bornstein, 2004).

Examples of social entrepreneurs are many and they can be found through time. While not necessarily viewed as social entrepreneurs at the time, many well-known historical figures have worked tirelessly to change minds and systems. Florence Nightingale established nursing as a profession, while at the same time transforming how hospitals operate. John Muir fought to preserve and conserve nature's beauty by changing the way people experienced it, leading to the establishment of the U.S. National Park System and helping to found The Sierra Club. More recently, Nobel Prize winner Muhammad Yunnus showed that lending small amounts of money to impoverished people, against all instincts of traditional banks, can spur sustainable small businesses, lifting families out of hopelessness through enterprise and full economic citizenship, rather than through charity or government services. These visionaries represent only a small sampling of true social entrepreneurs.

Many foundations have realized the role of social entrepreneurs as mass mobilizers of citizens and as good value for the initial investment, leveraging dollars with the promise of high impact. There are now more than a dozen major grant-making organizations devoted specifically to the support of social entrepreneurs, in addition to Ashoka's focus on institution building and transformation of the entire citizen sector in entrepreneurial directions. Efforts to provide sustainable energy and clean water in remote villages, to conserve rainforests, to ensure quality education for all, to dismantle landmines, and to teach children living in poverty how to become self-sufficient are just some of the challenges being supported by social entrepreneur organizations. Efforts in the not-for-profit sector are complemented by those in the private sector, which has also taken up the mantle of social entrepreneurship. Bill Gates, Jr., is in every sense a business entrepreneur, but through the Bill and Melinda Gates Foundation he is working for the global public good through a range of innovative, large-scale programs. Furthermore, the for-profit sector has increasingly incorporated concepts of social entrepreneurship into corporate strategies: FastCompany and others track the social entrepreneurial movement embedded within the private sector.[8] Between 1990 and 2005, FastCompany trends show that for-profit companies involved in social entrepreneurship grew from about a dozen to more than 30. And more sophisticated approaches toward social entrepreneurship are being developed and tested. For example, Ashoka is now developing new approaches for combining multiple social entrepreneurs and innovative ideas in creative forums in order to refine and accelerate the most promising ideas and to spark entirely new insights based on the dynamic engagement of multiple entrepreneurs working together.

Clean Water and Sanitation— Consideration of Two Social Entrepreneurial Approaches

Boxes 4-2 and 4-3 provide snapshots of two Ashoka Fellows in Indonesia, each of whom approached the clean water challenge differently.[9] Both cases represent operational strategies related to collection and use of water data or to provision and marketing of sanitation services. In one case, river water was collected and analyzed from multiple locations across the span of several years to show the burden of pollution to nearby inhabitants. The volume of credible data collected by ordinary citizens, including school children, in a methodical and persistent way was used as a platform to engage journalists and policy makers, leading to reform. The second example focuses on delivery of sanitation measures in an urban slum. Using a business model and training for a local family, the program led to sustained used of toilet facilities for a small fee. Anecdotal evidence suggests that health was improved due to reductions of raw sewage in the streets.

There are many strengths in the social entrepreneur approach as currently deployed in both cases in Indonesia. First, both models depend on a firm understanding of local communities, cultures, and norms. Ashoka Fellows, who themselves are from the same countries in which they conduct their work, spend time with community members to define project parameters, gain trust, and learn essential insights for successful implementation. Second, both models turn on its head the expectation of whose job it is to solve social problems, engaging community members as "experts," not just for the purpose of consultation but to invest community members with decision-making authority. Project development is informed by the community experts and then executed in close partnership with them. Both programs valued community members from the start, to be consulted and respected and vested with responsibility instead of treating them as uninformed future beneficiaries. At least in part because of this approach,

[8] See http://www.fastcompany.org.

[9] For additional details on each Fellow's work, see www.ashoka.org.

BOX 4-2
Communities Develop Evidence Base for Water Quality Policies

Ashoka Fellow Prigi Arasandi works in 51 Indonesian villages with the goal of cleaning the water of the Surabaya River, a large drainage from the interior of East Java which is important to fisheries, more than 600 industrial sites, and the health of more than 5 million people. His approach, now 10 years in, is based on strong community involvement and regular communication with journalists (who are in turn interested in the broad-based community action). The citizen base, including children and adults, conduct water quality monitoring and testing. Using the data over several years from the community base, coupled with support from the media, policy changes through the legal system have become a reality. Specifically, the first ever lawsuit of its kind was won recently to require the governor to set a total maximum daily load for industrial pollutants and to implement a system to monitor that load.

Working as a bridge, efforts now between the Ashoka Fellow, the communities, and the governments are moving toward implementation and maintenance of new clean water standards by reducing untreated industrial and residential wastewater emissions into the river.

BOX 4-3
Community Leadership to Improve Sanitation

Ashoka Fellow Hamzah Harum Al'Rasyid works in poor urban areas in Indonesia to provide "community-based sanitation centers." Mr. Al'Rasyid works with community members to identify a family willing (and sometimes eager) to learn how to operate a portable toilet facility as a small business. Before making the match, he listens to communities about how the technology would be accepted, and how it might be adapted to meet local needs. For example, the toilets for one particular facility were refitted in keeping with community comments. Raising awareness to the health benefits of reduced sewage on the streets is one of Mr. Al'Rasyid's main objectives. If a community ultimately decides to adopt the technology, and if a family can be identified for training, the match is made. Initial experience shows success in reduced sewage in the streets, increased awareness of good sanitation practices, and sustained efforts. Furthermore, the network of facilities Mr. Al'Rasyid is developing offers promise for introducing other community-based services in the future, such as reverse osmosis clean water systems and local production of biogas for practical use in the communities.

important cobenefits were realized: The citizen base developed feelings of ownership for the river and its quality in the first case, which led to further direct river stewardship actions and to mobilization as a political constituency when needed. In the second case, cobenefits included awareness of and pride in, and sometimes substantial revenue from, the health benefits of good sanitation measures. Mr. Al'Rasyid's project, along with a handful of other social entrepreneurs in other countries,[10] also shows that very poor people are willing to pay to use community toilets. Because of the delivery methodology used, locals report feelings of ownership and interest in helping to keep the facilities clean and well maintained.

Third, solid data is the foundation of both programs. In Mr. Arisandi's work, schools along the river were identified and teachers were trained to teach children about hydrology, water use and sources, and environmental connectedness through classroom work and field trips. This was complemented by additional outreach to those not associated with schools. Indeed, one of the foundations to the success of the whole effort was in convincing citizens, often uneducated adults or children, to rigorously test water quality at more than 50 sites for several years and in a way which is credible (using bioindicators). With large volumes of credible data, efforts to engage policy makers were facilitated. And, not incidentally, the process of collecting those data from the citizen base also created an important political constituency that then became influential in the public policy process.

Fourth, in the case of the river clean-up, the engagement of journalists to help spread information and awareness about the effort was an integral and successful component. Mr. Arisandi provided weekly information updates to 14 journalists who then collaborated with him on pieces for a range of outlets. Such outreach paved the way for policy maker involvement, leading directly to new laws requiring the monitoring of the river's water quality.

Finally, both cases point to the benefits of systems thinking. What changed in each case was not so much the data collected or the service provided but rather the ways in which the business was conducted. This comprehensive approach places the effort on a broader backdrop of activity, thereby integrating it more fully into the fabric of the community. It successfully puts the financial, educational, and other incentives directly in front of the people whose behavior needs to change in order to achieve the goals. Furthermore, once the work is understood in this way, both systemic models can be replicated with appropriate modification in other locales.

Rigorous evaluation of the programs has not yet been completed. While anecdotal evidence points to success, evaluation may provide insights that could further strengthen the programs and suggest future actions.

---

[10]See Ashoka Fellows David Kuria in Kenjya and Isaac Durojaiye in Nigeria, www.ashoka.org.

Clean Water—Consideration of Two Health Research Approaches

Many of the qualities ascribed to social entrepreneurs may be attributed to those health researchers who have worked over long periods to achieve success on a single goal. As social entrepreneurs, medical researchers can best be described as "driven," "willing to work across confines of fields and geography," and having "long-term vision," as just some examples of the similarities. The work of two such "driven" researchers is outlined in Boxes 4-4 and 4-5. Dr. Rita Colwell has contributed to the knowledge base to understand the life cycle of the pathogenic agent for cholera, *Vibrio cholerae*, and its copepod vector. Over decades, she and her colleagues showed that a variety of characteristics of water, including temperature and salinity, affect the spread of the cholera bacterium. She also showed that the copepod vector could be effectively filtered out using readily available sari cloth in poor villages, leading to reduced illness.

Dr. Joseph Graziano and his colleagues provided critical insights into the health impacts of arsenic, including the molecular mechanisms underlying disease. Recent work (Gamble et al., 2006) has led to interventional studies focused on folic acid as a means to assist subsets of the population in metabolizing arsenic to inactive forms. Moving from the bench back to the field, Dr. Graziano and his colleague Dr. Alexander van Green worked with a community-based effort to localize wells with high- or low-arsenic levels. With thousands of participants and attendant data, it was found that within a single village both high- and low-arsenic wells can exist. More recent studies have shown that, even when presented with information on arsenic impacts, other factors come into play when determining whether to use water from high- or low-arsenic wells. While roughly half of residents with high-arsenic wells switched to low-arsenic wells, others did not, largely because the distances to the safer wells were too long. The "durability" of well-switching over time is not clear, and other options such as the provision of a deep, low-arsenic community well appear to have much greater promise as a longer term solution.

Both of these examples show the impact and import of sustained hypothesis-driven research efforts over time. With the dedicated support and vision of these medical researchers over decades, new knowledge was gained to underpin intervention efforts related to cholera and arsenic-related disease. Both examples also show the ability of the researchers to engage communities as part of the interventional research. In the case of the folded sari intervention for cholera, the community played a leadership role in ensuring that the interventional pilot study was conducted. Some in the research community were concerned that the pilot would fail since it was believed that men would not drink water that had been filtered through sari cloth. It was discovered, however, that men had already been using the same sari cloth as part of the process for their local production of fermented drinks. In this case, the community voice allowed an intervention to be tested that might otherwise have been left off the drawing board.

BOX 4-4
A Multidisciplinary Approach to Prevent Cholera

Dr. Rita Colwell, now at the University of Maryland, and her colleagues have studied transmission dynamics of cholera in the Bangladeshi villages for more than 20 years (see, for example, Colwell et al., 2003). While surveillance figures for cholera are not exact due to under-reporting by many countries, lack of standard case definitions, and lack of surveillance technologies, there were at least 131,943 cholera cases in 52 countries in 2005, resulting in over 2,000 deaths (WHO, 2006). Cholera emergence and reemergence continues to challenge health authorities, particularly in African nations. The 2008 outbreak of cholera in Zimbabwe led to over 26,000 documented cases including 1,518 deaths (WHO, 2008). Over the past two decades, the knowledge base has been developed to show that one of the key components of the transmission cycle for the cholera bacterium is a planktonic copepod. Specifically, the copepod plays a major role in the multiplication, survival, and transmission of the *Vibrio cholerae* bacterium.

Dr. Colwell and her team devised and tested a simple intervention to filter out particles larger than 20 microns, including copepods, from drinking water. The intervention relied on folded sari cloth, which is used for women's attire and is readily available in all villages. Pilot studies showed the best filtering occurred by folding a particular mesh of sari cloth four or more times. Between September 1999 and July 2002, 65 villages of rural Bangladesh were introduced to the filtering technique. More than 130,000 individuals took part in the study, which resulted in a 48 percent reduction of cholera compared with the control (Colwell et al., 2003).

Dr. Colwell recounts (personal communication to author, April 2008) that health researchers were skeptical about uptake of the filtering technique during the study itself since the expectation was that men would not accept water filtered through women's sari cloth. In partnership with social scientists, the team soon discovered that men had already devised a similar filtering strategy as part of their own process in making fermented drinks. Anecdotal evidence shows that there were no obstacles in social acceptance of the filtered water.

It might be expected that this simple intervention would be widely and sustainably in place in the 65 villages, and that it would have been adopted by others. In fact this was the case, in that the "control" villages were found to filter their water when a study of sustainability of the method was done five years later. It was found that the filtration frequently employed less than the required four or more folds. If a continuing education and training effort had been provided, reduction in illness and even deaths would have been enhanced. Even so, it was observed that a "herd effect" occurred in that those not filtering their water benefited by a lower cholera rate if they lived within a village where filtration was practiced.

BOX 4-5
Basic and Behavioral Science to Reduce Arsenic Exposures

Dr. Joseph Graziano and his colleagues have worked over many years to understand the dose-response relationships between exposure to arsenic and human disease. It is estimated that between 35 and 77 million people are at risk of drinking arsenic-contaminated water in Bangladesh alone (Khan et al., 1997). Over long periods of time, such consumption leads to skin lesions, cancer, and in some cases death.

Working in a 25 km^2 region in Bangladesh, the Graziano team blends a spectrum of disciplines, including environmental health, geochemistry, hydrology, and social science. This interdisciplinary approach grew naturally due to a number of factors, one of which was the startling discovery that tube wells put in place in the 1970s led to unhealthy levels of arsenic in drinking water over much of Bangladesh and South Asia. With accomplishments in lead toxicology already, Dr. Graziano turned attention to the issue of arsenic and manganese in drinking water. Basic research led to discoveries about the cellular mechanisms of toxicity and interventions aimed at reducing impact of arsenic on specific subsets of individuals.

In 2000, Graziano and colleagues van Geen and Ahsan expanded their studies to try to understand more fully water usage patterns and preferences for mitigation should local wells be found to exceed accepted limits of arsenic (see Mead, 2005, for overview). With help from local villagers, handheld Global Positioning System devices were used to map the location of each well in an individual village. Later, 12,000 village residents were recruited for a study to determine arsenic levels in urine compared to that found in wells. The team determined that high-arsenic wells and low-arsenic wells could be found in the same village, and the distribution of arsenic in one village did not necessarily correlate with that in the next. This finding suggests that mitigation strategies should be tailored at the subvillage level.

More recent work has shown that, even when presented with information on the health impacts of arsenic, other factors come into play in decisions about whether to use water from high- or low-arsenic wells. Distance to the low-arsenic well appears to be one of the factors involved: if the distance is too far, some will choose to use water from the high-arsenic well over the low-arsenic well.

In terms of scale-up and long-term adoption of the interventions, both examples illustrate the challenges in moving from a basic research understanding of a problem to a populations-based intervention strategy. In the case of the folded sari intervention, despite its clear effectiveness in filtering out the cholera vector, and demonstrated reduction in illness after filtering, uptake of the intervention was sustainable in that filtration continued after the project was completed, but the details, namely, the number of folds effective for filtering out the plankton were not adhered to in every case. Similarly, there was a lack of permanent switching to the use of low-arsenic wells. Two challenges present themselves. First, the incomplete uptake of the intervention may be due to a lack of local reinforcement of the links between clean water and better health within the community. Continued efforts to raise awareness are needed. Second, given the number of donors working in the region, questions may also be raised as to why the simple sari intervention or well-switching interventions were not incorporated into larger development programs. Placing these simple interventions into broader development perspective bears examination.

Discussion

The global burden of illness and death directly related to lack of clean water and sanitation demands that we consider all possible strategies. As various groups take on this challenge, new ideas and programs will be put in place. By merging the best features of different approaches, even better strategies may surface.

This paper juxtaposes two approaches to improving clean water and sanitation in resource-poor settings. The social entrepreneur approach focuses on delivery of interventions or gathering of information for policy purposes. The medical research approach focuses on understanding the links between toxins or microbes and attendant ill health, then working to mitigate exposures. Both approaches have yielded immense data sets and new knowledge that has led to improved health.

There are similarities and differences between the approaches. Some of the key features of the two approaches are outlined in Table 4-9. Closer examination of two features in particular provide insights into potential future actions. First, how community members were included in the work had an impact on the overall outcomes. The social entrepreneurship model embraces community perspectives as true experts, thereby ensuring that voices heard from the earliest of stages are those from ultimate beneficiaries, and that the work witnesses and subsequently accounts for the incentives faced by those whose social behavior it seeks to change. Interventions based on community involvement had a high degree of uptake. This particular approach leads to feelings of "ownership" of the work, driven by educational or economic incentives, and efforts to ensure its sustainability. The medical research model includes important elements of community engagement, particularly during data gathering for pilot efforts and as part of educational efforts related to scale-up of interventions. Long-term sustainability of effective interventions might improve if community engagement were viewed through another lens. Increased community involvement, including consults with local social entrepreneurs, may lead to more effective and long-lasting uptake of interventions, both geographically and on a long-term timescale. One can easily imagine the impact that the folded sari intervention might have in the hands of a social entrepreneur. By the same token, linking information about low- and high-arsenic wells in particular villages to motivated community members or social entrepreneurs might yield new insights on critical operational aspects.

TABLE 4-9 Key Features of the Two Approaches on Clean Water and Sanitation

Social Entrepreneurship	Medical Research
Vision of public health goal from the outset	
Community viewed as experts and as responsible for solving the problem.	Community plays a critical role in informing approaches to scientific project development, in gathering of data, and in education and training related to scale-up.
Funding from wherever it can be found, often single nongovernmental sources, multiple sources, or parallel small private enterprises.	Funding via government grants, multiple sources over span of years.
Science is one of several, equal facets needed to get the job done. Other facets include business, education, social, technology. Each is relevant insofar as it provides incentives for changes in social behavior.	Scientific and public health dimensions are primary considerations.
Public policy goals may be front and center in guiding the work.	Public policy goals to be informed by the outcomes of the scientific efforts.
Evaluations important but take place more on an ad hoc basis.	Rigorous and regular evaluations required to sustain long-term research support.

Second, the type of funding for the two approaches dictates in many ways the range of activities that may be supported. Medical research grants supported by governments tend to limit activities to those that lead to new knowledge about population-based health risks, outcomes due to exposures, including cellular and subcellular responses, and effectiveness of clinical interventions, for example. The development of policy-relevant data has not been a traditional focus of medical research grants. Given the priorities, funding cycles tend to be on the four- to five-year range. The researchers described in the present text pieced together long-term funding strategies over time using multiple sources. This is in contrast to the Ashoka model, which provides support with relatively few conditions for a critical phase of start-up activity intended to lead to population-based policy or intervention results and intended to get the entrepreneurial work to a solid enough footing that more traditional organizations will step in to support and/or replicate. Looking at the two models, it could be speculated that some medical researchers would benefit from a granting system that provided longer term support, perhaps on a 10-year cycle, which included a mid-term review timed to moving basic science knowledge into practice. Such an approach could capture the best elements of both systems.

Increasing the dialogue between the medical research community and the social entrepreneur community would likely enhance operations on both sides. Formal and informal means could be identified for exchange of views and expertise. Such exchanges might lead to exciting new actions and programs. Possible mechanisms to bolster communication are many and include, first, nominations of social entrepreneurs to public slots on governmental research agency advisory boards and, second, linking social entrepreneurs on the ground with researchers funded through medical research councils or other national or international grant-making bodies. In the latter case, ambassadors, aid mission directors, and other senior government officials, including military, agricultural, and health attachés, could play a role in brokering and facilitating the exchange of ideas between the two communities on the ground. Third, efforts to raise awareness within the medical research community about the social entrepreneurship movement through conferences and lectures would spark ideas for action, particularly from those already attuned to this wave—the next generation. Finally, medical researchers, public health professionals, students, and others should be made aware of new tools such as Changemakers.com, a partnership supported by Ashoka, the Robert Wood Johnson Foundation, and the Global Water Challenge. Through that tool, ideas and entrepreneurial approaches to providing clean water and other seemingly intractable problems, including strengthening of health-care systems, have been identified. Encouraging deposits of good ideas and mining the site for good ideas would be useful undertakings. This would in fact be in line with the next-generation vision of Ashoka, "Everyone is a Changemaker."

Acknowledgments

I am very grateful to Rita Colwell and Joseph Graziano for allowing me to tell their stories in this context, and also for their generosity in reviewing and editing the draft paper. David Strelneck of Ashoka provided invaluable insights at every step of the way. He and former Ashoka colleague Carol Grodzins reviewed the draft and made extremely helpful suggestions. Appreciation also goes to NIH colleagues Patricia Mabry (OBSSR) and Claudia Thompson (NIEHS) for their guidance and perspectives early on in the framing of this effort.

OVERVIEW REFERENCES

Clasen, T. F. 2008. *Scaling up household water treatment: looking back, seeing forward*. Geneva: WHO.

Clasen, T. F., W. P. Schmidt, T. Rabie, I. Roberts, and S. Cairncross. 2007a. Interventions to improve water quality for preventing diarrhoea: systematic review and meta-analysis. *British Medical Journal* 334(7597):782.

Clasen, T. F., L. Haller, D. Walker, J. Bartram, and S. Cairncross. 2007b. Cost-effectiveness of water quality interventions for preventing diarrhoeal disease in developing countries. *Journal of Water and Health* 5(4):599-608.

Clasen, T., C. McLaughlin, N. Nayaar, S. Boisson, R. Gupta, D. Desai, and N. Shah. 2008. Microbiological effectiveness and cost of disinfecting water by boiling in semi-urban India. *American Journal of Tropical Medicine and Hygiene* 79(3):407-413.

Marcotullio, P. J., and G. McGranahan, eds. 2007. *Scaling urban environmental challenges: from local to global and back*. London: Earthscan Publications.

WHO (World Health Organization). 2008. *International network to promote household water treatment and safe storage*. Geneva: WHO.

SINGER REFERENCES

AWWA (American Water Works Association). 2006. *Water chlorination/chloramination practices and principles*, M20. Denver, CO: AWWA.

Boyer, T. H., and P. C. Singer. 2006. A pilot-scale evaluation of magnetic ion exchange treatment for removal of natural organic material and inorganic anions. *Water Research* 40(15):2865-2876.

Letterman, R., ed. 1999. *Water quality and treatment*, 5th edition. Denver, Co: AWWA.

MWH. 2005. *Water treatment principles and design*, 2nd edition. Hoboken, NJ: John Wiley & Sons.

CARAVATI ET AL. REFERENCES

ABB Ltd. 2007. *A2Z of H2O guide to water cycle*, http://www.tagteam.com/tagteam/client/detail.asp?dbid=539&siteid=836048&dataid=118594 (accessed February 17, 2009).

AWWA (American Water Works Association). 1999. *Water quality and treatment: a handbook of community public water supplies*, 5th ed. New York: McGraw-Hill.

Bayliss, K., and T. McKinley. 2007. Privatising basic utilities in sub-Saharan Africa: the MDG impact. *United Nations Development Programme and International Policy Center Policy Research Brief No. 3*, http://www.undp-povertycentre.org/pub/IPCPolicyResearchBrief003.pdf (accessed February 17, 2009).

Black, M. 2008. Sanitation: creating a stink about the world's wastewater. *The Guardian,* http://www.guardian.co.uk/environment/2008/aug/20/water.waste (accessed February 17, 2009).

Cardone, R., and C. Fonseca. 2006. Experiences with innovative financing: small town water supply and sanitation service delivery. Background Paper prepared for *UN-HABITAT publication, Meeting Development Goals in Small Urban Centres: Water and Sanitation in the World's Cities 2006*, http://www.irc.nl/content/download/26959/289739/file/UN_Habitat_Innovative_Finan.pdf (accessed February 17, 2009).

Crites, R., and G. Tchobanoglous. 1998. *Small and decentralized wastewater management systems*. New York: McGraw-Hill.

EPA (Environmental Protection Agency). 2006. *New York City watershed partnership*, http://www.epa.gov/NCEI/collaboration/nyc.pdf (accessed February 17, 2009).

———. 2009. *Benefits of water efficiency*, http://www.epa.gov/watersense/water/benefits.htm (accessed February 17, 2009).

George, R. 2008. *The big necessity: the unmentionable world of human waste and why it matters*. New York: Metropolitan Books.

Gleick, P. H. 2000. The changing water paradigm: a look at twenty-first century water resources development. *Water International* 25(1):127-138.

———. 2006. *The world's water 2006-2007: the biennial report on fresh water resources*. Washington, DC: Island Press.

Hightower, M., and S. A. Pierce. 2008. The energy challenge. *Nature* 452(7185):285-286.

Hillie, T., M. Munasinghe, M. Hope, and Y. Deraniyagala. 2006. *Nanotechnology, water and development*, http://www.merid.org/nano/waterpaper/NanoWaterPaperFinal.pdf (accessed February 17, 2009).

Hutton, G., L. Haller, and J. Bertram. 2006. Economic and health effects of increasing coverage of low cost water and sanitation interventions. *Human Development Report Office Occasional Paper*, http://hdr.undp.org/en/reports/global/hdr2006/papers/who.pdf (accessed February 17, 2009).

IEA (International Energy Agency). 2008. *World energy outlook 2008*. Paris: IEA.

J. P. Morgan and Consultative Group to Assist the Poor. 2009. *Microfinance: shedding light on microfinance equity valuation: past and present*, http://www.jpmorgan.com/pages/jpmorgan/investbk/research/mfi (accessed February 17, 2009).

Lantagne, D. S., R. Quick, and E. Mintz. 2006. *Household water treatment andsafe storage options in developing countries: a review of current implementation practices.* Wilson Center Environmental Change and Security Program, http://www.irc.nl/page/37316 (accessed February 17, 2009).

Logan, B. E., and J. M. Regan. 2006. Microbial fuel cells: challenges and applications. *Environmental Science and Technology* 40(17):5172-5180.

Lovley, D. R. 2006. Microbial fuel cells: novel microbial physiologies and engineering approaches. *Current Opinion in Biotechnology* 17(3):327-332.

McKinsey Global Institute. 2007. *Curbing global energy demand growth: the energy productivity opportunity*, http://www.mckinsey.com/mgi/publications/Curbing_Global_Energy/index.asp (accessed February 17, 2009).

Morgan Stanley. 2007. *Blue orchard loans for development 2007-1 ("BOLD 2") prices*, http://www.morganstanley.com/about/press/articles/4977.html (accessed February 17, 2009).

NWP (Netherlands Water Partnership). 2007. *Microfinance for water, sanitation and hygiene*, http://www.irc.nl/content/download/128991/353215/file/Microfinance_%20India.pdf (accessed February 17, 2009).

OECD (Organisation for Economic Co-operation and Development). 2008. *OECD Environmental Outlook to 2030*, http://www.oecd.org/dataoecd/29/33/40200582.pdf (accessed February 17, 2009).

Platz, D., and F. Schroeder. 2007. Moving beyond the privatization debate. *Dialogue on Globalization Occasional Papers* 34. New York: Friedrich-Ebert-Stiftung.

Revenga, C. 2009. *The next big ideas in conservation: paying water's real costs*, http://www.nature.org/tncscience/bigideas/people/art23907.html (accessed February 17, 2009).

Revenga, C., S. Murray, J. Abramowitz, and A. Hammond. 1998. *Watersheds of the world: ecological value and vulnerability*. Washington, DC: World Resources Institute.

Schnoor, J. 2008. Living with a changing water environment. *The Bridge* 38(3):46-54.

Shannon, M. A., P. W. Bohn, M. Elimelech, J. G. Georgiadis, B. J. Marinas, and A. M. Mayes. 2008. Science and technology for water purification in the coming decades. *Nature* 452(7185): 301-310.

Tchobanoglous, G., F. L. Burton, and H. D. Stensel. 2003. *Wastewater engineering: treatment and reuse*, 4th edition. New York: McGraw-Hill.

USGS (United States Geological Survey). 2005. *Estimated use of water in the United States in 2000*, http://pubs.usgs.gov/fs/2005/3051/pdf/fs2005-3051.pdf (accessed February 17, 2009).

———. 2008. *The water cycle*, http://ga.water.usgs.gov/edu/watercycle.html (accessed February 17, 2009).

Viessman, W., and M. J. Hammer. 1985. *Water supply and pollution control,* 4th edition. New York: Harper and Row.

WHO (World Health Organization). 2008. *Access to improved drinking-water sources and to improved sanitation (percentage)*, http://www.who.int/whosis/indicators/compendium/2008/2wst/en/ (accessed February 18, 2009).

World Bank. 2008. *Private participation in infrastructure database,* http://ppi.worldbank.org/ (accessed February 17, 2009).

World Water Assessment Programme. 2003. *Water for people, water for life*. New York: Berghahn Books; Paris: UNESCO.

———. 2006. *Water, a shared responsibility*. New York: Berghahn Books; Paris: UNESCO.

Zimmerman, J. B., J. R. Mihelcic, and J. Smith. 2008. Global stressors on water quality and quantity. *Environmental Science and Technology* 42(12):4247-4254.

BOSTOEN ET AL. REFERENCES

Bern, C., Martines, J., Zoysa, I. D. and Glass, R. I. (1992) 'The magnitude of the global problem of diarrhoeal disease: A 10-year update', *Bulletin of WHO*, vol 70, no 6, pp705–714

BHS (1998) 'International Conference on Hydrology in a Changing Environment', University of Exeter, British Hydrological Society, www.hydrology.org.uk/exeter.html

Briscoe, J., Feachem, R. G. and Rahman, M. (1985) *Measuring the Impact of Water Supply and Sanitation Facilities on Diarrhoea Morbidity; Prospects for Case-control Methods*, World Health Organization (WHO), Environmental Health Division,Geneva

Cairncross, S. (1989) 'Water supply and sanitation: An agenda for research', *Journal of Tropical Medicine and Hygiene*, vol 92, pp301–314

Cairncross, S. (1990) 'Health impacts in developing countries: New evidence and new prospects', *Journal of the Institution of Water and Environmental Management*, vol 4, no 6, pp571–577

Cairncross, S. (1995) *Water Quality, Quantity and Health. Safe Water Environments*, Swedish International Development Cooperation Agency (Sida), Eldoret

Cairncross, S. (1999) *Measuring the Health Impact of Water and Sanitation*, WEDC, London School of Hygiene and Tropical Medicine (LSHTM), London, p2, www.worldbank.org/watsan/pdf/tn02.pdf, www.lboro.ac.uk/well/resources/fact-sheets/fact-sheets-htm/mthiws.htm

Cairncross, S., Blumenthal, U., Kolsky, P., Moraes, L. and Tayeh, A. (1996) 'The public and domestic domains in the transmission of disease', *Tropical Medicine and International Health*, vol 1, no 1, pp27–34

Cairncross, S. and Feachem, R. G. (1993) *Environmental Health Engineering in the Tropics*, John Wiley & Sons, Chichester

Clasen, T. (2006) 'Household water treatment for the prevention of diarrhoeal disease. Department of Infectious Diseases', PhD thesis, LSHTM, London, p271

Clasen, T. F. and Bastable, A. (2003) 'Faecal contamination of drinking water during collection and household storage: The need to extend protection to the point of use', *Journal of Water and Health*, vol 1, no 3, pp109–115

Clasen, T. F. and Cairncross, S. (2004) 'Editorial: Household water management: Refine the dominant paradigm', *Tropical Medicine and International Health*, vol 9, no 2, pp187–191

Clasen, T., Roberts, I., Rabie, T. and Cairncross, S. (2004) *Interventions to Improve Water Quality for Preventing Infectious Diarrhoea*, Cochrane Library, Infectious Diseases Group, John Wiley & Sons, Chichester

Conroy, R. M., Meegan, M. E., Joyce, T., McGuigan, K. and Barnes, J. (2001) 'Solar disinfection of drinking water protects against cholera in children under 6 years of age', *Archives for Diseases in Childhood*, vol 85, no 4, pp293–295

Curtis, V., Cairncross, S. and Yonli, R. (2000a) 'Domestic hygiene and diarrhoea – pinpointing the problem', *Tropical Medicine and International Health*, vol 5, no 1, pp22–23

Curtis, V., Cairncross, S. and Yonli, R. (2000b) 'Domestic hygiene and diarrhoea – pinpointing the problem', *Tropical Medicine and International Health*, vol 5, no 1, pp22–32

Deb, B. C., Sircar, B. K., Sengupta, P. G., De, S. P., Mondal, S. K., Gupta, D. N., Saha, N. C., Ghosh, S., Mitra, U. and Pal, S. C. (1986) 'Studies on interventions to prevent el tor cholera transmission in urban slums', *Bulletin of the World Health Organization*, vol 64, no 1, pp127–131

Esrey, S. A., Feachem, R. G. and Hughes, J. M. (1985) 'Interventions for the control of diarrhoeal diseases among young children: Improving water supplies and excreta disposal facilities', *Bulletin of the World Health Organization*, vol 63, no 4, pp757–772

Esrey, S. A., Potash, J. B., Roberts, L. and Shiff, C. (1991) 'Effects of improved water supply and sanitation on ascariasis, diarrhoea, dracunculiasis, hookworm infection, schistosomiasis and trachoma', *Bulletin of the World Health Organization*, vol 69, no 5, pp609–621

Falkenmark, M., Lundqvist, J. and Widstrand, C. (1989) 'Macro-scale water scarcity requires micro-scale approaches: Aspects of vulnerability in semi-arid development', *Natural Resources Forum*, vol 13, no 4, pp258–267

Feachem, R. G., Bradley, D. J., Garelick, H. and Mara, D. D. (1983a) *Sanitation and Disease; Health Aspects of Excreta and Wastewater Management*, John Wiley & Sons, Chichester

Feachem, R. G., McGarry, M. and Mara, D. (eds) (1977) *Water, Waste and Health in Hot Climates*, John Wiley & Sons, Chichester

Feachem, R. G., Hogan, R. C. and Merson, M. H. (1983b) 'Diarrhoeal disease control: Reviews of potential interventions', *Bulletin of the World Health Organization*, vol 61, no 4, pp637–640

Fewtrell, L., Kaufmann, R. B., Kay, D., Enanoria, W., Haller, L. and Colford, J. M., Jr (2005) 'Water, sanitation, and hygiene interventions to reduce diarrhoea in less developed countries: A systematic review and meta-analysis', *Lancet Infectious Diseases*, vol 5, no 1, pp42–52

Handy, C. B. (1985) *Understanding Organizations*, Penguin Books, London

Hardin, G. (1968) 'The tragedy of the commons', *Science*, vol 162, pp1243–1248

Hellard, M. E., Sinclair, M. I., Forbes, A. B. and Fairley, C. K. (2001) 'A randomized, blinded, controlled trial investigating the gastrointestinal health effects of drinking water quality', *Environmental Health Perspectives*, vol 109, no 8, pp773–778

Hinrichsen, D., Robey, B. and Upadhyay, U. D. (1997) *Solution for a Water-Short World*, Johns Hopkins School of Public Health, Population Program, Baltimore, MA, www.infoforhealth.org/pr/m14edsum.shtml

Iijima, Y., Karama, M., Oundo, J. O. and Honda, T. (2001) 'Prevention of bacterial diarrhea by pasteurization of drinking water in Kenya', *Microbiology and Immunology*, vol 45, no 6, pp413–416

Jagals, P., Grabow, W. O. K. and Williams, E. (1997) 'The effects of supplied water quality on human health in an urban development with limited basic subsistence facilities', *Water South Africa*, vol 23, no 4, pp373–378

Jonnalagadda, P. R. and Bhat, R. V. (1995) 'Parasitic contamination of stored water used for drinking/cooking in Hyderbad', *South-East Asian Journal of Tropical Medicine and Public Health*, vol 26, no 4, pp789–794

Kirchhoff, L. V., McClelland, K. E., Do Carmo Pinho, M., Araujo, J. G., De Sousa, M. A. and Guerrant, R. L. (1985) 'Feasibility and efficacy of in-home water chlorination in rural North-eastern Brazil', *Journal of Hygiene*, vol 94, no 2, pp173–180

Kolsky, P. J. (1993) 'Diarrhoeal disease: Current concepts and future challenges. Water', *Transactions of the Royal Society of Tropical Medicine and Hygiene*, vol 87 (supplement no 3), pp43–46

Kolsky, P. (2002) 'Water, health and cities: Concepts and examples', Paper presented at an international Workshop on Planning for Sustainable Development: Cities and Natural Resource Systems in Developing Countries, University of Wales, Cardiff, 13–17 July, 1992

Lewin, S., Stephens, C. and Cairncross, S. (1996) 'Health impacts of environmental improvements in Cuttack and Cochin, India', Review prepared for the Overseas Development Administration by LSHTM, London

McGranahan, G., Jacobi, P., Songsore, J., Surjadi, C. and Kjellen, M. (eds) (2001) *The Citizens at Risk: From Urban Sanitation to Sustainable Cities*, Earthscan, London

Mintz, E. D., Reiff, F. M. and Tauxe, R. V. (1995) 'Safe water treatment and storage in the home. A practical new strategy to prevent waterborne disease', *Journal of the American Medical Association*, vol 273, no 12, pp948–953

Murray, C. J. L. and Lopez, A. D. (eds) (1996) *The Global Burden of Disease: A Comprehensive Assessment of Mortality and Disability from Diseases, Injuries, and Risk Factors in 1990 and Projected to 2020*, Harvard University Press, Harvard, MA

Okun, D. A. (1978) *The Regionalization of Water Management: A Revolution in England and Wales*, Applied Science Publishers, London

Quick, R. E., Venczel, L. V., Mintz, E. D., Soleto, L., Aparicio, J., Gironaz, M., Hutwagnar, L., Greene, K., Bopp, C., Maloney, K., Chavez, D., Sobsey, M. and Tauxe, R. (1999) 'Diarrhoea prevention in Bolivia through point of use water treatment and safe storage: A promising new strategy', *Epidemiology and Infection*, vol 122, pp83–90

Ruff, L. E. (1970) 'The economic common sense of pollution', *The Public Interest*, vol 19, Spring, pp69–85

Thompson, J., Porras, I. T., Tumwine, J. K., Mujwahuzi, M. R., Katui-Katua, M., Johnstone, N. and Wood, L. (2002) *Drawers of Water: 30 Y ears of Change in Domestic Water Use and Environmental Health – Summary*, Earthprint, www.iied.org/sarl/pubs/drofwater.html#9049IIED?to=9049IIED; www.iied.org/sarl/dow

UNEP (United Nations Environment Programme) (2002) *Global Environmental Outlook 3*, Earthscan, London

UN-HABITAT (United Nations Human Settlements Programme) (2003) *Water and Sanitation in the World's Cities*, Earthscan, London

VanDerslice, J. and Briscoe, J. (1993) 'All coliforms are not created equal: A comparison of the effects of water source and in-house water contamination on infantile diarrhoeal disease', *Water Resources Research*, vol 29, no 7, pp1983–1993

Wagner, E. G. and Lanoix, J. N. (1958) 'Excreta disposal for rural areas and small communities', *WHO monograph series*, Geneva, p39

White, G. F., Bradley, D. J. and White, A. U. (1972) *Drawers of Water*, University of Chicago Press, Chicago, IL

WHO (1983) *Minimum Evaluation Procedure (MEP) for Water Sanitation Projects*, World Health Organization, Geneva, p52

WHO (2001) *The World Health Report 2001; Mental Health: New Understanding, New Hope*, World Health Organization, Geneva, www.who.int/whr/

WHO (2004) The World Health Report 2004 – Changing History, WHO, Geneva, www.who.int/whr/2004/en/index.html

WHO/UNICEF (United Nations Children's Fund) (2000) *Global Water Supply and Sanitation Assessment 2000 Report*, www.unicef.org/programme/wes/pubs/global/global.htm

WHO/UNICEF (2005) *Water for Life. Making it Happen*, World Health Organization, Geneva, www.who.int/water_sanitation_health/waterforlife.pdf

WHO/UNICEF (2006) *Meeting the MDG Drinking Water and Sanitation Target: The Urban and Rural Challenge of the Decade*, www.childinfo.org/areas/water/pdfs/ jmp06final.pdf

World Bank (1976) *Measurement of the Health Benefits of Investments in Water Supply*, The World Bank, Washington, DC

HRYNKOW REFERENCES

Ashoka. 2009a. *Bill Drayton: nothing more powerful*, http://www.ashoka.org/video/4108 (accessed February 12, 2009).

———. 2009b. *What is a social entrepreneur?* http://www.ashoka.org/social_entrepreneur (accessed February 12, 2009).

Bornstein, D. 2004. *How to change the world: social entrepreneurship and the power of new ideas*. New York: Oxford Univesity Press.

Colwell, R. C., A. Hug, M. S. Islam, K. M. Aziz, M. Yunus, N. H. Khan, A. Mahmud, R. B. Sack, G. B. Nair, J. Chakraborty, D. A. Sack, and E. Russek-Cohen. 2003. Reduction of cholera in Bangladeshi villages by simple filtration. *Proceedings of the National Academy of Sciences* 100(3):1051-1055.

Gamble, M. V., X. Liu, H. Ahsan, J. R. Pilsner, V. Ilievski, V. Slavkovich, F. Parvez, Y. Chen, D. Levy, P. Factor-Litvak, and J. H. Graziano. 2006. Folate and arsenic metabolism: a double-blinded, placebo-controlled folic acid-supplementation trial in Bangladesh. *American Journal of Clinical Nutrition* 84(5):1093-1101.

Khan, A. W., S. K. Akhtar Ahmad, M. H. S. U. Sayed, S. K. A. Hadi, M. H. Khan, M. A. Jalil, R. Ahmed, and M. H. Faruquee. 1997. Arsenic contamination in groundwater and its effect on human health with particular reference to Bangladesh. *Journal of Preventive and Social Medicine* 16(1):65-73.

Mead, M. N. 2005. Columbia programs digs deeper into arsenic dilemma. *Environmental Health Perspectives* 113(6):A374-377.

NIH (National Institutes of Health). 2008. *Summary of the FY 2009 President's Budget*, http://officeofbudget.od.nih.gov/ui/2008/Summary%20of%20FY%202009%20Budget-Press%20Release.pdf (accessed February 12, 2009).

Prüss-Üstün, A., R. Bos, F. Gore, and J. Bartram. 2008. *Safer water, better health: costs, benefits and sustainability of interventions to protect and promote health*. Geneva: World Health Organization.

WHO (World Health Organization). 2006. *Weekly epidemiological record*, www.who.int/wer/2006/wer8131.pdf (accessed April 6, 2009).

———. 2008. *Cholera in Zimbabwe—update*, http://www.who.int/csr/don/2008_12_26/en/index.html (accessed April 6, 2009).

Appendix A

Agenda

Global Issues in Water, Sanitation, and Health

September 23-24, 2008
The National Academies
500 Fifth Street, NW—Room 100
Washington, DC

DAY 1: SEPTEMBER 23, 2008

8:45-9:15 Registration and continental breakfast

9:15-9:45 Welcoming remarks
David Relman, M.D., Chair
Margaret "Peggy" A. Hamburg, M.D., Vice-Chair

9:45-10:15 *Running Dry*—19 minute version
Followed by discussion with Jim Thebaut, writer, producer, and director

10:15-11:00 KEYNOTE REMARKS:
Improving water, sanitation, and health at the grassroots
Donald Hopkins, M.D., M.P.H.
The Carter Center

11:00-11:45 Discussion

11:45-12:45 Lunch

Session I
Models of Disease Emergence and Transmission

Moderator: David A. Relman, Stanford University

12:45-1:15 The spectrum of water-related disease transmission processes
David Bradley, Ph.D.
London School of Hygiene and Tropical Medicine

1:15-1:45 Disease prevention strategy that starts with clean water: Safer water, safer hands, and safer food
Robert Tauxe, M.D., M.P.H.
Centers for Disease Control and Prevention

1:45-2:15 Discussion

2:15-2:30 Break

Session II
Infrastructure Vulnerabilities—Water Distribution and Metrics for Measuring Water Quality

Moderator: Margaret "Peggy" A. Hamburg, M.D., Nuclear Threat Initiative

2:30-3:00 The changing epidemiology of waterborne disease outbreaks in the United States: Implications for system infrastructure and future planning
Michael Beach, Ph.D.
Centers for Disease Control and Prevention

3:00-3:30 Climate change and water quality
Joan Rose, Ph.D.
Michigan State University

3:30-4:00 Quantitative microbial risk assessment: State of the art
Kelly Reynolds, Ph.D., M.S.P.H.
University of Arizona

4:00-4:30 Break

4:30-5:00 Measures of water quality impacting disinfection
Philip Singer, Ph.D.
University of North Carolina

5:00-5:30 Testing methodology: Lab and field
Mark Sobsey, Ph.D.
University of North Carolina

5:30-6:15 Discussion of Session II

6:15 Conclusion of Day 1

7:00-9:30 Executive Session Working Dinner

DAY 2: SEPTEMBER 24, 2008

8:45-9:15 Continental breakfast

9:15-9:30 Summary of Day 1
Jim Hughes, M.D.
Emory University

Session III
Relationships Between Human Demographics, Land Use, Infrastructure, and Disease: Lessons from Waterborne Disease Outbreaks

Moderator: Jim Hughes, M.D.

9:30-10:00 Cholera in Peru: 1991, the impact of the water in the extension of the epidemic
Eduardo Gotuzzo, M.D.
Universidad Peruana Cayetano Heredia, Peru

10:00-10:30 Cryptosporidiosis (Milwaukee, 1993)
Jeff Davis, M.D.
Wisconsin Department of Health

10:30-11:00 Prevention is painfully easy in hindsight—fatal *E. coli* O157:H7 and *Campylobacter* outbreak in Walkerton, Canada, 2000
Steve Hrudey, Ph.D.
University of Alberta

11:00-11:45 Discussion

11:45-12:45 Lunch and continuation of Day 2 morning discussion

Session IV
Interventions to Improve Water Accessibility, Availability, and Sanitation

Moderator: Jerry Keusch, M.D., Boston University

12:45-1:15 Household water treatment to prevent diarrheal disease: Effectiveness, cost-effectiveness, and the challenge of scaling up
Thomas Clasen, J.D., Ph.D.
London School of Hygiene and Tropical Medicine

1:15-1:45 Civil infrastructure for water, sanitation, and improved health: Existing technology, barriers, and the need for innovation
Joseph Hughes, Ph.D., P.E.
Georgia Institute of Technology

1:45-2:15 Social entrepreneurship meets medical research: Lessons in clean water
Sharon Hrynkow, Ph.D.
National Institute of Environmental Health Sciences

2:15-2:45 Implementation issues
Vahid Alavian, Ph.D., and Pete Kolsky, Ph.D.
The World Bank

2:45-3:15 Discussion

3:15-3:45 Open discussion of Day 2

3:45-4:00 Concluding remarks/Meeting adjourns

Appendix B

Acronyms

AGI	acute gastrointestinal illness
AIDS	acquired immune deficiency syndrome
BOD	biochemical oxygen demand
CDC	Centers for Disease Control and Prevention
CSTE	Council for State and Territorial Epidemiologists
CWS	community water systems
DALY	disability-adjusted life year
DFBMD	Division of Foodborne, Bacterial, and Mycotic Diseases
DOC	dissolved organic carbon
DPD	diethyl-*p*-phenylenediamine
DPH	Wisconsin Division of Public Health
ENSO	El Niño–Southern Oscillation
EPA	Environmental Protection Agency
FDA	Food and Drug Administration
GIS	geographic information sciences
GPM	gallons per minute
HIV	human immunodeficiency virus
HPC	heterotrophic plate count

HUS	hemolytic uremic syndrome
ICC	integrated cell culture
IPCC	Intergovernmental Panel on Climate Change
JICA	Japan International Cooperation Agency
JMP	Joint Monitoring Programme for Water Supply and Sanitation
MDG	Millennium Development Goal
MFC	microbial fuel cell
MHD	Milwaukee Health Department
MOE	Ministry of Environment (Canada)
MOH	Medical Officer of Health (Canada)
MWW	Milwaukee Water Works
NCZVED	National Center for Zoonotic, Vector-borne and Enteric Diseases
NGO	nongovernmental organization
NIEHS	National Institute of Environmental Health Sciences
NIH	National Institutes of Health
NRC	National Research Council
NTU	nephelometric turbidity unit
OECD	Organization for Economic Cooperation and Development
PAHO	Pan American Health Organization
PCR	polymerase chain reaction
PET	polyethylene terephthalate
PUC	Walkerton Public Utilities Commission
SAC	Spills Action Centre (Canada)
SCADA	supervisory control and data acquisition system
SCID	severe combined immunodeficient
TCR	Total Coliform Rule
TCRDSAC	Total Coliform Rule/Distribution System Advisory Committee
UNICEF	United Nations Children's Fund
UV	ultraviolet
WBDOSS	Waterborne Disease and Outbreak Surveillance System
WHO	World Health Organization

Appendix C

Glossary

Anthroponotic: Transmission from human to human and potentially from human to animal.

Conjunctiva: The thin, transparent tissue that covers the outer surface of the eye. It begins at the outer edge of the cornea, covering the visible part of the sclera, and lining the inside of the eyelids. It is nourished by tiny blood vessels that are nearly invisible to the naked eye (http://www.stlukeseye.com/Anatomy/Conjunctiva.asp [accessed June 3, 2009]).

Copepod: A type of crustacean which may live in both salt and freshwater and is one of the most abundant animals on the planet (http://jaffeweb.ucsd.edu/pages/celeste/Intro/ index.html [accessed June 3, 2009]).

Cornea: The transparent, dome-shaped window covering the front of the eye (http://www.stlukeseye.com/anatomy/cornea.asp [accessed June 3, 2009]).

CT measurement: The product of free chlorine residual (C) and contact time (T) required for disinfection which measures the effectiveness of free chlorine in inactivating microorganisms.

Dracunculiasis: Dracunculiasis, or Guinea worm disease, is caused by the parasite *Dracunculus medinensis*. The disease affects poor communities in remote parts of Africa that do not have safe water to drink. There is no treatment for Guinea worm disease, yet removal of the worm as it emerges from the infected person's skin is curative or surgical removal by a trained doctor (http://www.cdc.gov/ncidod/dpd/parasites/dracunculiasis/ default.htm [accessed June 3, 2009]).

Evaporation: The conversion of liquid to gas.

Flocculation: The separation of a solution. Most commonly, flocculation is used to describe the removal of a sediment from a fluid. In addition to occurring naturally, flocculation can also be forced through agitation or the addition of flocculating agents. Numerous manufacturing industries use flocculation as part of their processing techniques, and it is also extensively employed in water treatment. The technique is also widely used in the medical world to analyze various fluids (http://www.wisegeek.com/what-is-flocculation.htm [accessed June 3, 2009]).

Fomite: Inanimate objects or substances that can transmit infectious organisms from one host to another (IOM. 1993. *Indoor allergens: assessing and controlling adverse health effects*. Washington, DC: National Academy Press).

Megacity: Urban concentration with more than 10 million inhabitants (wwap.unesco.org/ev.php [accessed June 4, 2009]).

Recreational water: That which is used for water-based activities in marine, freshwater, hot tubs, spas and swimming pools (Pond, K. 2005. *Water recreation and disease. plausibility of associated infections: acute effects, sequelae and mortality*. London: IWA Publishing on behalf of the World Health Organization).

Sullage: Sullage (or grey water) is dirty water from the laundry, kitchen, and bathroom. Grey water contains chemicals such as dish detergent and soap as well as fats, grease, and whatever washes off our body while bathing. Sullage does not usually contain sewage but can be equally contaminated and can cause infections (http://www.nt.gov.au/health/healthdev/health_promotion/bushbook/volume2/chap2/intro.htm [accessed May 18, 2009]).

Superfund: The U.S. government program to clean up the nation's uncontrolled hazardous waste sites.

Appendix D

Forum Member Biographies

David A. Relman, M.D. (*Chair*), is professor of medicine (infectious diseases and geographic medicine) and of microbiology and immunology at Stanford University School of Medicine, and chief of the infectious disease section at the Veterans Affairs (VA) Palo Alto Health Care System. Dr. Relman received his B.S. in biology from the Massachusetts Institute of Technology and his M.D. from Harvard Medical School. He completed his residency in internal medicine and a clinical fellowship in infectious diseases at Massachusetts General Hospital, Boston, after which he moved to Stanford for a postdoctoral fellowship in 1986 and joined the faculty there in 1994. His research focus is on understanding the structure and role of the human indigenous microbial communities in health and disease. This work brings together approaches from ecology, population biology, environmental microbiology, genomics, and clinical medicine. A second area of investigation explores the classification structure of humans and nonhuman primates with systemic infectious diseases, based on patterns of genome-wide gene transcript abundance in blood and other tissues. The goals of this work are to understand mechanisms of host-pathogen interaction, as well as predict clinical outcome early in the disease process. His scientific achievements include the description of a novel approach for identifying previously unknown pathogens; the characterization of a number of new human microbial pathogens, including the agent of Whipple's disease; and some of the most in-depth analyses to date of human indigenous microbial communities. Among his other activities, Dr. Relman currently serves as chair of the Board of Scientific Counselors of the National Institutes of Health (NIH) National Institute of Dental and Craniofacial Research, is a member of the National Science Advisory Board for Biosecurity, and advises a number of U.S. government departments and agencies on matters related to pathogen diversity, the future life sciences landscape, and the nature of present and future biological threats. He was cochair of the Committee on Advances in Technology and the Prevention of Their Application to Next Generation Biowarfare Threats for the National Academy of Sciences (NAS). He received the Squibb Award from the Infectious Diseases Society of America (IDSA) in 2001, the Senior Scholar Award in Global Infectious Diseases from the Ellison Medical Foundation in 2002, an NIH Director's Pioneer Award in 2006, and a Doris Duke Distinguished Clinical Scientist Award in 2006. He is also a fellow of the American Academy of Microbiology.

Margaret A. Hamburg, M.D. (*Vice Chair*),[1] was the founding vice president, Biological Programs, at the Nuclear Threat Initiative, a charitable organization working to reduce the global threat from nuclear, biological, and chemical weapons, and ran the program for many years. She currently serves as senior scientist for the organization. She completed her internship and residency in internal medicine at the New York Hospital-Cornell University Medical Center and is certified by the American Board of Internal Medicine. Dr. Hamburg is a graduate of Harvard College and Harvard Medical School. Before taking on her current position, she was the assistant secretary for planning and evaluation, U.S. Department of Health and Human Services (HHS), serving as a principal policy adviser to the secretary of health and human services, with responsibilities including policy formulation and analysis, the development and review of regulations and legislation, budget analysis, strategic planning, and the conduct and coordination of policy research and program evaluation. Prior to this, she served for nearly six years as the commissioner of health for the City of New York. As chief health officer in the nation's largest city, her many accomplishments included the design and implementation of an internationally recognized tuberculosis control program that produced dramatic declines in tuberculosis cases, the development of initiatives that raised childhood immunization rates to record levels, and the creation of the first public health bioterrorism preparedness program in the nation. She currently serves on the Harvard University Board of Overseers. She has been elected to membership in the Institute of Medicine (IOM), the New York Academy of Medicine, and the Council on Foreign Relations and is a fellow of the American Association for the Advancement of Science (AAAS) and the American College of Physicians.

David W. K. Acheson, M.D., F.R.C.P., is assistant commissioner for food protection in the U.S. Food and Drug Administration (FDA). Dr. Acheson graduated from the University of London Medical School in 1980 and, following training in internal medicine and infectious diseases in the United Kingdom, moved to the

[1]Until June 9, 2009. Dr. Hamburg is currently the Commissioner of the Food and Drug Administration.

New England Medical Center and Tufts University in Boston in 1987. As an associate professor at Tufts University, he undertook basic molecular pathogenesis research on foodborne pathogens, especially Shiga toxin-producing *Escherichia coli*. In 2001, Dr. Acheson moved his laboratory to the University of Maryland Medical School in Baltimore to continue research on foodborne pathogens. In September 2002, Dr. Acheson accepted a position as chief medical officer at the FDA Center for Food Safety and Applied Nutrition (CFSAN). In January 2004, he also became the director of CFSAN's Food Safety and Security Staff, and in January 2005, the staff was expanded to become the Office of Food Safety, Defense, and Outreach. In January 2007, the office was further expanded to become the Office of Food Defense, Communication, and Emergency Response. On May 1, 2007, Dr. Acheson assumed the position of FDA assistant commissioner for food protection to provide advice and counsel to the commissioner on strategic and substantive food safety and food defense matters. Dr. Acheson has published extensively and is internationally recognized both for his public health expertise in food safety and for his research in infectious diseases. Additionally, Dr. Acheson is a fellow of both the Royal College of Physicians (London) and the IDSA.

Ruth L. Berkelman, M.D., is the Rollins Professor and director of the Center for Public Health Preparedness and Research at the Rollins School of Public Health, Emory University, in Atlanta. She received her A.B. from Princeton University and her M.D. from Harvard Medical School. Board certified in pediatrics and internal medicine, she began her career at the Centers for Disease Control and Prevention (CDC) in 1980 and later became deputy director of the National Center for Infectious Diseases (NCID). She also served as a senior adviser to the director of CDC and as assistant surgeon general in the U.S. Public Health Service. In 2001 she came to her current position at Emory University, directing a center focused on emerging infectious diseases and other urgent threats to health, including terrorism. She has also consulted with the biologic program of the Nuclear Threat Initiative and is most recognized for her work in infectious diseases and disease surveillance. She was elected to the IOM in 2004. Currently a member of the Board on Life Sciences of the National Academies, she also chairs the Board of Public and Scientific Affairs at the American Society of Microbiology (ASM).

Enriqueta C. Bond, Ph.D., is president Emeritus of the Burroughs Wellcome Fund. She received her undergraduate degree from Wellesley College, her M.A. from the University of Virginia, and her Ph.D. in molecular biology and biochemical genetics from Georgetown University. She is a member of the IOM, the AAAS, the ASM, and the American Public Health Association. Dr. Bond chairs the Academies' Board on African Science Academy Development and serves on the Report Review Committee for the Academies. She serves on the board and executive committee of the Hamner Institute, the board of the Health Effects Institute, the board of the James B. Hunt Jr. Institute for Educational Leadership and Policy, the council of the National Institute of Child Health and Human Development and the NIH Council of Councils. In addition Dr. Bond serves on a scientific advisory committee for the World Health Organization (WHO) Tropical Disease Research Program. Prior to being named president of the Burroughs Wellcome Fund in 1994, Dr. Bond served on the staff of the IOM beginning in 1979, becoming its executive officer in 1989.

Roger G. Breeze, Ph.D., received his veterinary degree in 1968 and his Ph.D. in veterinary pathology in 1973, both from the University of Glasgow, Scotland. He was engaged in teaching, diagnostic pathology, and research on respiratory and cardiovascular diseases at the University of Glasgow Veterinary School from 1968 to 1977 and at Washington State University College of Veterinary Medicine from 1977 to 1987, where he was professor and chair of the Department of Microbiology and Pathology. From 1984 to 1987 he was deputy director of the Washington Technology Center, the state's high-technology sciences initiative, based in the College of Engineering at the University of Washington. In 1987, he was appointed director of the U.S. Department of Agriculture's (USDA's) Plum Island Animal Disease Center, a Biosafety Level 3 facility for research and diagnosis of the world's most dangerous livestock diseases. In that role he initiated research into the genomic and functional genomic basis of disease pathogenesis, diagnosis, and control of livestock RNA and DNA virus infections. This work became the basis of U.S. defense against natural and deliberate infection with these agents and led to his involvement in the early 1990s in biological weapons defense and proliferation prevention. From 1995 to 1998, he directed research programs in 20 laboratories in the Southeast for the USDA Agricultural Research Service before going to Washington, DC, to establish biological weapons defense research programs for USDA. He received the Distinguished Executive Award from President Clinton in 1998 for his work at Plum Island and in biodefense. Since 2004 he has been chief executive officer of Centaur Science Group, which provides consulting services in biodefense. His main commitment is to the Defense Threat Reduction Agency's Biological Weapons Proliferation Prevention Program in Europe, the Caucasus, and Central Asia.

Steven J. Brickner, Ph.D., is an independent consultant based in southeastern Connecticut. He received his Ph.D. in organic chemistry from Cornell University, and completed an NIH postdoctoral research fellowship at the University of Wisconsin-Madison. He is co-inventor of Zyvox® (linezolid), a leading antibiotic with annual worldwide sales first exceeding US$1 billion in 2008. He initiated the oxazolidinone research program at Upjohn and led the team that discovered linezolid and an earlier clinical candidate, eperezolid. Linezolid is the first member of any entirely new class of antibiotic to reach the market in the more than

35 years since the discovery of the first quinolone. Dr. Brickner is a corecipient of the Pharmaceutical Research and Manufacturers of America (PhRMA) 2007 Discoverers Award, and the 2007 American Chemical Society Award for Team Innovation. He was named the 2002-2003 Outstanding Alumni Lecturer, College of Arts and Science, Miami University (Ohio). Dr. Brickner is a synthetic organic/ medicinal chemist with over 25 years of research experience focused entirely on the discovery of novel antibacterial agents during his prior tenure at Upjohn, Pharmacia & Upjohn, and Pfizer. He is an inventor or co-inventor on 21 U.S. patents and has published over 30 peer-reviewed scientific papers, particularly on the oxazolidinones and novel azetidinones. An internationally recognized drug discoverer with over 20 invited speaker presentations, he has been a member of the IOM Forum on Microbial Threats since 1997 and is on the Editorial Advisory Board of *Current Pharmaceutical Design* and the Faculty of 1000 Biology. In February 2009, he established SJ Brickner Consulting, LLC, which primarily offers consulting services on all aspects of medicinal chemistry and drug design related to the discovery and development of new antibiotics.

John E. Burris, Ph.D., became president of the Burroughs Wellcome Fund in July 2008. He is the former president of Beloit College. Prior to his appointment at Beloit in 2000, Dr. Burris served for eight years as director and CEO of the Marine Biological Laboratory in Woods Hole, Massachusetts. From 1984 to 1992 he was at the National Research Council/National Academies, where he served as the executive director of the Commission on Life Sciences. A native of Wisconsin, he received an A.B. in biology from Harvard University in 1971, attended the University of Wisconsin–Madison in an M.D.-Ph.D. program, and received a Ph.D. in marine biology from the Scripps Institution of Oceanography at the University of California–San Diego in 1976. A professor of biology at the Pennsylvania State University from 1976 to 1985, he held an adjunct appointment there until coming to Beloit. His research interests were in the areas of marine and terrestrial plant physiology and ecology. He has served as president of the American Institute of Biological Sciences and is or has been a member of a number of distinguished scientific boards and advisory committees including the Grass Foundation; the Stazione Zoologica "Anton Dohrn" in Naples, Italy; the AAAS; and the Radiation Effects Research Foundation in Hiroshima, Japan. He has also served as a consultant to the National Conference of Catholic Bishops' Committee on Science and Human Values.

Gail H. Cassell, Ph.D., is currently vice president, Scientific Affairs, and Distinguished Lilly Research Scholar for Infectious Diseases, Eli Lilly and Company, in Indianapolis, Indiana. She is the former Charles H. McCauley Professor and chairman of the Department of Microbiology at the University of Alabama Schools of Medicine and Dentistry at Birmingham, a department that ranked first in research funding from NIH during her decade of leadership. She obtained her

B.S. from the University of Alabama in Tuscaloosa and in 1993 was selected as one of the top 31 female graduates of the twentieth century. She obtained her Ph.D. in microbiology from the University of Alabama at Birmingham and was selected as its 2003 Distinguished Alumnus. She is a past president of the ASM (the oldest and single-largest life sciences organization, with a membership of more than 42,000). She was a member of the NIH Director's Advisory Committee and a member of the Advisory Council of the National Institute of Allergy and Infectious Diseases (NIAID) of NIH. She was named to the original Board of Scientific Councilors of the CDC Center for Infectious Diseases and served as chair of the board. She recently served a three-year term on the Advisory Board of the director of the CDC and as a member of the HHS secretary's Advisory Council of Public Health Preparedness. Currently she is a member of the Science Board of the FDA Advisory Committee to the Commissioner. Since 1996 she has been a member of the U.S.–Japan Cooperative Medical Science Program responsible for advising the respective governments on joint research agendas (U.S. State Department–Japan Ministry of Foreign Affairs). She has served on several editorial boards of scientific journals and has authored more than 250 articles and book chapters. Dr. Cassell has received national and international awards and an honorary degree for her research in infectious diseases. She is a member of the IOM and is currently serving a three-year term on the IOM Council, its governing board. Dr. Cassell has been intimately involved in the establishment of science policy and legislation related to biomedical research and public health. For nine years she was chairman of the Public and Scientific Affairs Board of the ASM; she has served as an adviser on infectious diseases and indirect costs of research to the White House Office of Science and Technology Policy (OSTP); and she has been an invited participant in numerous congressional hearings and briefings related to infectious diseases, antimicrobial resistance, and biomedical research. She has served two terms on the Liaison Committee for Medical Education (LCME), the accrediting body for U.S. medical schools, as well as other national committees involved in establishing policies for training in the biomedical sciences. She has just completed a term on the Leadership Council of the School of Public Health of Harvard University. Currently she is a member of the Executive Committee of the Board of Visitors of Columbia University School of Medicine, the Board of Directors of the Burroughs Wellcome Fund, and the Advisory Council of the School of Nursing of Johns Hopkins.

Mark Feinberg, M.D., Ph.D., is vice president for medical affairs and policy in global vaccine and infectious diseases at Merck & Co., Inc., and is responsible for global efforts to implement vaccines to achieve the greatest health benefits, including efforts to expand access to new vaccines in the developing world. Dr. Feinberg received a bachelor's degree magna cum laude from the University of Pennsylvania in 1978 and his M.D. and Ph.D. degrees from Stanford University School of Medicine in 1987. His Ph.D. research at Stanford was supervised by Dr. Irving Weissman and included time spent studying the molecular biology of the human retroviruses—HTLV-I (human T-cell lymphotrophic virus, type I) and HIV—as a visiting scientist in the laboratory of Dr. Robert Gallo at the National Cancer Institute. From 1985 to 1986, Dr. Feinberg served as a project officer for the IOM Committee on a National Strategy for AIDS. After receiving his M.D. and Ph.D. degrees, Dr. Feinberg pursued postgraduate residency training in internal medicine at the Brigham and Women's Hospital of Harvard Medical School and postdoctoral fellowship research in the laboratory of Dr. David Baltimore at the Whitehead Institute for Biomedical Research. From 1991 to 1995, Dr. Feinberg was an assistant professor of medicine and microbiology and immunology at the University of California, San Francisco (UCSF), where he also served as an attending physician in the AIDS-oncology division and as director of the virology research laboratory at San Francisco General Hospital. From 1995 to 1997, Dr. Feinberg was a medical officer in the Office of AIDS Research in the Office of the Director of the NIH, the chair of the NIH Coordinating Committee on AIDS Etiology and Pathogenesis Research, and an attending physician at the NIH Clinical Center. During this period, he also served as executive secretary of the NIH Panel to Define Principles of Therapy of HIV Infection. Prior to joining Merck in 2004, Dr. Feinberg served as professor of medicine and microbiology and immunology at the Emory University School of Medicine, as an investigator at the Emory Vaccine Center, and as an attending physician at Grady Memorial Hospital. At UCSF and Emory, Dr. Feinberg and colleagues were engaged in the preclinical development and evaluation of novel vaccines for HIV and other infectious diseases and in basic research studies focused on revealing fundamental aspects of the pathogenesis of AIDS. Dr. Feinberg also founded and served as the medical director of the Hope Clinic of the Emory Vaccine Center—a clinical research facility devoted to the clinical evaluation of novel vaccines and to translational research studies of human immune system biology. In addition to his other professional roles, Dr. Feinberg has also served as a consultant to, and a member of, several IOM and NAS committees. Dr. Feinberg currently serves as a member of the National Vaccine Advisory Committee and is a member of the Board of Trustees of the National Foundation for Infectious Diseases. Dr. Feinberg has earned board certification in internal medicine; he is a fellow of the American College of Physicians, a member of the Association of American Physicians, and the recipient of an Elizabeth Glaser Scientist Award from the Pediatric AIDS Foundation and an Innovation in Clinical Research Award from the Doris Duke Charitable Foundation.

Capt. Darrell R. Galloway, M.S.C., Ph.D., is chief of the Medical Science and Technology Division for the Chemical and Biological Defense Directorate at the Defense Threat Reduction Agency. He received his baccalaureate degree in microbiology from California State University in Los Angeles in 1973. After completing military service in the U.S. Army as a medical corpsman from 1969 to

1972, Captain Galloway entered graduate school and completed a doctoral degree in biochemistry in 1978 from the University of California, followed by two years of postgraduate training in immunochemistry as a fellow of the National Cancer Institute (NCI) at the Scripps Clinic and Research Foundation in La Jolla, California. Captain Galloway began his Navy career at the Naval Medical Research Institute in Bethesda, Maryland, where he served as a research scientist working on vaccine development from 1980 to 1984. In late 1984, Captain Galloway left active service to pursue an academic appointment at Ohio State University, where he is now a tenured faculty member in the Department of Microbiology. He also holds appointments at the University of Maryland Biotechnology Institute and the Uniformed Services University of the Health Sciences. He has an international reputation in the area of bacterial toxin research and has published more than 50 research papers on various studies of bacterial toxins. In recent years, Captain Galloway's research has concentrated on anthrax and the development of DNA-based vaccine technology. His laboratory has contributed substantially to the development of a new DNA-based vaccine against anthrax that has completed the first phase of clinical trials. Captain Galloway is a member of the ASM and has served as president of the Ohio branch of that organization. He received an NIH Research Career Development Award. In 2005, Captain Galloway was awarded the Joel M. Dalrymple Award for significant contributions to biodefense vaccine development.

S. Elizabeth George, Ph.D., is deputy director, Biological Countermeasures Portfolio Science and Technology Directorate, Department of Homeland Security. Until it merged into the new department in 2003, she was program manager of the Chemical and Biological National Security Program in the Department of Energy's National Nuclear Security Administration's Office of Nonproliferation Research and Engineering. Significant accomplishments include the design and deployment of BioWatch, the nation's first civilian biological threat agent monitoring system, and PROTECT, the first civilian operational chemical detection and response capability deployed in the Washington, DC, area subway system. Previously, she spent 16 years at the U.S. Environmental Protection Agency (EPA), Office of Research and Development, National Health and Ecological Effects Research Laboratory, Environmental Carcinogenesis Division, where she was branch chief of the Molecular and Cellular Toxicology Branch. She received her B.S. in biology in 1977 from Virginia Polytechnic Institute and State University and her M.S. and Ph.D. in microbiology in 1979 and 1984, respectively, from North Carolina State University. From 1984 to 1986, she was a National Research Council (NRC) fellow in the laboratory of Dr. Larry Claxton at EPA. Dr. George is the 2005 chair of the Chemical and Biological Terrorism Defense Gordon Research Conference. She has served as councillor for the Environmental Mutagen Society and president and secretary of the Genotoxicity and Environmental Mutagen Society. She holds memberships in the ASM and the AAAS and is an adjunct faculty member in the School of Rural Public Health, Texas A&M University. She is a recipient of the EPA Bronze Medal and Scientific and Technological Achievement Awards and the Department of Homeland Security (DHS) Under Secretary's Award for Science and Technology. She is the author of numerous journal articles and has presented her research at national and international meetings.

Jesse L. Goodman, M.D., M.P.H., is director of the FDA's Center for Biologics Evaluation and Research, which oversees medical, public health, and policy activities concerning the development and assessment of vaccines, blood products, tissues, and related devices and novel therapeutics, including cellular and gene therapies. He moved to the FDA full-time in 2001 from the University of Minnesota, where he was professor of medicine and director of the Division of Infectious Diseases. A graduate of Harvard College, he received his M.D. from the Albert Einstein College of Medicine; did residency and fellowship training at the Hospital of the University of Pennsylvania and at the University of California, Los Angeles (UCLA), where he was also chief medical resident; and is board certified in internal medicine, oncology, and infectious diseases. He trained in the virology laboratory of Jack Stevens at UCLA and has had an active laboratory program in the molecular pathogenesis of infectious diseases. In 1995, his laboratory isolated the etiologic agent of human granulocytic ehrlichiosis and subsequently characterized fundamental events involved in the infection of leukocytes, including their cellular receptors. He is editor of the book *Tick Borne Diseases of Humans* published by ASM Press in 2005, and is a staff physician and infectious diseases consultant at the NIH Clinical Center and the National Naval Medical Center-Walter Reed Army Medical Center, as well as adjunct professor of medicine at the University of Minnesota. He is active in a wide variety of clinical, public health, and product development issues, including pandemic and emerging infectious disease threats; bioterrorism preparedness and response; and blood, tissue, and vaccine safety and availability. In these activities, he has worked closely with CDC, NIH, and other HHS components, academia, and the private sector, and he has put into place an interactive team approach to emerging threats. This model was used in the collaborative development and rapid implementation of nationwide donor screening of the U.S. blood supply for West Nile virus. He has been elected to the American Society for Clinical Investigation (ASCI) and to the IOM.

Eduardo Gotuzzo, M.D., is principal professor and director at the Instituto de Medicina Tropical Alexander von Humbolt, Universidad Peruana Cayetan Heredia in Lima, Peru, as well as chief of the Department of Infectious and Tropical Diseases at the Cayetano Heredia Hospital. He is also an adjunct professor of medicine at the University of Alabama, Birmingham, School of Medicine. Dr. Gotuzzo is an active member of numerous international societies and has been president of the Latin America Society of Tropical Disease (2000-2003), the IDSA Scientific Program (2000-2003), the International Organizing Committee of the International Congress of Infectious Diseases (1994 to present), president-elect of the International Society for Infectious Diseases (1996-1998), and president of the Peruvian Society of Internal Medicine (1991-1992). He has published more than 230 articles and chapters as well as six manuals and one book. Recent honors and awards include being named an honorary member of the American Society of Tropical Medicine and Hygiene in 2002, an associate member of the National Academy of Medicine in 2002, an honorary member of the Society of Internal Medicine in 2000, and a distinguished visitor at the Faculty of Medical Sciences, University of Cordoba, Argentina, in 1999. In 1988 he received the Golden Medal for Outstanding Contribution in the Field of Infectious Diseases awarded by Trnava University, Slovakia.

Jo Handelsman, Ph.D., is a Howard Hughes Medical Institute professor in the Departments of Bacteriology and Plant Pathology and chair of the Department of Bacteriology at the University of Wisconsin (UW)–Madison. She received her Ph.D. in Molecular Biology from the UW–Madison in 1984 and joined the faculty of UW–Madison in 1985. Her research focuses on the genetic and functional diversity of microorganisms in soil and insect gut communities. She is one of the pioneers of functional metagenomics, an approach to accessing the genetic potential of unculturable bacteria in environmental samples. In addition to her research program, Dr. Handelsman is nationally known for her efforts to improve science education and increase the participation of women and minorities in science at the university level. She cofounded the Women in Science and Engineering Leadership Institute at UW–Madison, which has designed and evaluated interventions intended to enhance the participation of women in science. Her leadership in women in science led to her appointment as the first President of the Rosalind Franklin Society and her service on the National Academies' panel that wrote the 2006 report, *Beyond Bias and Barriers: Fulfilling the Potential of Women in Academic Science and Engineering*, which documented the issues of women in science and recommended changes to universities and federal funding agencies. In addition to more than 100 scientific research publications, Dr. Handelsman is coauthor of two books about teaching: *Entering Mentoring* and *Scientific Teaching*. Dr. Handelsman is the editor-in-chief of *DNA and Cell Biology* and the series, Controversies in Science and Technology, and a member of the National Academy of Sciences Board on Life Sciences and the IOM Forum on Microbial Threats. She is a National Academies Mentor in the Life Sciences, a fellow in the American Academy of Microbiology and the AAAS, Director of the Wisconsin Program for Scientific Teaching, and codirector of the National Academies Summer Institute on Undergraduate Education in Biology. In 2008 she received the Alice Evans Award from the ASM in recognition of her mentoring, and in 2009 she received the Carski Award from the ASM in recognition of her teaching contributions, and in 2009, Seed Magazine named her "A Revolutionary Mind" in recognition of her unorthodox ideas.

Carole A. Heilman, Ph.D., is the director of the Division of Microbiology and Infectious Diseases (DMID) at NIAID, a component of NIH-HHS. As director of DMID she has responsibility for scientific direction, oversight, and management of all extramural research programs on infectious diseases (except AIDS) within NIH. In addition, since 2001 Dr. Heilman has played a critical role in launching and directing NIAID's extramural biodefense research program. Previously, Dr. Heilman served as deputy director of NIAID's Division of AIDS for three years. Dr. Heilman has a Ph.D. in microbiology from Rutgers University. She did her postdoctoral work in molecular virology at the National Cancer Institute (NCI) and continued at the NCI as a senior staff fellow in molecular oncology. She moved into health science administration in 1986, focusing on respiratory pathogens, particularly vaccine development. She has received numerous awards for scientific management and leadership, including three HHS Secretary's Awards for Distinguished Service for her contributions to developing pertussis, biodefense, and AIDS vaccines.

David L. Heymann, M.D., is currently chair of the Health Protection Agency, United Kingdom, and head of the Global Health Security Programme at Chatham House, London. Until April 2009, he was assistant director-general for Health Security Environment and Representative of the director-general for Polio Eradication at WHO. Prior to that, from July 1998 until July 2003, he was executive director of the WHO Communicable Diseases Cluster, which included WHO's programmes on infectious and tropical diseases, and from which the public health response to SARS was mounted in 2003. From October 1995 to July 1998, he was director of the WHO Programme on Emerging and other Communicable Diseases, and prior to that was the chief of research activities in the WHO Global Programme on AIDS. Dr. Heymann has worked in the area of public health for the past 35 years, 25 of which were on various assignments from the U.S. Centers for Disease Control and Prevention (CDC), and 10 of which have been with WHO. Before joining WHO, Dr. Heymann worked for 13 years as a medical epidemiologist in sub-Saharan Africa (Cameroon, Côte d'Ivoire, Malawi, and the Democratic Republic of Congo, formerly Zaire) on assignment from the CDC in CDC-supported activities. These activities aimed at strengthening capacity in surveillance of infectious diseases and their control, with special emphasis on the childhood immunizable diseases including measles and polio, African hemorrhagic fevers, poxviruses and malaria. While based in Africa, Dr. Heymann participated in the investigation of the first outbreak of Ebola in Yambuku (former Zaire) in 1976, then again investigated the second outbreak of Ebola in 1977 in Tandala, and in 1995 directed the international response to the Ebola outbreak in Kikwit for WHO. Prior to assignments in Africa he was assigned for two years to

India as a medical epidemiologist in the WHO Smallpox Eradication Programme. Dr. Heymann's educational qualifications include a B.A. from the Pennsylvania State University, an M.D. from Wake Forest University, a Diploma in Tropical Medicine and Hygiene from the London School of Hygiene and Tropical Medicine, and practical epidemiology training in the two-year Epidemic Intelligence Service (EIS) of CDC. He is a member of the IOM; and has been awarded the 2004 Award for Excellence of American Public Health Association, the 2005 Donald Mackay Award from the American Society for Tropical Medicine and Hygiene, and the 2007 Heinz Award on the Human Condition. Dr. Heymann has been visiting professor at Stanford University, the University of Southern California, and the George Washington University School of Public Health; has published over 145 scientific articles on infectious diseases and related issues in peer-reviewed medical and scientific journals; and has authored several chapters on infectious diseases in medical textbooks. He is currently the editor of the 19th edition of the *Control of Communicable Diseases Manual*, a joint publication of the American Public Health Association and WHO.

Phil Hosbach is vice president, New Products and Immunization Policy, at sanofi pasteur. The areas under his supervision are new product marketing, state and federal government policy, business intelligence, bids and contracts, medical communications, public health sales, and public health marketing. His current responsibilities include oversight of immunization policy development. He acts as sanofi pasteur's principal liaison with CDC. Mr. Hosbach graduated from Lafayette College in 1984 with a degree in biology. He has 20 years of pharmaceutical industry experience, including the past 17 years focused solely on vaccines. He began his career at American Home Products in clinical research in 1984. He joined Aventis Pasteur (then Connaught Labs) in 1987 as clinical research coordinator and has held research and development positions of increasing responsibility, including clinical research manager and director of clinical operations. Mr. Hosbach also served as project manager for the development and licensure of Tripedia, the first diphtheria, tetanus, and acellular pertussis (DTaP) vaccine approved by the FDA for use in U.S. infants. During his clinical research career at Aventis Pasteur, he contributed to the development and licensure of seven vaccines, and he has authored or coauthored several clinical research articles. From 2000 through 2002, Mr. Hosbach served on the board of directors for Pocono Medical Center in East Stroudsburg, Pennsylvania. Since 2003 he has served on the board of directors of Pocono Health Systems, which includes Pocono Medical Center.

James M. Hughes, M.D.,[2] is professor of medicine and public health at Emory University's School of Medicine and Rollins School of Public Health, serving as

[2]Current Vice Chair.

director of the Emory Program in Global Infectious Diseases, associate director of the Southeastern Center for Emerging Biological Threats, and senior adviser to the Emory Center for Global Safe Water. He also serves as senior scientific adviser for infectious diseases to the International Association of National Public Health Institutes funded by the Bill and Melinda Gates Foundation. Prior to joining Emory in June 2005, Dr. Hughes served as director of the NCID at the CDC. Dr. Hughes received his B.A. and M.D. degrees from Stanford University and completed postgraduate training in internal medicine at the University of Washington, infectious diseases at the University of Virginia, and preventive medicine at the CDC. After joining the CDC as an EIS officer in 1973, Dr. Hughes worked initially on foodborne and waterborne diseases and subsequently on infection control in health-care settings. He served as director of CDC's Hospital Infections Program from 1983 to 1988, as deputy director of NCID from 1988 to 1992, and as director of NCID from 1992 to 2005. A major focus of Dr. Hughes' career has been on building partnerships among the clinical, research, public health, and veterinary communities to prevent and respond to infectious diseases at the national and global levels. His research interests include emerging and reemerging infectious diseases; antimicrobial resistance; foodborne diseases; health-care-associated infections; vectorborne and zoonotic diseases; rapid detection of and response to infectious diseases and bioterrorism; strengthening public health capacity at the local, national, and global levels; and prevention of water-related diseases in the developing world. Dr. Hughes is a fellow of the AAAS, the American College of Physicians, and the IDSA, a member of IOM, and a councillor of the American Society of Tropical Medicine and Hygiene.

Stephen A. Johnston, Ph.D., is currently director of the Center for Innovations in Medicine in the Biodesign Institute at Arizona State University. His center focuses on formulating and implementing disruptive technologies for basic problems in health care. The center has three divisions: Genomes to Vaccines, Cancer Eradication, and DocInBox. Genomes to Vaccines has developed high-throughput systems to screen for vaccine candidates and is applying them to predict and produce chemical vaccines. The Cancer Eradication group is working on formulating a universal prophylactic vaccine for cancer. DocInBox is developing technologies to facilitate presymptomatic diagnosis. Dr. Johnston founded the Center for Biomedical Inventions (also known as the Center for Translation Research) at the University of Texas–Southwestern, the first center of its kind in the medical arena. He and his colleagues have developed numerous inventions and innovations, including the gene gun, genetic immunization, TEV (tobacco etch virus) protease system, organelle transformation, digital optical chemistry arrays, expression library immunization, linear expression elements, and others. He also was involved in transcription research for years, first cloning *Gal4* and later discovering functional domains in transcription factors and the connection of the proteasome to transcription. He has been professor at the University of

Texas Southwestern Medical Center at Dallas and associate and assistant professor at Duke University. He has been involved in several capacities as an adviser on biosecurity since 1996 and is a member of the WRCE SAB and a founding member of BioChem 20/20.

Gerald T. Keusch, M.D., is associate provost and associate dean for global health at Boston University and Boston University School of Public Health. He is a graduate of Columbia College (1958) and Harvard Medical School (1963). After completing a residency in internal medicine, fellowship training in infectious diseases, and two years as an NIH research associate at the Southeast Asia Treaty Organization (SEATO) Medical Research Laboratory in Bangkok, Thailand, Dr. Keusch joined the faculty of the Mt. Sinai School of Medicine in 1970, where he established a laboratory to study the pathogenesis of bacillary dysentery and the biology and biochemistry of Shiga toxin. In 1979 he moved to Tufts Medical School and New England Medical Center in Boston to found the Division of Geographic Medicine, which focused on the molecular and cellular biology of tropical infectious diseases. In 1986 he integrated the clinical infectious diseases program into the Division of Geographic Medicine and Infectious Diseases, continuing as division chief until 1998. He has worked in the laboratory and in the field in Latin America, Africa, and Asia on basic and clinical infectious diseases and HIV/AIDS research. From 1998 to 2003, he was associate director for international research and director of the Fogarty International Center at NIH. Dr. Keusch is a member of ASCI, the Association of American Physicians, the ASM, and the IDSA. He has received the Squibb (1981), Finland (1997), and Bristol (2002) awards of the IDSA. In 2002 he was elected to the IOM.

Rima F. Khabbaz, M.D., is director of the National Center for Preparedness, Detection, and Control of Infectious Diseases at CDC. She became director of NCID at CDC in December 2005 and led its transition to the current centers. She is a graduate of the American University of Beirut, Lebanon, where she obtained both her bachelor's degree in science and her medical doctorate degree. She trained in internal medicine and completed a fellowship in infectious diseases at the University of Maryland in Baltimore. She is also a clinical associate professor of medicine (infectious diseases) at Emory University. She began her CDC career in 1980 as an epidemic intelligence service officer in the Hospital Infections Program. She later served as a medical epidemiologist in CDC's Retrovirus Diseases Branch, where she made major contributions to defining the epidemiology of non-HIV retroviruses (HTLV-I and II) in the United States and developing guidance for counseling HTLV-infected persons. Following the hantavirus pulmonary syndrome outbreak in the southwestern United States in 1993, she led CDC's efforts to set up national surveillance for the syndrome. Prior to becoming director of NCID, she was acting deputy director and, before that, associate director for epidemiologic science, NCID. Additional positions held at

CDC include associate director for science and deputy director of the Division of Viral and Rickettsial Diseases. She played a leading role in developing CDC's blood safety programs and its food safety programs related to viral diseases. She also had a key role in CDC's responses to outbreaks of new and/or reemerging viral infections including Nipah, Ebola, West Nile, SARS, and monkeypox. She led CDC's field team to the nation's capital during the public health response to the anthrax attack of 2001. She is a fellow of the IDSA, a member of the American Epidemiologic Society, the ASM, and the Council of State and Territorial Epidemiologists. She served on FDA's Blood Product Advisory Committee and on its Transmissible Spongiform Encephalopathy Advisory Committee. She also served on IDSA's Annual Meeting Scientific Program Committee and serves on the society's National and Global Public Health Committee. She is a graduate of the National Preparedness Leadership Initiative at Harvard University and of the Public Health Leadership Institute at the University of North Carolina.

Lonnie J. King, D.V.M., is currently director of CDC's new National Center for Zoonotic, Vector-Borne, and Enteric Diseases (NCZVED). Dr. King leads the center's activities for surveillance, diagnostics, disease investigations, epidemiology, research, public education, policy development, and disease prevention and control programs. NCZVED also focuses on waterborne, foodborne, vectorborne, and zoonotic diseases of public health concern, which also include most of CDC's select and bioterrorism agents, neglected tropical diseases, and emerging zoonoses. Before serving as director, he was the first chief of the agency's Office of Strategy and Innovation. In 1996, Dr. King was appointed dean of the College of Veterinary Medicine, Michigan State University. He served for 10 years as dean of the college. As dean, he was the chief executive officer for academic programs, research, the teaching hospital, diagnostic center for population and animal health, basic and clinical science departments, and outreach and continuing education programs. As dean and professor of large animal clinical sciences, Dr. King was instrumental in obtaining funds for construction of the $60 million Diagnostic Center for Population and Animal Health, initiated the Center for Emerging Infectious Diseases in the college, served as the campus leader in food safety, and had oversight for the National Food Safety and Toxicology Center. He brought the Center for Integrative Toxicology to the college and was the university's designated leader for counterbioterrorism activities for his college. Prior to this, Dr. King was administrator for USDA's Animal and Plant Health Inspection Service. Dr. King served as the country's chief veterinary officer for five years and worked extensively in global trade agreements within the North American Free Trade Agreement and the World Trade Organization. Before beginning his government career in 1977, he was in private veterinary practice for seven years in Dayton, Ohio, and in Atlanta, Georgia. He received his B.S. and D.V.M. from Ohio State University in 1966 and 1970, respectively. He earned his M.S. in epidemiology from the University of Minnesota while on special assignment with USDA in 1980. He received his master's in public administration from the American University in Washington, DC, in 1991. Dr. King has a broad knowledge of animal agriculture and the veterinary profession through his work with other government agencies, universities, major livestock and poultry groups, and private practitioners. Dr. King is a board-certified member of the American College of Veterinary Preventive Medicine and has completed the senior executive fellowship program at Harvard University. He served as president of the Association of American Veterinary Medical Colleges from 1999 to 2000 and was vice chair for the National Commission on Veterinary Economic Issues from 2000 to 2004. Dr. King helped start the National Alliance for Food Safety, served on the Governor's Task Force on Chronic Wasting Disease for the State of Michigan, and was a member of four NAS committees; most recently, he chaired the National Academies Committee on Assessing the Nation's Framework for Addressing Animal Diseases. Dr. King is one of the developers of the Science, Politics, and Animal Health Policy Fellowship Program, and he lectures extensively on the future of animal health, emerging zoonoses, and veterinary medicine. He served as a consultant and member of the Board of Scientific Counselors to CDC's NCID and is a member of the IOM's Forum on Microbial Threats. Dr. King was an editor for the OIE (World Organisation for Animal Health) *Scientific Review on Emerging Zoonoses*, is a current member of FDA's Board of Scientific Advisors, and is president of the American Veterinary Epidemiology Society. Dr. King was elected to the IOM in 2004.

Col. George W. Korch, Ph.D.,[3] is commander, U.S. Army Medical Research Institute for Infectious Diseases, Ft. Detrick, Maryland. Dr. Korch attended Boston University and earned a B.S. in biology in 1974, followed by postgraduate study in mammalian ecology at the University of Kansas from 1975 to 1978. He earned his Ph.D. from the Johns Hopkins School of Hygiene and Public Health in immunology and infectious diseases in 1985, followed by postdoctoral experience at Johns Hopkins from 1985 to 1986. His areas of training and specialty are the epidemiology of zoonotic viral pathogens and medical entomology. For the past 15 years, he has also been engaged in research and program management for medical defense against biological pathogens used in terrorism or warfare.

Stanley M. Lemon, M.D., is the John Sealy Distinguished University Chair and director of the Institute for Human Infections and Immunity at the University of Texas Medical Branch (UTMB) at Galveston. He received his undergraduate A.B. degree in biochemical sciences from Princeton University summa cum laude and his M.D. with honors from the University of Rochester. He com-

[3]Until January 16, 2009. Kent Kester, M.D., Commander of Walter Reed Army Institute of Research, is the current U.S. Army Medical Research and Materiel Command representative on the Forum.

pleted postgraduate training in internal medicine and infectious diseases at the University of North Carolina at Chapel Hill and is board certified in both. From 1977 to 1983 he served with the U.S. Army Medical Research and Development Command, followed by a 14-year period on the faculty of the University of North Carolina School of Medicine. He moved to UTMB in 1997, serving first as chair of the Department of Microbiology and Immunology, then as dean of the School of Medicine from 1999 to 2004. Dr. Lemon's research interests relate to the molecular virology and pathogenesis of the positive-stranded RNA viruses responsible for hepatitis. He has had a long-standing interest in antiviral and vaccine development and has served as chair of FDA's Anti-Infective Drugs Advisory Committee. He is the past chair of the Steering Committee on Hepatitis and Poliomyelitis of the WHO Programme on Vaccine Development. He is past chair of the NCID-CDC Board of Scientific Counselors and currently serves as a member of the U.S. Delegation to the U.S.–Japan Cooperative Medical Sciences Program. He was cochair of the NAS Committee on Advances in Technology and the Prevention of Their Application to Next Generation Biowarfare Threats, and he recently chaired an IOM study committee related to vaccines for the protection of the military against naturally occurring infectious disease threats.

George Ludwig, Ph.D., is the Civilian Deputy Principal Assistant for Research and Technology with the U.S. Army Medical Research and Materiel Command, MEDCOM, where he is responsible for developing and implementing medical research policy, facilitating strategic partnerships, and coordinating medical research and development intellectual capital and physical infrastructure. Dr. Ludwig plays an integral role in the planning, programming, budgeting, and execution processes for the science and technology components of a $2 billion per year medical RDT&E effort. Previously he served as science director at the U.S. Army Research Institute of Infectious Diseases (USAMRIID) where he helped USAMRIID meet the challenges of a changing national and international biodefense landscape. Dr. Ludwig also served as chief of the Diagnostic Systems Division at USAMRIID, where he coordinated a program for development of advanced diagnostics capable of identifying potential biological weapons and other high-hazard infectious and noninfectious disease agents. Dr. Ludwig also worked extensively on vaccine development and traveled for the military while serving as a team leader for disease outbreak investigations in the former Zaire (Ebola virus), Colombia (Venezuelan equine encephalitis virus), and the southwestern United States (Sin Nombre virus). Dr. Ludwig received his Ph.D. from the University of Wisconsin in 1990 and is the author of nearly 70 manuscripts, technical reports, book chapters, and other publications written during 25 years of relevant experience.

Edward McSweegan, Ph.D., is a program officer at NIAID. He graduated from Boston College with a B.S. in biology in 1978. He has an M.S. in microbiology from the University of New Hampshire and a Ph.D. in microbiology from the

University of Rhode Island. He was an NRC associate from 1984 to 1986 and did postdoctoral research at the Naval Medical Research Institute in Bethesda, Maryland. Dr. McSweegan served as an AAAS diplomacy fellow in the U.S. State Department from 1986 to 1988, where he helped to negotiate science and technology agreements with Poland, Hungary, and the former Soviet Union. After moving to NIH, he continued to work on international health and infectious disease projects in Egypt, Israel, India, and Russia. Currently, he manages NIAID's bilateral program with India, the Indo–U.S. Vaccine Action Program, and he represents NIAID in the HHS Biotechnology Engagement Program with Russia and related countries. He is a member of AAAS, the ASM, and the National Association of Science Writers. He is the author of numerous journal and freelance articles.

Stephen S. Morse, Ph.D., is professor of epidemiology and founding director of the Center for Public Health Preparedness at the Mailman School of Public Health of Columbia University. He returned to Columbia in 2000 after four years in government service as program manager at the Defense Advanced Research Projects Agency, where he codirected the Pathogen Countermeasures Program and subsequently directed the Advanced Diagnostics Program. Before coming to Columbia, he was assistant professor of virology at the Rockefeller University in New York, where he remains an adjunct faculty member. He is the editor of two books, *Emerging Viruses* (Oxford University Press, 1993; paperback, 1996), which was selected by American Scientist for its list of 100 Top Science Books of the 20th Century, and *The Evolutionary Biology of Viruses* (Raven Press, 1994). He was a founding section editor of the CDC journal *Emerging Infectious Diseases* and was formerly an editor-in-chief of the Pasteur Institute's journal *Research in Virology*. Dr. Morse was chair and principal organizer of the 1989 NIAID-NIH Conference on Emerging Viruses, for which he originated the term and concept of emerging viruses/infections. He has served as a member of the IOM-NAS Committee on Emerging Microbial Threats to Health, chaired its Task Force on Viruses, and was a contributor to the resulting report *Emerging Infections* (1992). He was a member of the IOM Committee on Xenograft Transplantation. Dr. Morse also served as an adviser to WHO and several government agencies. He is a fellow of the New York Academy of Sciences and a past chair of its microbiology section, a fellow of the American Academy of Microbiology of the American College of Epidemiology, and an elected life member of the Council on Foreign Relations. He was the founding chair of ProMED, the nonprofit international Program to Monitor Emerging Diseases, and was one of the originators of ProMED-mail, an international network inaugurated by ProMED in 1994 for outbreak reporting and disease monitoring using the Internet. Dr. Morse received his Ph.D. from the University of Wisconsin–Madison.

Michael T. Osterholm, Ph.D., M.P.H., is director of the Center for Infectious Disease Research and Policy and director of the NIH-sponsored Minnesota Center for Excellence in Influenza Research and Surveillance at the University of Minnesota. He is also professor at the School of Public Health and adjunct professor at the Medical School. Previously, Dr. Osterholm was the state epidemiologist and chief of the acute disease epidemiology section for the Minnesota Department of Health. He has received numerous research awards from NIAID and CDC. He served as principal investigator for the CDC-sponsored Emerging Infections Program in Minnesota. He has published more than 300 articles and abstracts on various emerging infectious disease problems and is the author of the best-selling book *Living Terrors: What America Needs to Know to Survive the Coming Bioterrorist Catastrophe*. He is past president of the Council of State and Territorial Epidemiologists. He currently serves on the IOM Forum on Microbial Threats. He has also served on the IOM Committee to Ensure Safe Food from Production to Consumption, and on the IOM Committee on the Department of Defense Persian Gulf Syndrome Comprehensive Clinical Evaluation Program, and as a reviewer for the IOM report *Chemical and Biological Terrorism: Research and Development to Improve Civilian Medical Response*.

George Poste, Ph.D., D.V.M., is director of the Biodesign Institute and Del E. Webb Distinguished Professor of Biology at Arizona State University. From 1992 to 1999, he was chief science and technology officer and president, Research and Development, of SmithKline Beecham (SB). During his tenure at SB, he was associated with the successful registration of 29 drug, vaccine, and diagnostic products. He is chairman of Orchid Cellmark. He serves on the board of directors of Monsanto and Exelixis. He is a distinguished fellow at the Hoover Institution at Stanford University. He is a member of the Defense Science Board of the U.S. Department of Defense and of the IOM Forum on Microbial Threats. Dr. Poste is a board-certified pathologist, a fellow of the Royal Society, and a fellow of the Academy of Medical Sciences. He was awarded the rank of Commander of the British Empire by Queen Elizabeth II in 1999 for services to medicine and for the advancement of biotechnology. He has published more than 350 scientific papers; has coedited 15 books on cancer, biotechnology, and infectious diseases; and serves on the editorial board of several technical journals.

John C. Pottage, Jr., M.D., has been vice president for Global Clinical Development in the Infectious Disease Medicine Development Center at GlaxoSmithKline since 2007. Previously he was senior vice president and chief medical officer at Achillion Pharmaceuticals in New Haven, Connecticut. Achillion is a small biotechnology company devoted to the discovery and development of medicines for HIV, hepatitis C virus (HCV), and resistant antibiotics. Dr. Pottage initially joined Achillion in May 2002. Prior to Achillion, Dr. Pottage was medical director of

Antivirals at Vertex Pharmaceuticals. During this time he also served as an associate attending physician at the Tufts New England Medical Center in Boston. From 1984 to 1998, Dr. Pottage was a faculty member at Rush Medical College in Chicago, where he held the position of associate professor, and also served as the medical director of the Outpatient HIV Clinic at Rush-Presbyterian-St. Luke's Medical Center. While at Rush, Dr. Pottage was the recipient of several teaching awards and is a member of the Mark Lepper Society. Dr. Pottage is a graduate of St. Louis University School of Medicine and Colgate University.

Gary A. Roselle, M.D., received his medical degree from the Ohio State University School of Medicine in 1973. He served his residency at the Northwestern University School of Medicine and his infectious diseases fellowship at the University of Cincinnati School of Medicine. He is program director for infectious diseases for the Department of Veterans Affairs Central Office in Washington, DC, as well as the chief of the medical service at the Cincinnati VA Medical Center. He is a professor of medicine in the Department of Internal Medicine, Division of Infectious Diseases, at the University of Cincinnati College of Medicine. Dr. Roselle serves on several national advisory committees. In addition, he is currently heading the Emerging Pathogens Initiative for the VA. He has received commendations from the under secretary for health for the VA and the secretary of VA for his work in the Infectious Diseases Program for the VA. He has been an invited speaker at several national and international meetings and has published more than 90 papers and several book chapters.

Kevin Russell, M.D., M.T.M.&H., F.I.D.S.A. CAPT MC USN, graduated from the University of Texas Health Science Center San Antonio Medical School in 1990; after a Family Practice internship he was accepted into the Navy Undersea Medicine program. He was stationed in Panama City, Florida, at the Experimental Diving Unit where he worked in diving medicine research from 1991 to 1995. After a Preventive Medicine Residency with a Masters in Tropical Medicine and Hygiene, he was transferred to Lima, Peru, where he became head of the Virology Laboratory. His portfolio included febrile illness (largely arboviral in origin) and HIV surveillance studies in eight different countries of South America, as well as prospective dengue transmission studies. In 2001, he moved back to the states and became the director of the Respiratory Disease Laboratory at the Naval Health Research Center in San Diego, California. Febrile respiratory illness surveillance in recruits of all services was expanded into shipboard populations, Mexican border populations, support for outbreaks, and deployed settings. Validation and integration of new and emerging advanced diagnostic capabilities, utilizing the archives of specimens maintained at the laboratory, became a priority. A BSL-3-Enhanced is currently nearing completion. Projects expanded in 2006 to clinical trials support as Dr. Russell became the Principal Investigator for the Navy site in the FDA Phase 3 adenovirus vaccines trial, and more recently to support the Phase 4 post-marketing trial of the recently FDA-approved ACAM2000 Smallpox vaccine. Dr. Russell recently became director of the Department of Defense Global Emerging Infections Surveillance and Response System (DoD-GEIS).

Janet Shoemaker is director of the American Society for Microbiology's Public Affairs Office, a position she has held since 1989. She is responsible for managing the legislative and regulatory affairs of this 42,000-member organization, the largest single biological science society in the world. Previously, she held positions as assistant director of public affairs for ASM; as ASM coordinator of the U.S.–U.S.S.R. Exchange Program in Microbiology, a program sponsored and coordinated by the NSF and the U.S. Department of State; and as a freelance editor and writer. She received her baccalaureate, cum laude, from the University of Massachusetts and is a graduate of the George Washington University programs in public policy and in editing and publications. She is a member of Women in Government Relations, the American Society of Association Executives, and AAAS. She has coauthored articles on research funding, biotechnology, biodefense, and public policy issues related to microbiology.

P. Frederick Sparling, M.D., is the J. Herbert Bate Professor Emeritus of Medicine, Microbiology, and Immunology at the University of North Carolina (UNC) at Chapel Hill, and professor of medicine, Duke University. He is director of the North Carolina Sexually Transmitted Infections Research Center and also the Southeast Regional Centers of Excellence in Biodefense and Emerging Infections. Previously he served as chair of the Department of Medicine and chair of the Department of Microbiology and Immunology at UNC. He was president of the Infectious Diseases Society of America from 1996 to 1997. He was also a member of the IOM Committee on Microbial Threats to Health (1990-1992) and the IOM Committee on Emerging Microbial Threats to Health in the 21st Century (2001-2003). Dr. Sparling's laboratory research has been on the molecular biology of bacterial outer membrane proteins involved in pathogenesis, with a major emphasis on gonococci and meningococci. His work helped to define the genetics of antibiotic resistance in gonococci and the role of iron-scavenging systems in the pathogenesis of human gonorrhea.

Terence Taylor is director of the Global Health and Security Initiative and president and director of the International Council for the Life Sciences (ICLS). He is responsible for the overall direction of the ICLS and its programs, which have the goal of enhancing global biosafety and biosecurity. From 1995 to 2005, he was assistant director of the International Institute for Strategic Studies (IISS), a leading independent international institute, and president and executive director of its U.S. office (2001-2005). He studies international security policy, risk analysis, and scientific and technological developments and their impact on political and economic stability worldwide. He was one of IISS's leading experts on issues associated with nuclear, biological, and chemical weapons and their means of delivery. In his previous appointments, he has had particular responsibilities for issues affecting public safety and security in relation to biological risks and advances in the life sciences. He was one of the commissioners to the United Nations Special Commission on Iraq, for which he also conducted missions as a chief inspector. He was a science fellow at the Center for International Security and Cooperation at Stanford University, where he carried out, among other subjects, studies of the implications for government and industry of the weapons of mass destruction treaties and agreements. He has also carried out consultancy work for the International Committee of the Red Cross (ICRC) on the implementation and development of the laws of armed conflict and serves as a member of the Editorial Board of the *ICRC Review*. He has served as chairman of the World Federation of Scientists' Permanent Monitoring Panel on Risk Analysis. He was a career officer in the British Army on operations in many parts of the world, including counterterrorist operations and United Nations peacekeeping. His publications include monographs, book chapters, and articles for, among others, Stanford University, the World Economic Forum, Stockholm International Peace Research Institute (SIPRI), the Crimes of War Project, the *International Herald Tribune*, the *Wall Street Journal*, the *International Defence Review*, the *Independent* (London), *Tiempo* (Madrid), the *International and Comparative Law Quarterly*, the *Washington Quarterly*, and other scholarly journals, including unsigned contributions to IISS publications.

Murray Trostle, Dr.P.H., is a foreign service officer with the U.S. Agency for International Development (USAID), presently serving as the deputy director of the Avian and Pandemic Influenza Preparedness and Response Unit. Dr. Trostle attended Yale University, where he received a master's in public health in 1978, focusing on health services administration. In 1990, he received his doctorate in public health from UCLA. His research involved household survival strategies during famine in Kenya. Dr. Trostle has worked in international health and development for approximately 38 years. He first worked overseas in the Malaysian national malaria eradication program in 1968 and has since focused on health development efforts in the former Soviet Union, Africa, and Southeast Asia. He began his career with USAID in 1992 as a postdoctoral fellow with AAAS. During his career he has worked with a number of development organizations such as the American Red Cross, Project Concern International, and the Center for Development and Population Activities. With USAID, Dr. Trostle has served as director of the child immunization cluster, where he was chairman of the European Immunization Interagency Coordinating Committee and the USAID representative to the Global Alliance on Vaccines and Immunization. Currently, Dr. Trostle leads the USAID Infectious Disease Surveillance Initiative as well as the Avian Influenza Unit.